GALLIUM ARSENIDE
Proceedings of the Second International Symposium

GALLIUM ARSENIDE

Proceedings of the Second International Symposium
organized by
Southern Methodist University
and
The Institute of Physics and The Physical Society
in co-operation with
The Avionics Laboratory of the U.S. Air Force

Dallas, Texas, October, 1968

Institute of Physics and Physical Society
Conference Series No. 7

The Second International Gallium Arsenide Symposium was held at Southern Methodist University, Dallas, Texas, on 16–18 October 1968.

Organizing Committee

Dr. H. Strack (Chairman)
Dr. W. Carr
Dr. J. Franks
Dr. J. Lamorte
Col. R. Runnels

Editor

C. I. Pedersen
   assisted by E. L. Dellow

Copyright © 1969 by The Institute of Physics and The Physical Society and individual contributors. All rights of reproduction, translation, and adaptation are reserved for all countries.

SBN 85498 002 4

Published by
THE INSTITUTE OF PHYSICS AND THE PHYSICAL SOCIETY
47 Belgrave Square, London, S.W.1
Editorial Department: 1 Lowther Gardens, Prince Consort Road, London, S.W 7

Printed in Great Britain by
Adlard & Son Ltd, Dorking, Surrey

Preface

It has been the intent of the sponsors of the 'International Conference on Gallium Arsenide' series to give specialists in the field of research on gallium arsenide devices and materials an opportunity of interchanging ideas and to provide an up-to-date survey of the gallium arsenide technology. Because of the similarity of the technology of gallium arsenide and that of its alloys with other III–V compound materials, the 1968 conference included papers on $Ga_xAl_{1-x}As$, $Ga_xIn_{1-x}As$ and $GaAs_xP_{1-x}$. It is of interest to compare the papers given at the 1966 and 1968 conferences to establish progress made within the last few years, to identify new materials, devices, and technologies, and to visualize trends in the gallium arsenide field.

The papers were divided into four major categories: materials, light emitters, microwave devices, and other devices. Based on the number of papers presented in each category, the interest in the materials area has considerably increased over the past two years, and the work on microwave devices has decreased. This might indicate that researchers improve the material first before expanding a large device effort. The light emitter work, both in the area of spontaneous and stimulated emission, dominates other gallium arsenide device work. In the section on 'Other Devices', the interest has shifted from bipolar and MIS field-effect transistors to Schottky barrier field-effect transistors. A new device, the gallium arsenide photocathode, was not reported on during the last conference.

In the 1966 conference no papers in the materials session dealt with liquid epitaxy, while a whole session was devoted to this subject in 1968. The most remarkable observation is that solution-grown gallium arsenide reached the purity level of vapour-phase material. Materials with a carrier concentration in the low 10^{13} cm^{-3} and with liquid nitrogen mobilities of 13,500 cm^2 v^{-1} s^{-1} have been obtained. One material was not even considered as a material for device fabrication at the time of the last conference. Now, GaAlAs might become an important material for fabrication of light emitters in the visible red wavelength region.

Other material papers discussed advanced research on vapour-phase growth. All aspects of this field such as doping of vapour-phase epitaxial material, the effect of the growth direction on growth rate and doping level, and the growth of high purity, high mobility material were discussed. Reports on bulk gallium arsenide covered subjects such as the properties of dislocation-free material and the behaviour of amphoteric dopants in gallium arsenide.

Most device papers concerned themselves with spontaneous and stimulated light emitters. Progress has been made both in understanding the light emission characteristics and in fabrication of devices with high quantum efficiency. The injection mechanism of diffused junctions has been thoroughly investigated, as well as the phenomena of unconventional light emitters such as filamentary emission in semi-insulating GaAs and two-photon absorption emitters. The laser field has reached a degree of maturity with most papers reporting on special aspects of lasers rather than on device fabrication. This is not the case in the microwave device area, where problems such as the influence of materials properties on microwave diode and Gunn oscillator performance, as well as design calculations for LSA oscillators were discussed.

In the session on 'Other Devices', several papers dealt with various types of gallium arsenide transistors. Though the gallium arsenide transistor was one of the first III–V compound devices and now has a ten-year history, a useful device has not been achieved yet. At the 1966 conference it was reported that bulk states limit the high frequency performance of bipolar devices and surface states limit MIS field-effect transistors. At the 1968 conference some of the problems of the newly developed Schottky barrier field-effect transistor seem to be related to interface states. Transistors developed for the other possible applications for gallium arsenide, the high temperature electronics field, suffer at the present time from high junction leakage currents. The remainder of the papers dealt with optoelectronic devices such as photodetectors and photocathodes. It is expected that III–V compound optoelectronic devices, including emitters and detectors, will see the strongest expansion in years to come.

The *Proceedings* will have achieved their purpose if the major trends in the gallium arsenide field are reflected and if stimulating ideas are given to researchers planning their work for the following years.

H. STRACK

October 1968

Contents

Page Paper

CHAPTER 1 LIQUID PHASE EPITAXIAL GROWTH

3   1   Derivation of the Ga–Al–As ternary phase diagram with applications to liquid phase epitaxy
M. Ilegems and G. L. Pearson

11 2   Factors influencing the electrical and physical properties of high quality solution grown GaAs
R. Solomon

18 3   Tin and tellurium doping characteristics in gallium arsenide epitaxial layers grown from Ga solution
C. S. Kang and P. E. Greene

22 4   Solution epitaxy of gallium arsenide with controlled doping
J. Kinoshita, W. W. Stein, G. F. Day, and J. B. Mooney

28 5   Electrical properties of solution grown GaAs layers
J. C. Carballès, D. Diguet and J. Lebailly

36 6   Liquid phase epitaxial growth of gallium arsenide
A. R. Goodwin, C. D. Dobson, and J. Franks

CHAPTER 2 VAPOUR PHASE EPITAXIAL GROWTH AND BULK MATERIAL

43 7   Tin doping of epitaxial gallium arsenide
C. M. Wolfe, G. E. Stillman, and W. T. Lindley

50 8   Influence of substrate orientation on GaAs epitaxial growth rates
Don W. Shaw

55 9   The origin of macroscopic surface imperfections in vapour-grown GaAs
J. J. Tietjen, M. S. Abrahams, A. B. Dreeban, and H. F. Gossenberger

59 10 Preparation of epitaxial gallium arsenide for microwave applications
F. J. Reid and L. B. Robinson

66 11 Site distribution of silicon in silicon-doped gallium arsenide
W. P. Allred, G. Cumming, J. Kung, and W. G. Spitzer

73 12 Correlation between diffusion and precipitation of impurities in dislocation-free GaAs
H. R. Winteler and A. Steinemann

77 13 Diffusion through and from solid layers into gallium arsenide
W. von Münch

CHAPTER 3 STIMULATED EMISSION

83 14 Correlation of GaAs junction laser thresholds with photo-luminescence measurements
C. J. Hwang and J. C. Dyment

91 15 Theory of Q-switching and time delays in GaAs junction lasers
J. E. Ripper

96 16 Doping profiles of solution grown GaAs injection lasers
H. Beneking and W. Vits

Page	Paper	
101	17	Properties of a GaAs laser coupled to an external cavity E. Mohn
110	18	Filamentary lasing and delay time in GaAs laser diodes H. B. Kim
116	19	Stimulated emission from $(Ga_{1-x}Al_x)As$ junctions Wataru Susaki

CHAPTER 4 SPONTANEOUS EMISSION

123	20	Investigation of liquid-epitaxial GaAs spontaneous light-emitting diodes K. L. Ashley and H. A. Strack
131	21	A visible light source utilizing a GaAs electroluminescent diode and a stepwise excitable phosphor S. V. Galginaitis and G. E. Fenner
136	22	Light emitting devices utilizing current filaments in semi-insulating GaAs A. M. Barnett, H. A. Jensen, V. F. Meikleham, and H. C. Bowers
141	23	Radiative tunnelling in GaAs: a comparison of theoretical and experimental properties H. C. Casey, Jr. and Donald J. Silversmith

CHAPTER 5 MICROWAVE DEVICES

153	24	Epitaxial GaAs Gunn effect oscillators: influence of material properties on device performance L. Cohen, F. Drago, B. Shortt, R. Socci, and M. Urban
160	25	Gallium Arsenide p–i–n diode as a microwave device G. R. Antell
167	26	Bulk GaAs travelling-wave amplifier J. Koyama, S. Ohara, S. Kawazura, and K. Kumabe
173	27	Design calculations for cw millimetre wave L.S.A. oscillators T. J. Riley

CHAPTER 6 OTHER DEVICES

181	28	A GaAs pn-junction FET and gate-controlled Gunn effect device R. Zuleeg
187	29	The Schottky barrier gallium arsenide field-effect transistor P. L. Hower, W. W. Hooper, D. A. Tremere, W. Lehrer, and C. A. Bittmann
195	30	Implications of carrier velocity saturation in a gallium arsenide field-effect transistor J. A. Turner and B. L. H. Wilson
205	31	Study of GaAs devices at high temperature F. H. Doerbeck, E. E. Harp, and H. A. Strack
213	32	Vapour-phase growth of large-area microplasma-free p–n junctions in GaAs and $GaAs_{1-x}P_x$ R. E. Enstrom and J. R. Appert
222	33	Characteristics of GaAs based heterojunction photodetectors T. L. Tansley

Page	Paper	
230	34	The GaAs photocathode
		L. W. James, J. L. Moll, and W. E. Spicer
238	35	Photon emission during avalanche breakdown in GaAs
		M. H. Pilkuhn and G. Schul
244		Author index

CHAPTER 1

Liquid phase epitaxial growth

Derivation of the Ga–Al–As ternary phase diagram with applications to liquid phase epitaxy[†]

M. ILEGEMS and G. L. PEARSON

Stanford Electronics Laboratories; Stanford University, Stanford, California 94305, U.S.A.

Abstract. The ternary liquidus-solidus Ga–Al–As phase diagram has been calculated under the assumptions that the Ga–As and Al–As binary systems are quasi-regular and that the mixed crystals form an ideal solid solution as suggested by the fact that the lattice constants of GaAs and AlAs are nearly identical. Excellent agreement is obtained with experimental liquidus and solidus isotherms obtained in the gallium rich region of the diagram. The liquidus lines were determined at 1000, 900, and 800°C using a weight-loss technique. The solidus composition measurements were made on epitaxial layers of $Ga_xAl_{1-x}As$ grown on $\langle 111 \rangle$ GaAs substrates out of gallium-rich melts using a vertical solution regrowth system. The growth temperature was 1000°C and compositions ranging from approximately 20 to 80% AlAs were obtained from melts containing between 0·4 and 4·6% Al.

1. Introduction

Solution grown $Ga_xAl_{1-x}As$ is a promising material for use in a wide variety of applications such as light-emitting diodes, injection lasers, and high voltage p–n junctions. While this mixed crystal compound can be grown from the liquid by essentially the same techniques as those used for GaAs, special problems arise from the addition of a third component to the melt. In particular the composition of the layers is a complex function of melt composition and growth temperature, thus making composition control and homogeneity difficult to achieve even when growing over a limited temperature range. In this paper experimental data obtained for several isotherms in the Ga-rich region of this system will be presented and these will be shown to be consistent with the theoretical analysis presented in section 3. The calculations are then extended to several other liquidus and solidus isotherms and are used to assess the composition variations to be expected during growth in which the growing solid interface is in equilibrium with the liquid phase. The experimental results are in good agreement with these reported recently by Panish and Sumski (1969).

2. Experimental crystal composition and solubility data

A vertical solution regrowth setup similar to that described by Rupprecht (1966) was used in this work. The crucible is alumina and the seed was clamped either between two quartz plates or between two plates of boron-titanium-nitride.

2.1. Crystal growth

In a typical growth experiment the crucible was loaded with 5 to 8 g of Ga, 20 to 200 mg of Al depending on the desired crystal composition and the exact amount of GaAs source material required to saturate the melt at the specified growth temperature. At the same time the substrate was mounted in the seed holder and positioned at the top end of the reaction tube. The substrates were slices of $\langle 111 \rangle$ oriented GaAs single crystal approximately 5 mm wide, 10 mm long and 0·5 mm thick which had been lapped and chemically polished. The system was heated under a purified stream of hydrogen to the temperature of growth, usually 1000°C, and kept there for a saturation period of 1 hour. Growth was started by

[†] This work was sponsored by the Advanced Research Projects Agency and the U.S. Army Mobility Equipment Research and Development Center.

lowering the seed holder very slowly into the hot zone of the furnace and dipping the seed into the melt. Cooling rates were usually around 15 to 20°C h^{-1}. It is assumed that under these conditions the growing solid surface is essentially in equilibrium with the melt. Layer thicknesses varying from 250 μm at low Al concentrations to about 50 μm at higher Al concentrations were obtained for a cooling interval of 100°C. The layers were not flat and had to be polished before optical and electrical measurements could be made. Room-temperature carrier concentrations, measured by the Schottky barrier technique, on undoped material were in the low 10^{16} cm^{-3} range. Since only high purity starting materials were used the crucible may have been the main source of contamination.

The crystal composition is determined by the composition of the melt and the temperature of growth. Since the layers are not homogeneous in depth, estimation of composition from optical transmission data is difficult and, at present, electron beam probing appears to be the only practical method.

The experimental composition data versus the Al concentration in the melt are shown in figure 1. The sample with the lowest AlAs concentration was grown by cooling from 1000°C over a 100°C interval, and the composition measured on the substrate side of the

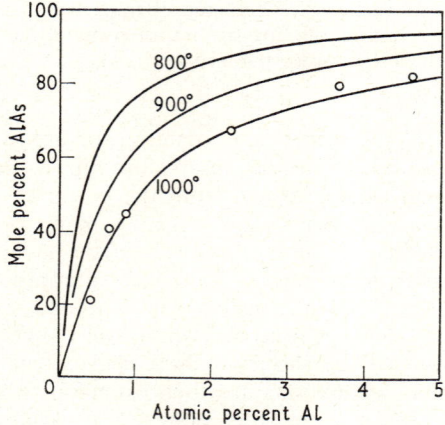

Figure 1. Crystal composition as measured by electron beam probe versus melt composition for samples grown at 1000°C. Solid lines are calculated.

layer after the substrate had been lapped away. All other samples were grown by cooling from 1000°C over a temperature interval less than 20°C, and compositions were measured directly on the growth side of the layer.

The electron beam probing was carried out by the Materials Analysis Company. The Ga and Al concentrations were both measured and the results agreed within the order of five per cent after corrections for absorption, so that the uncertainty in x was estimated at ±5%. Samples of high Al content were exposed to air for the shortest possible time and kept in a dry hydrogen stream or under vacuum without apparent decomposition. The material appears to be stable in air up to about 60% AlAs.

2.2. Liquidus composition measurements

In order to start crystal growth at the exact equilibrium temperature it was found advantageous to establish the equilibrium condition prior to each run by dipping a piece of polycrystalline GaAs source material in the melt for 20 to 30 minutes. By measuring the weight-loss after dipping and repeating the procedure for a series of different temperatures and compositions the liquidus isotherms were determined experimentally in the Ga-rich region of the diagram. The solubility data obtained at 1000, 900 and 800°C are plotted as a function of the Al/Ga weight ratio in the melt in figure 2. The melt was kept at the saturation temperature for approximately 30 minutes before dipping and was stirred frequently during dipping by rotating the seed holder.

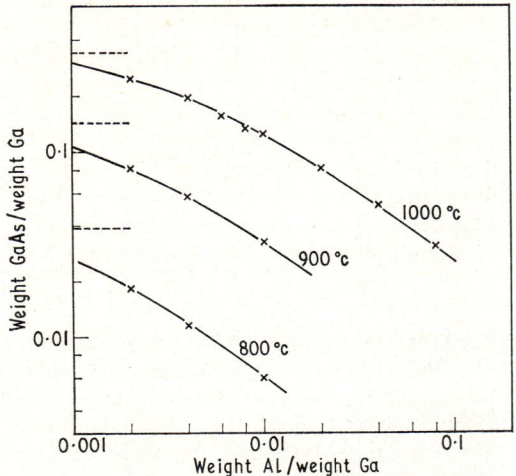

Figure 2. Solubility of GaAs versus Al concentration in the Ga–Al melt at 1000, 900, and 800 °C. The dashed lines correspond to the values in pure Ga.

As soon as sufficient GaAs has dissolved in the melt to reach the liquidus composition any further dissolution of GaAs will result in the formation of the mixed $Ga_xAl_{1-x}As$ crystals in equilibrium with the melt at that temperature. The mixed crystals will be formed in the region immediately surrounding the seed and apparently deposit as a thin layer on the seed thereby protecting it from further attack. That a true equilibrium situation was established in this manner was verified by repeating several measurements at longer dipping times (up to 90 minutes at 1000 °C) where it was found that the weight loss increased only very slightly as might be expected from temperature fluctuations and evaporation of As from the seed.

3. Interpretation of the data

The experimental results presented above can only be properly understood with reference to the conditions of chemical equilibrium between the mixed crystals and the melts from which they were grown. An analysis of these equilibrium conditions is presented next and shows that the experimental data are consistent with theory. The representation adopted for the three binary systems which form the boundaries of the ternary phase diagram will be discussed first and their deviation from ideality characterized in terms of the interaction energy $\alpha = \Delta G_m^e / N_A N_B = (\Delta H_m - T \Delta S_m^e)/N_A N_B$ in the liquid phase, where ΔG_m^e is the excess free energy of mixing. The equilibrium conditions will then be derived in terms of the chemical potentials μ, making the assumption that GaAs and AlAs form ideal solid solutions as suggested by the fact that their lattice constants are nearly identical.

3.1. Ga–Al system

The Ga–Al system was analysed using the method developed by Hiskes and Tiller (1968). Deviations from ideality were found to be fairly small and to have only a limited influence on the phase diagram calculations. In order to keep the activity coefficient relationships consistent with the Gibbs–Duhem equations, a strictly regular representation was adopted for this system and only the first term ($\alpha_{GaAl} = 104$ cal/mole) was kept in the expansion obtained for α.

3.2. Ga–As system

The phase equilibria in this system have been analysed by Thurmond (1965) and Arthur (1967). Calculating α from the liquidus data following Vieland (1963), it was found that α was a linear function of temperature so that the binary liquid could be described as quasi-regular. In our calculations this representation was adopted and the expression $\alpha_{GaAs} = -9.16T + 5160$ cal/mole was taken from Arthur (1967).

By analysing all III–V compound solutions for which experimental data are given and using published experimental values for the entropies of fusion when available, it was found that the quasi-regular approximation was valid in all cases within the uncertainty in the experimental data. It was observed, moreover, that all α versus T lines had approximately the same slope for a given group V element, indicating the same degree of association in the liquid, while the heat of mixing term decreased regularly as one goes down the column III elements. This similarity in behaviour enables reasonable estimates to be made in section 3.3 for the Al–As system even though few experimental data are available in the literature.

3.3. Al–As system

No experimental liquidus data have been reported on this system but from the preceding discussion it seemed reasonable to assume that the Al–As system is quasi-regular with a value for the excess entropy of mixing equal to that of Ga–As. With this assumption the α versus T line will be completely determined if the value of α at one particular temperature is known. This single value of α is derived here from the standard heat of formation of AlAs, $\Delta H_{298} = -27.8 \pm 0.5$ kcal/mole from Kischio (1964), by calculating the free energy of formation of the compound from the pure liquid components at the melting point. The result is $\Delta G_{AlAs} = -10.25 \pm 1.8$ kcal/mole at the melting point of 2013°K as reported by Kischio (1964). The derivation was carried out by (1) using the free energy functions for Al and As given by Stull and Sinke (1956), (2) taking the specific heat of liquid As equal to that of solid As at the melting point, (3) adopting the value for the heat of fusion of As selected by Thurmond (1965), and (4) assuming that the specific heat of AlAs is the same as that given for GaP by Thurmond (1965). The rather large error limit in ΔG_{AlAs} arises by assuming that any or all of the estimated specific heats may be in error by 10% over the entire temperature range and includes the 0.5 kcal/mole error in the standard heat of formation.

From $\Delta G_{AlAs}(2013)$ the value $\alpha_{AlAs}(2013) = -9.4$ kcal/mole is obtained so that in the quasi-regular approximation the Al–As binary system is described by

$$\alpha_{AlAs} = -9.16T + 9040 \text{ cal/mole.}$$

3.4. Equilibrium conditions between solid and liquid phases

The activity coefficients and the chemical potentials in the liquid phase can be derived from the α's given in the previous sections. Following Prigogine and Defay (1965), the relation for a three-component system is

$$RT \ln \gamma_{Ga} = \alpha_{GaAs} N_{As}^2 + \alpha_{GaAl} N_{Al}^2 + (\alpha_{GaAs} - \alpha_{AlAs} + \alpha_{GaAl}) N_{As} N_{Al}, \quad (1)$$

where γ_i and N_i designate respectively the activity coefficient and concentration of constituent i in solution. Only the formula for Ga has been given since the other two can be deduced by cyclic permutation of the indices. These equations obey the Gibbs–Duhem relationships and reduce to the appropriate binary form at the boundaries.

The chemical potentials in the liquid phase are given by equations of the form

$$\mu_{Ga}^{l}(T) = \mu_{Ga}^{ol}(T) + RT \ln \gamma_{Ga} N_{Ga}, \quad (2)$$

and in the solid phase by

$$\mu_{GaAs}^{c}(T) = \mu_{GaAs}^{oc}(T) + RT \ln x. \quad (3)$$

Here the superscripts o, l, and c refer to the pure state, the liquid phase and the crystalline phase respectively; x designates the GaAs fraction of the $(GaAs)_x (AlAs)_{1-x}$ mixed crystals which are assumed to form perfect solid solutions. A similar equation applies to AlAs.

The chemical potential of the pure compound which appears in equation (3) can be related to the chemical potentials of its constituents in the liquid phase by the following formula given by Vieland (1963):

$$\mu_{GaAs}^{oc}(T) = \mu_{Ga}^{sl}(T) + \mu_{As}^{sl}(T) - \Delta S_{GaAs}^{F}(T_{GaAs}^{F} - T)$$
$$- \Delta C_p \left(T_{GaAs}^{F} - T - T \ln \frac{T_{GaAs}^{F}}{T} \right), \quad (4)$$

where ΔS^F is the entropy of fusion of the compound, ΔC_p the difference in specific heat between the compound and its supercooled liquid and where the superscript sl refers to the stochiometric liquid. ΔS^F and ΔC_p are assumed independent of temperature.

At equilibrium the chemical potentials of GaAs and AlAs in the mixed crystals must be equal respectively to the sum of the chemical potentials of Ga and As and Al and As in the liquid:

$$\mu_{GaAs}^{c}(T) = \mu_{Ga}^{l}(T) + \mu_{As}^{l}(T) \quad (5a)$$

$$\mu_{AlAs}^{c}(T) = \mu_{Al}^{l}(T) + \mu_{As}^{l}(T). \quad (5b)$$

Substituting equations (2), (3), and (4) into the above equilibrium conditions and solving for x and $1-x$ it is found, neglecting the specific heat terms, that:

$$x = 4N_{Ga}N_{As} \frac{\gamma_{Ga}\gamma_{As}}{\gamma_{Ga}^{sl}\gamma_{As}^{sl}} \exp\left[\Delta S_{GaAs}^{F}(T_{GaAs}^{F} - T)/RT\right] \quad (6a)$$

and

$$1 - x = 4N_{Al}N_{As} \frac{\gamma_{Al}\gamma_{As}}{\gamma_{Al}^{sl}\gamma_{As}^{sl}} \exp\left[\Delta S_{AlAs}^{F}(T_{AlAs}^{F} - T)/RT\right]. \quad (6b)$$

Elimination of x between these two equations gives the liquidus surface while elimination of T gives the solidus lines. The complete phase diagram can therefore be obtained by calculating the liquidus equilibrium temperature T and the composition x of the mixed crystals in equilibrium with the melt for each set of values of N_{Ga}, N_{Al}, and N_{As}. Unfortunately, the resulting functions have no simple analytical expression and a computer solution must be used.

In the following calculations the value 16·64 cal/mole deg was adopted for the entropy of fusion of GaAs as quoted by Arthur (1967). The entropy of fusion of AlAs has not been reported in the literature but it was found that its value could be estimated from crystal composition measurements. Dividing equation (6b) by equation (6a) gives the following simple linear relationship between the entropies of fusion of the two compounds and the logarithm of the ratio of the distribution coefficients of Ga and Al:

$$\Delta S_{AlAs}^{F}(T_{AlAs}^{F} - T) - \Delta S_{GaAs}^{F}(T_{GaAs}^{F} - T) = RT \ln\left(\frac{1-x}{N_{Al}} \bigg/ \frac{x}{N_{Ga}}\right) + RT \ln A \quad (7)$$

with

$$RT \ln A = (\alpha_{AlAs} - \alpha_{GaAs})(1/2 - N_{As}) - \alpha_{GaAl}(N_{Ga} - N_{Al}), \quad (8)$$

as derived from the activity coefficient relationships in the quasi-regular approximation. Substituting for x, N_{Ga}, and N_{Al} from the six experimental composition points in figure 1 and solving for ΔS_{AlAs}^{F} as shown in table 1, values ranging from 21·3 to 23·3 e.u./mole are obtained. An average value of 22·8 e.u./mole was finally adopted for the entropy of fusion of AlAs.

All thermodynamic constants required in solving the phase equilibria equations are now available and tabulated in table 2. Upon substitution in equations (6a) and (6b) the solidus lines at 1000, 900, and 800°C were calculated and plotted in figure 1. In like manner the

Table 1. ΔS_{AlAs}^F from measured distribution coefficients at 1000°C

N_{Al} %	N_{Ga} %	x %	$\ln \dfrac{(1-x)/N_{Al}}{x/N_{Ga}}$	$\ln A$	ΔS_{AlAs}^F e.u./mole
0·416	90	79	4·05	0·60	21·3
0·637	90·7	59	4·6	0·60	23·1
0·86	91·4	55	4·46	0·61	22·7
2·26	92·6	32·5	4·45	0·65	22·8
3·65	92·4	20	4·61	0·66	23·3
4·56	92	18	4·52	0·68	22·9

Table 2. Thermodynamic constants used in the calculations

	Ga–As	Al–As	Ga–Al
$T^F(°K)$	1511	2013	—
ΔS^F (e.u./mole)	16·64	22·8	—
α (cal/mole)	$-9·16T+5160$	$-9·16T+9040$	104

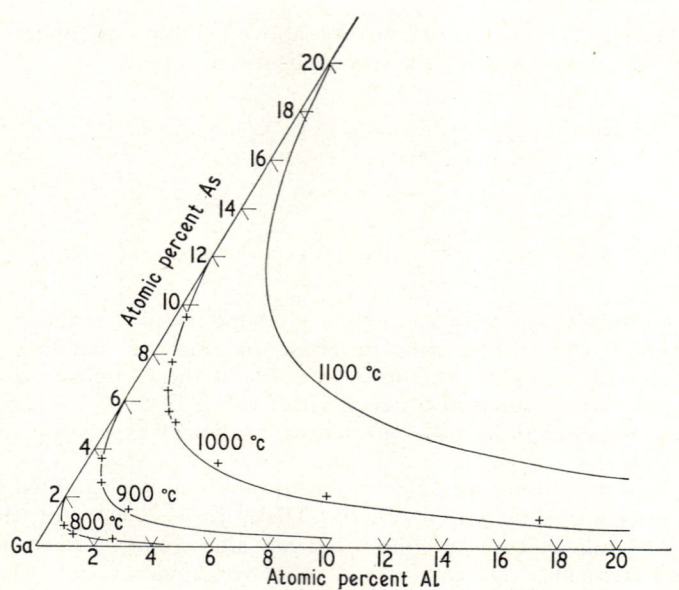

Figure 3. Calculated liquidus lines in the Ga rich region of the ternary diagram. Experimental data points are from figure 2.

liquidus lines in the Ga rich region of the diagram were calculated and plotted in figure 3 for several fixed temperatures. In each case the experimental data points are shown for comparison. The agreement is evident and demonstrates that the composition and solubility measurements are consistent with each other within the framework of the above analysis and that the assumption of ideal solid solutions is essentially correct.

The excellent agreement obtained may at first seem surprising in view of the large uncertainty that goes with the thermodynamic data and the approximations made in this work. However, the manner in which ΔS_{AlAs}^F was derived from the composition data makes it an adjustable parameter which compensates for the assumptions in the analysis so that the final liquidus curves are still reliable at least over the region in which composition measurements were made. With this limitation in mind, the liquidus and solidus lines covering the entire phase diagram can be generated from equations (6a) and (6b).

4. Applications to liquid phase epitaxy

4.1. *Comparison with GaAs solution growth*

Since the Al distribution coefficients are high, only small amounts of Al must be added to the melt to grow crystals throughout the entire composition range from pure GaAs to practically pure AlAs. For this reason the procedures developed for GaAs liquid phase epitaxy in a vertical system can be used with comparable success in the Ga–Al–As system. Due to the very high reactivity of molten Al, an oxide layer may form on the surface of the melt. Difficulties associated with the formation of this crust were minimized by working in a purified hydrogen atmosphere and by starting growth at the equilibrium temperature.

4.2. *Homogeneity of the layers*

A major drawback to the solution growth method arises from the fact that the melt becomes depleted in Al during the growth process so that the composition of the layers varies during the cooling cycle. Depletion of the melt is to some extent offset by the fact that the Al distribution coefficient increases with decreasing temperature as shown in figure 1. Figure 4 shows the combined effect of these two factors upon the composition of the mixed crystals while cooling under equilibrium conditions from 1000 to 800°C.

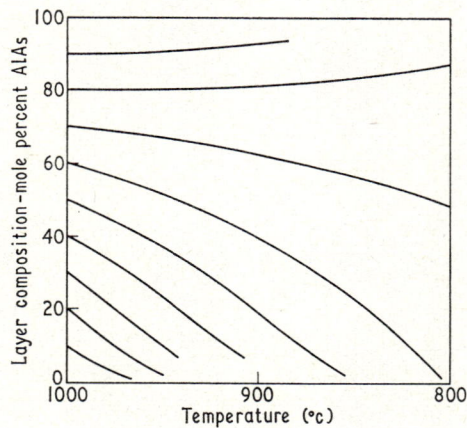

Figure 4. Variation in composition of layers grown during cooling from different initial compositions at 1000°C.

At rates faster than equilibrium the lines shown in figure 4 will tend to be lowered as the region surrounding the substrate becomes more depleted than the remainder of the melt. This indicates that it will be possible to grow homogeneous crystals by this method only in the region around 70 to 80% AlAs since the fall off with temperature is enhanced as the AlAs content decreases.

5. Conclusion

A theoretical analysis of the relationship between melt composition, crystal composition, and growth temperature for crystals grown from a ternary liquid has been presented. The experimental solidus and liquidus data are shown to be consistent with each other in the framework of this analysis and it is concluded that the calculations give a fairly accurate description of the Ga–Al–As system. Finally, although this discussion is in terms of one particular system, the general approach should prove useful when applied to the solution growth of other III–V compound mixed crystals.

References

ARTHUR, J. R., 1967, *J. Phys. Chem. Solids*, **28**, 2257.
HISKES, R., and TILLER, W. A., 1968, *Mater. Sci. Eng.*, **2**, 320.
KISCHIO, W., 1964, *Z. Anorg. Allg. Chemie*, **328**, 187.
PANISH, M. B., and SUMSKI, S., *J. Phys. Chem. Solids*, to be published.
PRIGOGINE, I., and DEFAY, R., 1965, *Chemical Thermodynamics* (London: Longmans), p. 257.
RUPPRECHT, H., 1966, *Proc. Int. Symp. on GaAs*, (London: The Institute of Physics and the Physical Society), p. 57.
STULL, D. R., and SINKE, G. C., 1956, *Thermodynamic Properties of the Elements*, (Washington, D.C.: American Chemical Society), p. 37, 44.
THURMOND, C. D., 1965, *J. Phys. Chem. Solids*, **26**, 785.
VIELAND, L. J., 1963, *Acta Met.*, **11**, 137.

Factors influencing the electrical and physical properties of high quality solution grown GaAs

R. SOLOMON

Fairchild Semiconductor Research and Development Laboratory,
Palo Alto, California 94304, U.S.A.

Abstract. Growth from gallium solutions is capable of yielding epitaxial GaAs of exceptionally high quality. In this paper, we present experimental results on the effect of several important variables on the electrical and physical properties of these layers.

Boat grown GaAs is used to saturate 7–9's pure Ga in a Pd-purified H_2 atmosphere. Growth temperatures in the range 600 to 900°C and cooling rates from 8 to 300°C h^{-1} have been investigated for several substrate orientations. Results are reported for both a horizontal tilt furnace and a vertical dip furnace.

It is found that oxygen introduces a shallow donor level in solution grown layers, and is probably the principal residual donor in these layers.

Some results are presented for tin doped layers. The mobility at 77°K is compared to theory and provided certain assumptions are made, the fit is reasonably good.

A discussion of the nucleation problem is presented for (100) oriented wafers.

By suitable adjustment of conditions, the technique has consistently yielded layers with net donor concentrations in the range 1×10^{12} to 5×10^{13} cm^{-3}, with LN mobilities in the range 100 000 to 130 000 cm^2 v^{-1} s^{-1}.

1. Introduction

Until recently, interest in liquid phase epitaxy (LPE) has centred around heavily doped layers for ohmic contacts and electroluminescent applications. Within the past year or two, however, there has been growing awareness that LPE is also capable of yielding high resistivity, high mobility layers, as has been demonstrated by Kang and Green (1967).

In this paper we will consider some of the factors which influence the electrical and physical properties of high resistivity layers. It will be shown that oxygen contamination is a serious obstacle to achieving high quality results.

2. Experimental technique

In these experiments both a tilt furnace, similar to that used by Nelson (1963), and a vertical dip furnace have been used. In the latter technique a wafer, held in a graphite chuck is dipped into a graphite crucible containing gallium and GaAs. For both furnaces gallium of 7–9's purity is saturated at temperature with boat grown GaAs having a background impurity concentration of $1-3 \times 10^{16}$ cm^{-3}. The furnaces are cooled at programmed rates which vary from 10 to 100°C h^{-1}. The ambient in both furnace arrangements is Pd-purified hydrogen.

The major advantage of the dip furnace is that it is much easier to grow crystals over small temperature intervals, and surfaces tend to be much smoother. On the other hand, the electrical parameters are somewhat inferior to the tilt furnace.

Although the temperature range 600 to 900°C has been explored, most of the work reported here has been performed at lower temperatures, i.e., 600–750°C, than have been previously reported. Our selection of this range stems from our interest in growing relatively thin layers, e.g., from 2 to 15 μ. Although the electrical properties can be relatively independent of temperature over a wide range, low temperature growth does impose special problems with surface smoothness and thickness uniformity.

Most of the effort has been placed on the (100) orientation; however, the (111) B and (110) orientations have also been studied.

3. The role of oxygen

In the early phases of this work, when moderately heavy n-type layers were being obtained, it became increasingly clear that oxygen contamination significantly influenced results. In order to study this problem, a tilt furnace was designed to reduce oxygen contamination to very low levels, so that oxygen could be re-introduced quantitatively. In the first technique, a small flow of wet H_2 was bled into the main Pd-purified H_2 stream for varying lengths of time. Table 1 shows the electrical properties for two different

Table 1. The dependence of electrical properties on the quantity of water vapour passed over the gallium

Moles H_2O	$N_d - N_a$	$\mu_H(77)$
0	$5 \cdot 0 \times 10^{12}$	120 000
$1 \cdot 7 \times 10^{-4}$	$4 \cdot 0 \times 10^{14}$	90 000
$1 \cdot 8 \times 10^{-3}$	$1 \cdot 2 \times 10^{15}$	64 000

quantities of water vapour. The number of moles of H_2O passed over the gallium was calculated from the measured flow rate and the partial pressure of water at room temperature. It is seen that a relationship appears to be established. However, for longer flows than shown in table 1, an annoying experimental problem developed—a crust was formed on the gallium which prevented it from flowing onto the wafer. This technique was therefore abandoned in favour of adding gallium sesquioxide to the melt.

After performing a calibration run which turned out in the low 10^{12} cm^{-3} range, a series of runs were made in which 5-9's pure Ga_2O_3 was added to the melt. Figure 1 shows the

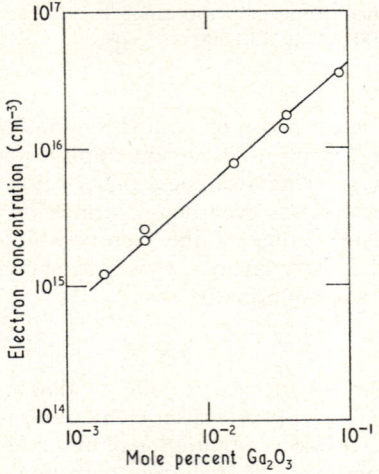

Figure 1. The electron concentration versus the concentration of Ga_2O_3 added to gallium. Wafer is (100) orientation.

dependence of the electron concentration in the layer on the mole percent of Ga_2O_3. The relationship is approximately linear over almost two decades in electron concentration. Assuming 100% doping efficiency for oxygen, a segregation coefficient of approximately 3×10^{-4} is calculated from the slope.

When a comparison is made between these data and the wet H_2 experiments, it is noted that the latter is a much less efficient technique. For a 10 g Ga charge, it requires about 10^{-5} equivalents of oxygen, in the form of Ga_2O_3, to create a donor density of 10^{15} cm^{-3}. From the wet H_2 experiments, however, 10^{-3} oxygen equivalents are needed to produce the same concentration. Hence only about 1% of the water vapour goes into solution. Since we have observed the formation of a thin skin of an apparently insoluble form of the oxide, it is probable that this impedes further penetration of water vapour.

4. Electrical evaluation

The van der Pauw method (1958) has been used for routine evaluation of all LPE layers at room temperature and 77°K. In addition, Hall-versus-temperature measurements were performed for several oxygen doped layers in the range 9×10^{13} to 1×10^{15} cm^{-3}. A magnetic field of 5·5 kG was used and the temperature interval from 6·5 to 600°K was covered.

A computer model, similar to that used by Fuller, et al. (1967), was developed to fit the experimental data. The model takes into account the temperature dependence of the band gap (Ehrenreich, 1960), thermal generation across the gap, and intervalley scattering. The degeneracy factors, and the temperature dependence of the impurity levels are adjustable parameters to be chosen on a best fit basis.

Although this work is still in its preliminary stages, several facts have already become apparent. First, oxygen introduces a very shallow donor level. A preliminary comparison with tin doped layers indicates that both levels are closely similar, with energy levels only several mev below the conduction band. Second, no other energy level was observed in the temperature range investigated. Although the Fermi level did not quite reach the centre of the gap at the highest temperature measured, the data suggest that if the deep oxygen level is present it must have a fairly low density. Thus it appears that oxygen behaves differently in LPE layers grown at low temperatures, than in boat grown material, for example. In the latter method, oxygen is usually added to suppress silicon doping, and at high concentrations it introduces a deep donor which may cause it to go semi-insulating (Woods and Ainslie 1963). The behaviour of oxygen in LPE layers, therefore, is quite remarkable.

There are several mechanisms by which oxygen might introduce shallow donors in LPE layers:

1. Oxygen could influence the incorporation of another impurity. The two most likely possibilities are a reaction with the graphite boat, and a wall reaction between a volatile oxide species and the quartz tube. Considering graphite first, the following reaction is possible.

$$Ga_2O_3 + 2C \rightleftharpoons Ga_2O + 2CO$$

Cochran and Foster (1962) have calculated the equilibrium constant versus temperature from thermodynamic data. At 700°C, the initial growth temperature for the series of runs shown in figure 1, they obtain a value for log K of approximately -10. Thus, the concentration of CO in gallium should be extremely small at this temperature, provided equilibrium is approached for this reaction. It is conceivable that a substantial loss of one of the reaction products could drive the reaction far to the right; however, we have not been able to observe any evidence of attack of the graphite boat, nor is there any evidence that gallium even wets graphite.

The reaction with silica can be summarized in the following set of reactions.

$$Ga_2O_3 + 4Ga \rightleftharpoons 3Ga_2O \text{ (v)}$$
$$Ga_2O \text{ (at quartz wall)} + H_2 \rightleftharpoons Ga \text{ (wall)} + H_2O$$
$$2Ga \text{ (wall)} + SiO_2 \rightleftharpoons SiO \text{ (v)} + Ga_2O$$
$$SiO \text{ (v)} + Ga \text{ (l)} \rightleftharpoons Si \text{ (l)} + Ga_2O$$

A detailed calculation of the overall reaction is difficult because of the cyclic character of the reactions, involving mass transport of several reactants between wall and boat. However, the rate-limiting step will clearly be the third reaction. Referring again to Cochran and Foster (1962), we find that log $K \simeq -20$ at 700°C. Hence it is unlikely that any significant quantity of SiO is formed at this temperature. In order to check this result, we have made runs with a small quantity of gallium in contact with the quartz wall and have not been able to find any deterioration of the electrical properties of the films. Finally, it should be pointed out that silicon is amphoteric in GaAs, and that the conductivity type is strongly temperature and concentration dependent. In particular, low temperatures, i.e. high Ga to As ratios, and high silicon concentrations should favour p-type conductivity. Similar considerations ought also to apply to carbon, although much less information is available for this element.

2. Oxygen could behave like a hydrogenic impurity. It is, after all, a group VI element with chemical properties similar to sulphur, and it is possible that in the presence of a large gallium excess it could be induced to substitute on arsenic sites.

3. Oxygen might introduce a complex, for example, an oxygen–vacancy or oxygen–oxygen pair. If the binding energy of the complex is relatively small, then it might be stable at relatively low growth temperatures, but unstable at higher growth temperatures—thus tending to reconcile some of the apparently conflicting evidence with respect to oxygen.

Although the first possibility discussed above seems unlikely on the basis of thermodynamic and other considerations, we cannot entirely exclude it, and we are currently undertaking mass spectroscopic analyses of the Ga and GaAs for a wide range of conditions. It is clear that the correct model must not only explain why oxygen introduces a shallow donor, but also why it does not appear to introduce a deep level in LPE layers. Unfortunately, very little is known about the nature of the deep oxygen level in GaAs. A possible model which might fit most of the observations is that the deep level arises when oxygen occupies gallium sites. Growth at very high Ga to As ratios could force oxygen to occupy arsenic sites, giving rise to shallow donor levels, while at the same time preventing the occupation of gallium sites.

5. Tin doped layers

When precautions are taken to reduce oxygen contamination to very low levels, the layers are usually of very high resistivity, nearly semi-insulating in the dark. Because of difficulties in evaluating these layers, it has proved useful in the study of a number of variables to backdope with a well-behaved n-type dopant. Tin has proved to be a very reproducible and easy to handle material. Figure 2 shows the dependence of the electron

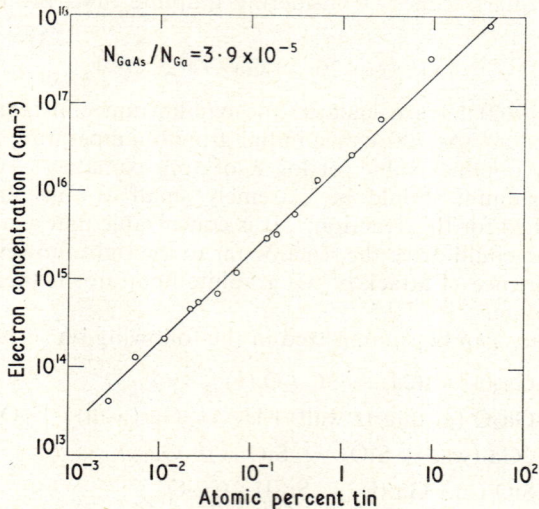

Figure 2. The electron concentration versus concentration of tin in gallium. Wafer is (100) orientation, initial growth temperature is 700°C.

concentration on the concentration of tin in solution. Each point is an average of 2 to 5 van der Pauw samples. The dependence is linear over more than four decades in electron concentration, with a segregation coefficient of $3 \cdot 9 \times 10^{-5}$. The segregation coefficient is approximately constant over the range 600 to 700°C. It is interesting to note that our value is more than three orders of magnitude smaller than a value of 0·08 obtained by Willardson and Allred (1966) for the stoichiometric composition at the melting point of GaAs.

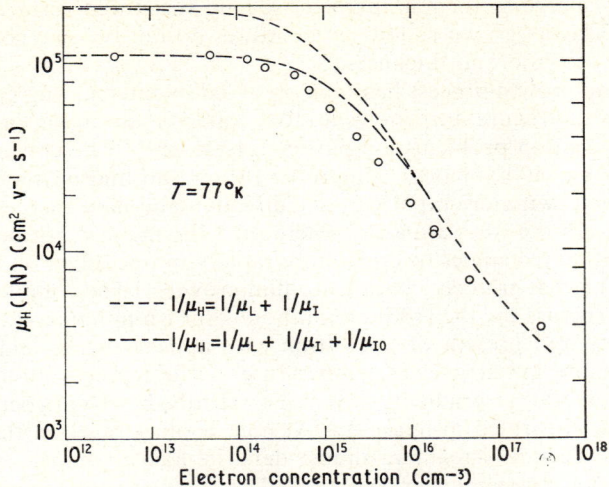

Figure 3. The dependence of mobility on electron concentration for tin doped layers. $T=77°K$.

The mobility as a function of the electron concentration at 77°K is shown in figure 3. The measured mobility saturates below 2×10^{14} cm^{-3} at a value, in this case of 110 000 cm^2 v^{-1} s^{-1}. The value of the saturation mobility can be varied by varying the experimental conditions.

The two curves are calculated by assuming a lattice mobility, calculated by Bolger *et al.* (1966), of 200 000 cm^2 v^{-1} s^{-1}. The impurity mobility is calculated from the Conwell–Weiskoff theory, assuming an effective mass of 0·064 (DeMers, 1965), and a static dielectric constant of 12·5 (Hambleton *et al.*, 1961). The dashed curve is calculated on the assumption that only tin donors are present, i.e.,

$$N_d + N_a = N_d \text{ (Sn)}$$

The lower curve (dash-dot curve) has been calculated on the more reasonable assumption that the smaller saturation value for the measured mobility is due to the presence of residual ionized impurities, i.e.,

$$N_d + N_a = N_d \text{ (Sn)} + N_0$$

where N_0 is the sum of all residual ionized impurites. A best fit has been obtained for a residual impurity concentration of 9×10^{14} cm^{-3}.

The value of N_0 obtained from figure 3 may be too high by a factor of three or more. Although the model of Bolger *et al.* (1966) provides a reasonably good fit to the experimental data, the fit could be improved either by choosing a smaller value for μ_L, or an appropriate modification of the C–W theory. Either approach would have the effect of reducing N_0. A critical evaluation of both theory and experiment has been initiated with the aim of improving the fit to the data.

The saturation mobility obtained from backdoping experiments has proved to be a useful tool for the evaluation of a number of variables. We regularly use it to keep a constant check on our system, and it was by this means that we noted a small but distinct dependence of background impurities on different gallium lots. A similar dependence also appears for different GaAs ingots; however, the data in this case are much less complete.

6. Nucleation problems

Liquid phase growth at low temperatures undoubtedly imposes difficult problems with thickness uniformity and surface smoothness. In solution growth there is a strong tendency to form a relatively small number of nucleation sites on the substrate surface. Because of the high mobility of molecules in solution, these islands grow at the expense of later forming sites, and gradually merge to form a characteristically structured surface. Under normal circumstances, LPE layers grown at low temperatures do not become continuous until the layer approaches 1 μ or more in thickness.

The nucleation and growth process has been observed by interrupting growth in very small temperature intervals. Figure 4(a) shows a (100) wafer for an initial growth temperature of 650 °C, where nucleation problems are particularly severe. The crystallites are oriented rectangles bounded by ⟨100⟩ planes. Figure 4(b) shows an intermediate stage of growth. The rectangles have grown more rapidly in one direction and have merged into rows. The darker regions between the rows are areas where the substrate has not been covered. The tendency for nucleating rectangles to grow more rapidly in one direction is, we believe, due to the presence of thermal gradients parallel to film growth. The completed layer showing the characteristic structure for the (100) orientation is shown in figure 4(c).

The initial density and growth of nucleation sites appears to depend on a number of variables, of which the initial growth temperature, surface preparation, rate of cooling, orientation, and temperature gradients have been identified. The dependence on cooling rate, for example, is illustrated in figure 5. At high cooling rates, in this case 100 °C h^{-1}, the crystallites have become irregular, almost dendrite-like in shape. These grow much more rapidly than the rectangular crystallites. The symmetry of the fingers suggest that twinning may occur and that growth may be similar to the dendritic growth discussed by Bennet and Longini (1959). The completed film exhibits a very rough surface consisting of flat plateaus scoured by deep troughs. Similar behaviour occurs on the (110) orientation; however, the (111) B orientation, because it is the slow-growing direction can tolerate substantially higher growth rates and still yield smooth surfaces.

Although it cannot be claimed that the complex conditions required for control over nucleation are fully mastered, growth at low temperature is capable of giving smooth, uniform layers. Figure 6 shows a layer grown on the (100) orientation in which conditions have been adjusted to give a high density of initial nucleating sites. The surface appears nearly mirror smooth to the eye, although very weak structure can be seen at 100× magnification.

7. Conclusions

It has been demonstrated that oxygen introduces a shallow donor level in GaAs grown from the liquid phase, and is probably the principal residual donor in these layers. When oxygen contamination is reduced to low levels, high-resistivity, high-mobility films can be grown at comparatively low temperatures. We have consistently been able to grow films in the range 1×10^{12} to 5×10^{13} cm^{-3}, with LN mobilities from 100 000 to 130 000 cm^2 v^{-1} s^{-1}.

Acknowledgment

I would like to thank my colleague, Dr. Megha Shyam, for his help with the Hall versus temperature measurements and for a number of valuable discussions.

Question on paper 2
J. P. Spratt (Philco-Ford Corp.)

At the highest mobilities reported ($\mu_H > 10^5$ cm^2 v^{-1} s^{-1}) simple Hall effect theory breaks down for $B \geqslant 10^3$ gauss ($\mu_H B \rightarrow 1$). Since your measurements were done at $5 \cdot 5 \times 10^3$ gauss, did you notice any effect of B on either σ or R_H?

R. Solomon

We have measured R_H over the range from about 1·0 to 17 kG, and to first order it is approximately constant. Closer examination of the data, however, indicates that it increases slightly with increasing field, peaks at about 6 kG, and then decreases. The total change is very small, of the order of two per cent. We have not examined σ as carefully, but my impression is that it, too, is approximately constant.

Despite this result I quite agree with you that the Hall measurement of high mobility layers needs more careful investigation. Our own investigation, in fact, was prompted by an anomaly that we have observed in high mobility vapour and liquid epi-layers. We have found that the apparent electron concentration decreases on going from 77° to 300°K. The decrease is in the range from 10 to 40%. We had presumed that the effect was probably due to the fact that the scattering coefficient, r, was a function of temperature.

References

BENNET, A. I., and LONGINI, R. L., 1959, *Phys. Rev.*, **116**, 53.
BOLGER, D. E., FRANKS, J., GORDON, J., and WHITAKER, J., 1966, *Proc. Int. Symp. on GaAs*, p. 16.
COCHRAN, C. N., and FOSTER, L. M., 1962, *J. Elect. Chem. Soc.*, **109**, 144.
DE MERS, W. M., 1965, *Tech. Rep. No. HP*-15, Gordon McKay Lab., Harvard Univ.
EHRENREICH, H., 1960, *Phys. Rev.*, **120**, 1951.
FULLER, C. S., WOLFSTERN, K. B., and ALBISON, H. W., 1967, *J. Appl. Phys.*, **38**, 2873.
HAMBLETON, K. G., HILSUM, C., and HOLEMAN, B. R., 1961, *Proc. Phys. Soc.*, **77**, 1197.
KANG, C. S., and GREEN, P. E., 1967, *Appl. Phys. Letters*, **11**, 171.
NELSON, H., 1963, *RCA Rev.*, **24**, 603.
VAN DER PAUW, L. J., 1958, *Phillips Res. Repts.*, **13**, No. 1.
WILLARDSON, R. K., and ALLRED, W. P., 1966, *Proc. Int. Symp. on GaAs*, p. 35.
WOODS, J. R., and AINSLIE, N. G., 1963, *J. Appl. Phys.*, **34**, 1469.

Tin and tellurium doping characteristics in gallium arsenide epitaxial layers grown from Ga solution

C. S. KANG and P. E. GREENE

Hewlett-Packard Laboratories, Palo Alto, California 94304, U.S.A.

Abstract. A marked impurity gradient was observed in the epitaxial layers grown by cooling GaAs saturated Ga solution from 850°C to 650°C. Doping of the solution by Sn has resulted in homogeneous impurity distribution in 1×10^{15} cm^{-3} carrier concentration range. In order to study the doping behaviour of Sn at constant temperatures the doping experiments were done with a steady-state growth system in which crystals were grown by maintaining a temperature gradient across the source GaAs and the substrate. The results of the experiments show a significant dependence of Sn doping behaviour on growth temperature. Similar experiments were done with Te as a dopant. Te also showed a marked temperature dependence in its doping behaviour in GaAs. The observed effects are attributed to the doping dependence of GaAs by these elements on the As activity in Ga solution.

1. Introduction

Recent experimental findings on the high quality of GaAs and p–n junctions grown by solution growth techniques has stirred a strong interest in this technique (Kang, *et al.* 1967; Nelson, 1967), as evidenced by the presentation of a number of papers on this topic in this symposium. In the fabrication of devices, it is important that one has a good control of doping to obtain desired electrical or optical characteristics. Although data have been published on the distribution coefficients of various important electrically active impurities in GaAs, these are mostly for the growth of GaAs from the stoichiometric melt at the melting point. Little information has been available on the doping characteristics of various important doping elements in GaAs when the crystals are grown from Ga-rich solution. We have initiated a study of doping behaviour of some of the important dopants in the solution growth of GaAs.

Some of the observations on the impurity distributions in the crystals grown by cooling saturated solutions and the temperature dependence of the doping behaviour of Sn and Te are reported in this paper.

2. Experiment

Two types of solution growth systems were used in the experiment. One was a horizontal tilt tube system in which epitaxial growth is achieved by cooling a GaAs saturated Ga solution. Details of this system have been reported earlier (Kang *et al.* 1967). In the other system, as shown in figure 1, a crystal is grown at a constant temperature. After saturating the Ga solution with a GaAs source at a desired temperature, the GaAs substrate is lowered and brought in contact with the solution. Subsequently, a thermal gradient is maintained by slowly increasing the source temperature a few degrees above that of the substrate. Due to the increased solubility of GaAs in Ga at higher temperatures, a concentration gradient is established across the Ga solution and this gradient results in a transport of As from the source to the substrate. Epitaxial layers can be grown at a constant temperature in this manner. The principle is the same as in the case of the travelling solvent method (Mlavsky *et al.* 1963), except that far greater amounts of Ga solution (30 g) are used in the present case.

Desired amounts of Sn and Te were added to the Ga solution for the doping studies.

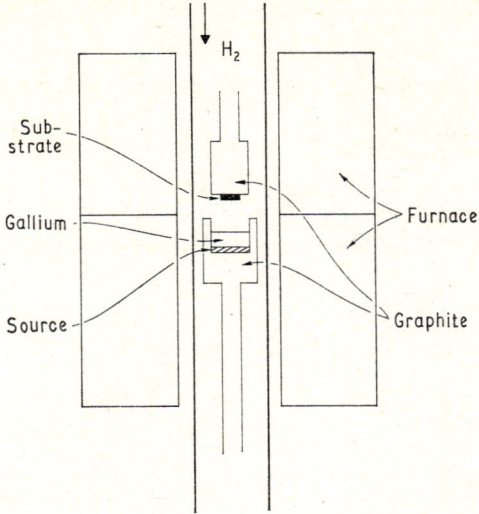

Figure 1. Schematic drawing of vertical steady-state system.

The epitaxial layers were evaluated by Hall effect measurements and the differential capacitance measurements on Schottky barriers fabricated by evaporating gold dots on GaAs.

3. Results

When crystals are grown in the tilt tube system by cooling the solution without any deliberate doping, a marked impurity concentration gradient along the growth direction of the epitaxial layer has resulted. The crystal is grown between 850 and 650°C in this system. In a typical $N_D - N_A$ profile as revealed by the differential capacitance measurements on Schottky barriers, $N_D - N_A$ increased from low 10^{13} cm^{-3} near the substrate to high 10^{14} cm^{-3} at the surface of the epitaxial layer. The curve labelled as TE180 in figure 2 is the profile for the undoped case. This impurity profile is believed to be the result of the changing doping behaviour of the dominant residual impurity as the temperature is changed. The growth velocity was in the order of 10^{-6} cm sec^{-1} and was too slow to cause any significant impurity inhomogeneity by impurity pile up at the growth front according to our calculations.

When the solution was doped with Sn, the profile changed very markedly. With sufficient doping of Sn, the profile became practically flat. This occurred when $N_D - N_A$ was in the 1×10^{15} cm^{-3} range. Above approximately 2×10^{15} cm^{-3} range, the direction of the profile went to the opposite of that exhibited by the undoped case. These are shown in figure 2. Various Sn doping levels were tried in the tilt tube system. The results are shown in figure 3. There were significant dependences of doping level on the substrate orientations as shown in figure 3.

In order to study the doping behaviour at a fixed temperature, Sn doping experiments were repeated with the vertical steady-state system described in the previous section. The distance between the substrate and the source was 0·5 cm. The temperature difference between the source and the substrate was approximately 2°C. The carrier concentration of the epitaxial layers grown at different temperatures changed significantly as shown in figure 4. The slope gives the activation energy of approximately 0·8 ev. The ratio of the carrier concentration to the Sn concentration in the solution changes from $3·6 \times 10^{-4}$ to $0·7 \times 10^{-4}$ in the region of investigation. The mole fractions of Sn in the Ga solution were $2·4 \times 10^{-2}$ and $1·1 \times 10^{-2}$.

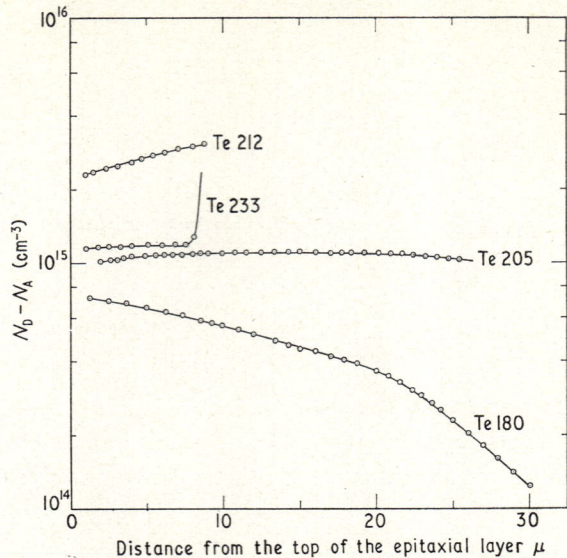

Figure 2. $N_D - N_A$ profile as revealed by differential capacitance measurements on Au–GaAs Schottky barriers.

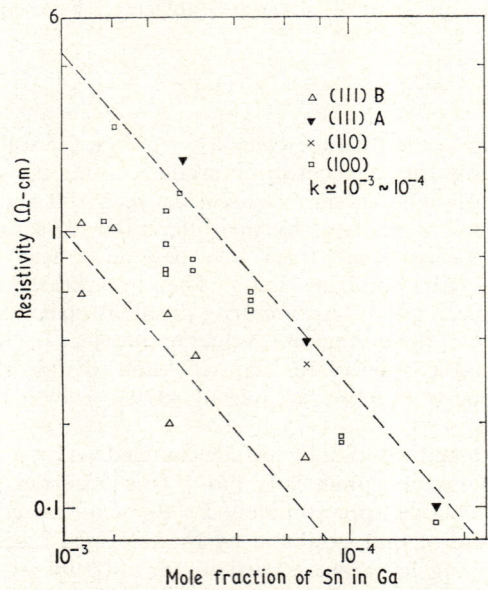

Figure 3. Resistivity of epitaxial layers vs. mole fraction of Sn in Ga.

The doping behaviour of Te was also studied with the steady-state system under growth conditions similar to those of the Sn case. The result is plotted in figure 4. One also notes a marked change in the carrier concentration with different growth temperatures. The ratio of the carrier concentration to the concentration of Te in the solution ranges from 1·1 to 5·9 as the temperature is changed from 850 to 700°C. The activation energy calculated from the slope of the semilog plot is approximately 1·0 ev. The mole fraction of Te in Ga solution was $3·5 \times 10^{-6}$.

It is speculated that the observed doping behaviours of Sn and Te in GaAs under the

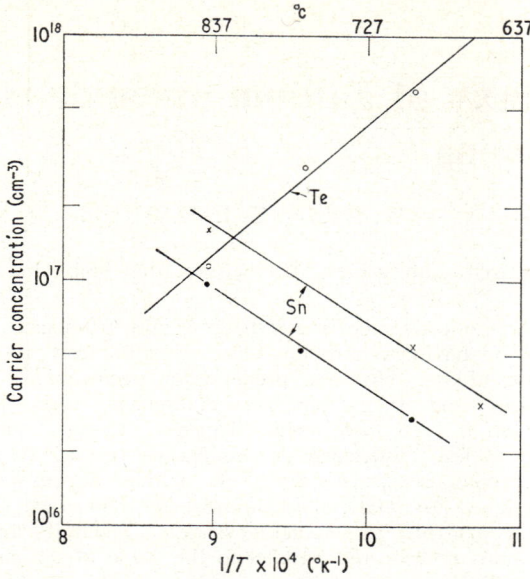

Figure 4. Temperature dependence of Sn and Te doping.

solution growth condition is the result of the strong dependence of the doping on the arsenic activity in the Ga solution. Rupprecht *et al.* (1966) have grown a p–n junction by growing GaAs out of Si doped Ga solution. The region grown at higher temperature became n-type and the region grown at lower temperature became p-type. They attribute this to the dependence of amphoteric doping behaviour of Si in GaAs on the As activity in the solution. Sn, being a group IV element as Si, can be expected to have similar doping characteristics to Si in GaAs when it is grown out of Ga solution.

The concentration of As in Ga changes one order of magnitude in the temperature range of the experiment. If Sn is an amphoteric impurity like Si in GaAs, the degree of preference of a Ga site over an As site by Sn can depend very much on the As activity in the growth environment. With a higher As activity at the higher temperature, the incorporation of Sn in an As site will be more difficult than at lower temperature where As activity is lower. This will produce the doping behaviour of Sn as observed. A similar argument can be presented for the doping behaviour of Te in GaAs grown out of Ga solution. Te becomes a donor in GaAs by substituting for As. The incorporation of Te in As site will be inhibited more at higher temperature where As concentration in the solution is higher, thus resulting in the doping dependence on the temperature as observed in this experiment.

When one compares the doping behaviour of Te to that of unknown residual impurity one notes the similarity in their temperature dependence. This suggests that perhaps the unknown residual impurity is a donor substituting As in GaAs. Group VI elements are the most likely to do this. Mass spectrographic analysis of the epitaxial layers showed the existence of a trace of sulphur but no other group VI elements such as Se and Te. The graphite used for the crucible was found to have approximately 0·2 p.p.m. of sulphur and it is believed that this was the main source of contamination.

Further study of doping characteristics of group II, IV, and VI elements in GaAs grown out of Ga solution will be conducted and the results will be reported.

References
KANG, C. S., and GREENE, P. E., 1967, *Appl. Phys. Let.*, **11**, 171.
MLAVSKY, A. I., and WEINSTEIN, M., 1963, *J. Appl. Phys.*, **34**, 2885.
NELSON, H., 1967, *Proc. IEEE*, **55**, 1415.
RUPPRECHT, H., WOODALL, J. M., KONNERTH, K., and PETTIT, D. G., 1966, *Appl. Phys. Let.*, **9**, 221.

Solution epitaxy of gallium arsenide with controlled doping

J. KINOSHITA, W. W. STEIN, G. F. DAY, and J. B. MOONEY

Central Research Laboratories, Varian Associates, Palo Alto, California, U.S.A.

Abstract. Gallium arsenide is grown epitaxially from gallium-tin solutions by the Nelson method having controlled donor densities from 10^{15} to 10^{19} cm^{-3}. Three general classes of layers have been prepared for specific microwave device applications. These devices and their material requirements are: 6-volt paramp diodes ($n = 2 \times 10^{17}$ cm^{-3}, $l = 2$ μm), high-frequency Gunn oscillators ($n = 1–3 \times 10^{15}$, $l < 20$ μm), and low-frequency Gunn oscillators ($n = 0.5–2 \times 10^{15}$, $l \simeq 100$ μm). The growth parameters used to produce layers for these devices are presented, and appropriate techniques for material evaluation are described. The reproducibility of doping density and thickness are demonstrated. The uniformity of net doping density in the directions normal and parallel to the plane of the substrate is controlled to better than ±20%.

1. Introduction

Solution epitaxy of GaAs was first reported by Nelson (1963) for light emitters, and more recently by Kang and Greene (1967) in the doping range useful for the Gunn effect. We have developed processes for the preparation of material for microwave devices through the control of doping density and profile by Sn addition to the melt.

Figure 1 summarizes the results of a study of the doping level of n-type GaAs grown from various Ga–Sn solutions in a graphite boat. The curves shown are for melt contact temperatures of 750°C and 650°C. The epitaxial layers were grown for 3 minutes at a cooling rate of about 8°C per minute. The net doping densities of the layers were determined at room temperature from measurements of Hall coefficient, Schottky barrier depletion capacitance, and infra-red reflectance.

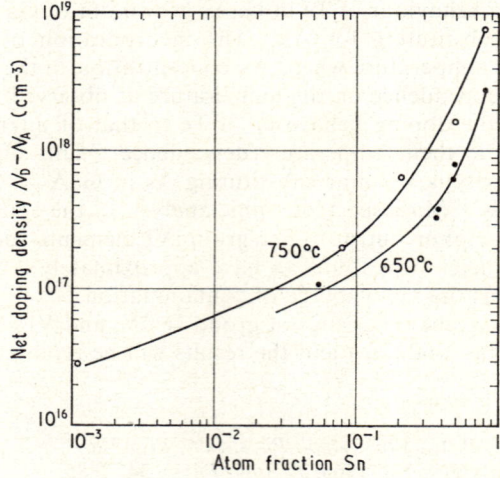

Figure 1. Net doping density vs. atom fraction of Sn.

2. Paramp diode material

The data presented in figure 1 were obtained during a study of the preparation of ohmic contacts by solution epitaxy. When it was clear that precise control of doping density was possible, a procedure for the preparation of GaAs for 6-v paramp diodes was developed. The first column of table 1 gives typical growth parameters. Table 2 lists the thickness and net doping density of 18 consecutive layers which were grown to evaluate the process and supply material for device development. Runs No. 21 and No. 32 were rejected due to the failure of the solution to contact the seed properly.

Table 1. Typical epitaxial layers and growth parameters

Application	Paramp diodes	10 GHz Gunn	1 GHz Gunn
Net doping $N_D - N_A$	10^{17} cm^{-3}	10^{15} cm^{-3}	10^{15} cm^{-3}
Thickness	3 μm	15 μm	100 μm
Melt composition	7·6 g Ga	6·0 g Ga	20 g Ga
	2·4 g GaAs	0·5 g GaAs	2 g GaAs
	2·4 g Sn		2 mg Sn
Starting temperature	720°C	770°C	850°C
Final temperature	630°C	600°C	750°C
Cool rate °C min^{-1}	9	7	0·3
Growth rate μm min^{-1}	0·25	1·0	0·5

Table 2. Reproducibility of net doping density at 10^{17} carriers cm^{-3}

Run No.	Thickness μm	$(N_D - N_A) \times 10^{-17}$, cm^{-3}
15	3·4	2·88
16	3·2	1·21
17	3·2	2·08
18	2·9	0·87
19	2·0	2·08
20	~1·2	1·84
22	1·6	2·21
23	2	1·96
24	1·8	3·17
25	1·5	1·84
26	1·5	3·56
27	2·5	1·17
28	3·0	1·36
29	2·8	2·40
30	1·5	2·33
31	3·0	2·88
33	2·5	2·46
34	2·7	2·79
Average	2·35	2·16
Standard deviation	0·69	0·70

The standard deviations shown at the bottom of table 2 in both net doping density and thickness demonstrate the control that is possible with solution epitaxy. Such reproducible epitaxial layers are required for high-yield production of devices.

3. High-efficiency Gunn effect material

The same general procedure was used but with parameters as shown in the second column of table 1 to produce more lightly doped material for the preparation of cw Gunn effect oscillators which operate in the X, K_u, and K-bands. Two profiles of net doping density are shown in figure 2. These data were taken by a series of Schottky barrier depletion capacitance measurements each of which can determine the net doping density at this level to a depth of about 10 μm. After each profile measurement the layer was etched to the

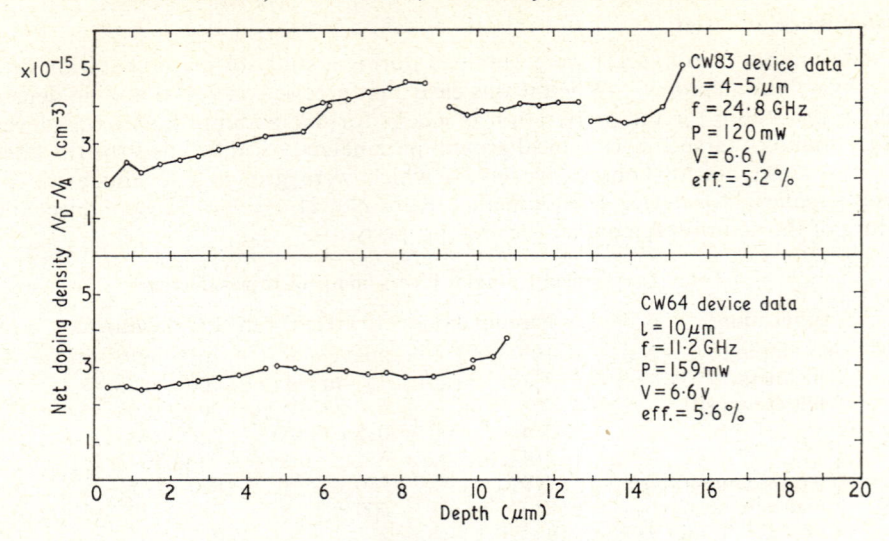

Figure 2. Doping density and device data.

next depth, diode contacts were applied by evaporation, and another depletion capacitance measurement of the profile made. The magnitudes of the errors involved in this analysis are indicated by the displacement between successive tests where the data overlap.

Figure 2 includes the microwave characteristics of oscillators which were prepared from these particular wafers. In the case of wafer CW83 the epitaxial layer was thinned to about 5 μm with $5H_2SO_4 : 1H_2O_2 : 1H_2O$ etch so that the level portion of the profile could be used for device fabrication. The resulting power and efficiency show that thorough characterization and careful selection of material leads to superior devices.

This lightly doped GaAs was further evaluated by the van der Pauw method for measuring resistivity and mobility. Layers were first deposited on Cr-doped semi-insulating GaAs substrates. Schottky barrier diodes were then evaporated onto the epitaxial surface and depletion capacitance measurements made. The profiles are shown in figure 3 for three

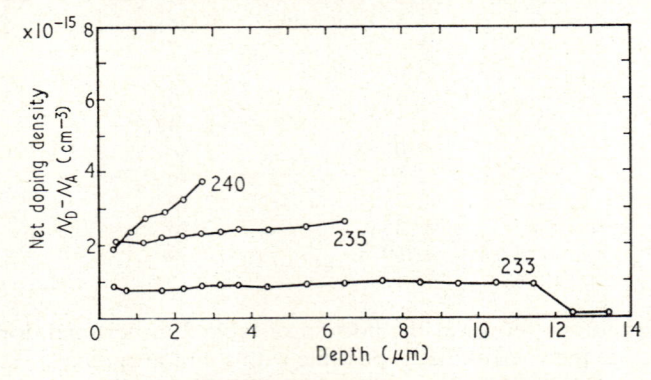

Figure 3. Depletion capacitance data from 10 GHz Gunn effect GaAs.

samples. The Schottky barrier diodes were removed, and van der Pauw samples were taken from the area of the wafer on which the depletion capacitance measurements were made. In the case of sample 235, there was exact spatial correspondence between the two measurements. The van der Pauw data, which were taken at 1000 G, are shown in table 3 for 300°K and 77°K.

Although the van der Pauw measurements give reliable averaged data on resistivity and mobility, we rely heavily upon the depletion capacitance measurements of the deposits on

Table 3. Comparison of depletion capacitance and van der Pauw data

No.	300°K			77°K		
	ρ	$n/10^{15}$	μ	ρ	$n/10^{15}$	μ
233	0·89	0·88	8030	0·13	0·78	60 500
235	0·34	2·35	7820	0·078	2·1	38 100
240	0·24	3·71	7040	0·077	3·01	26 000

ρ in ohm-cm; n in cm^{-3}; μ in cm^2 v^{-1} s^{-1}.

conducting substrates to determine suitability of a deposit for device preparation. A doping profile that is nearly flat or decreases slightly in the direction away from the substrate seems to be essential for Gunn oscillators in which the substrate is the anode contact; otherwise the devices are apt to be destroyed by avalanche breakdown. The cooling rate used in the present process results in such a profile.

4. Low-frequency Gunn effect material

The growth parameters for GaAs which is used for L-band, pulsed Gunn oscillators are listed in the third column of table 1. The depletion capacitance profiles shown in figure 4 were obtained by a series of controlled etches between measurements. The increasing net doping density in the growth direction is always present under these growth conditions. This doping gradient is opposite to that described for the thin layers discussed in section 3. The procedure is designed to produce an epitaxial layer of about 140 μm, but the surface is lapped and etched to 100 μm before analysis. This removes surface irregularities and the GaAs which was deposited after the boat was removed from the hot zone. The cooling rate was chosen to produce a 100 μm layer in a day without the growth imperfections that result from constitutional supercooling, as discussed by Hurle, Joner, and Mullin (1961) and Andre and LeDuc (1968).

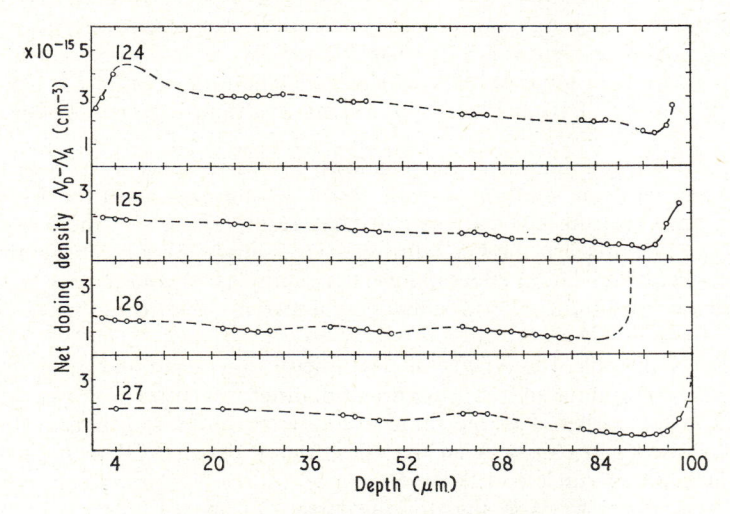

Figure 4. Doping density profiles.

The in-plane doping uniformity of an epitaxial layer is determined by measuring the capacitance at 0, 10, and 20 v bias of an array of 0·5 mm diameter Schottky barrier diodes on 1 mm centres over the entire surface of the wafer, or a representative sample of these diodes. The net doping for each diode and bias is calculated assuming homogeneity within the depleted region. The average net doping and its standard deviation are calculated for each wafer. A summary of these data at 20 v bias for 24 runs from a consecutive

Table 4. Layer homogeneity measured by depletion capacitance

Run No.	No. of diodes	$N_d/20$ v b	Standard deviation
103	80	4.38×10^{15}	0.389
103	10	4.42	0.658
104	21	1.75	0.256
105	10	0.66	0.081
106	7	4.42	0.275
107	16	2.32	0.123
108	11	0.59	0.077
109	16	1.55	0.415
110	8	5.02	0.371
111	18	2.49	0.166
114	14	0.72	0.418
116	17	2.59	0.159
117	15	0.42	0.142
118	14	3.96	1.13
119	16	1.46	0.130
120	12	3.16	0.304
121	7	0.89	0.200
122	11	21.66	3.97
123	10	9.32	1.32
124	12	3.58	0.515
125	14	2.61	0.080
126	23	1.90	0.269
127	14	2.08	0.153
129	16	2.68	0.200
130	20	2.43	0.329

series of 28 runs is shown in table 4. These measurements were all taken 100 μm from the substrate. Of the four missing runs, one did not deposit due to an experimental error, and three had doping densities below the 10^{14} cm^{-3} level.

There are two entries in table 4 for layer 103. The 80 diodes represent nearly the entire surface, while the group of 10 diodes is a representative sample similar to the others in the table.

Some of the layers listed in table 4 were found to have insulating layers near the substrate. This is not a problem if the solution is fresh, but if a solution is used for a second or third epitaxial layer, an insulating layer at the substrate results even with increased Sn content. Consequently, we now discard the Ga solution after a single layer has been grown. In the case of the high-frequency Gunn effect material, as many as 20 consecutive layers have been deposited from one solution. Tin is added as needed to keep the doping density in the desired range of $1-4 \times 10^{15}$ cm^{-3}.

The series 122–7 in table 4 is typical of the control that is possible with this procedure. Run No. 122 was contaminated in an unknown manner. The series was continued with a fresh Ga solution for each run in the same pyrolytic graphite boat and no other cleaning procedures were used. It can be seen that the doping density fell with each consecutive deposition. In fact, for run No. 127 the Sn concentration was increased from 5×10^{-3} to 7.5×10^{-3} wt.% in order to raise the doping density at the surface above 2×10^{15} cm^{-3}. Runs No. 129 and No. 130 had 1×10^{-2} wt.% Sn added for the same reason.

5. Conclusions

Processes have been developed for the epitaxial deposition of GaAs for three different microwave devices with the net doping density precisely controlled by the addition of Sn to the Ga solution. The reproducibility and uniformity of net doping density both normal and parallel to the substrate have been demonstrated. The significant remaining problem is a doping gradient in the direction normal to the substrate in 100 μm layers.

Acknowledgments

Helpful discussions with members of the Solid-State Electronics Laboratory at Stanford University are gratefully acknowledged, as well as the assistance of many of our associates at Varian.

This work was supported in part by the Air Force Avionics Laboratory under contract AF33615–67–C–1980.

References

ANDRE, E., and LEDUC, J. M., 1968, *Materials Research Bulletin*, **3**, 1–6.
HURLE, D. T. J., JONER, O., and MULLIN, J. B., 1961, *Solid State Electronics*, **3**, 317–20.
KANG, C. S., and GREENE, P. E., 1967, *Appl. Phys. Letters*, **11**, 171–3.
NELSON, H., 1963, *RCA Review*, **24**, 603–15.

Electrical properties of solution grown GaAs layers

J. C. CARBALLÈS, D. DIGUET, and J. LEBAILLY

Laboratoire de Développement Physico-Chimique, RTC La Radiotechnique Compelec, Caen, France

Abstract. The electrical properties of GaAs layers grown from Ga solution have been studied and compared to those of materials obtained by vapour phase epitaxy and grown by the horizontal Bridgman method. Two types of analyses have been used: Hall effect measurements and trapping measurements by capacitive methods. For undoped Ga-solution layers, Hall coefficient curves together with values of the mobility reveal residual acceptor concentrations $\sim 3 \times 10^{14}$ cm^{-3}, which are one order of magnitude lower than those of bulk crystals obtained from the melt. Studies of thermally stimulated variations of capacitance show that the good electronic properties of the layers are due to the relative absence of deep lattice defects and oxygen centres. More precisely, the deep centre concentrations are lower than 10^{14} cm^{-3}. Vapour phase epitaxy layers are shown to be rather comparable. The compensation mechanism of the layers is discussed.

1. Introduction

The electrical properties of undoped GaAs layers ($\rho \sim 0\cdot 1$–1 Ω cm) grown from Ga solution have been studied and compared to those of materials obtained by vapour phase epitaxy and grown by the horizontal Bridgman method.

Two types of analyses have been used: Hall effect measurements and trapping measurements by capacitive methods. Analyses of thermally stimulated variations of capacitance have been utilized to investigate the deep centre concentrations and to develop a compensation model. For the medium conductivity layers, it has been shown valid to consider a two-level compensation model. Under these conditions Hall coefficient variations and mobility values lead to reliable results, which have been confirmed by photoinduced detrapping measurements.

The layers were grown in a simple quartz crucible which could be rotated around its axis in a horizontal furnace. Bridgman grown GaAs was used as a source. The working temperature was fixed at 830°C, the growth being carried out in an atmosphere of hydrogen. When the temperature was stabilized, the substrate was immersed by rotating the crucible, and then the solution was cooled at approximately 1°C min^{-1}. In all the experiments the gallium solution contained about 5% indium, in order to avoid arsenic associations and to reduce the viscosity of the solution as discussed by Le Duc (André et al. 1968). The layers obtained under these conditions presented smooth surfaces in both [111] and [100] orientations and had thicknesses of about 50 μ. Studies of the electroluminescent spectra of the material have shown that the indium concentration in the crystals is negligible. Electron microprobe analysis has confirmed that it is lower than 1%.

Crystals which were studied for comparison were obtained in the following manner. Bulk material was grown by the horizontal Bridgman (H.B.) method in a gradient freeze apparatus (Fertin et al. 1966). Oxygen, or gallium oxide, or boric oxide was introduced into the vessel in order to reduce silicon contamination, and a quartz diaphragm was used to retard vapour diffusion between the hot and the cold parts of the vessel. In some cases, bulk crystals were annealed for 48 hours at 750° or 800°C (Woodall and Woods, 1966). Vapour phase epitaxy has been performed in a classical quartz apparatus, using the method of gallium transport by arsenic trichloride. Most of the layers have been doped with sulphur and grown on substrates orientated in the (100) orientation.

2. T.S.V.C. analyses

The principles of the capacitive measurements have been discussed previously by Carballès and Lebailly (1968). When the emptying of initially occupied centres is imposed by a thermal effect, it is simpler to observe the variations of the ionized centre concentrations by means of the capacitance value of a junction, rather than the variations of carrier concentrations.

To obtain information about the nature of the trapping level, defined by the values of the concentration, the energy depth and the capture cross section of the centres, it is necessary to change the system from an equilibrium to a non-equilibrium state and to observe transient phenomena. It is possible to vary several parameters: temperature, free carrier concentration (by illuminating), initial or final conditions (all traps filled or empty, initially or finally).

In this section, we shall only deal with thermally stimulated variations of capacitance (T.S.V.C.), which consists of a simple extension of the well-known method of analysis of thermally stimulated conductivity (T.S.C.).

To observe thermally stimulated variations of capacitance (T.S.V.C.), we first cool the sample to liquid-nitrogen temperature and saturate the traps by ultraviolet illumination. We then apply a reverse bias to the diode and heat the sample at a fixed heating rate.

The junctions are surface barriers made by metallizing with gold. The thickness of the metal is of the order of 300 Å so that it is partially transparent to ultraviolet and visible light.

Figure 1 shows typical T.S.V.C. curves of the different types of material studied. These curves exhibit several steps, due to different kinds of deep centres. The concentration, energy depth, and capture cross-section values of these centres have been previously studied in bulk material obtained by the horizontal Bridgman method (Carballès and Lebailly, 1968). These results are summarized in table 1.

The as-grown H.B. crystals contain oxygen centres and lattice defect centres in concentrations of about 3×10^{15} cm^{-3}. After equilibrium heat treatment, the concentration of deep centres attributed to lattice defects becomes one order of magnitude lower. However, the concentration of the acceptor centres related to oxygen remains about the same. In the

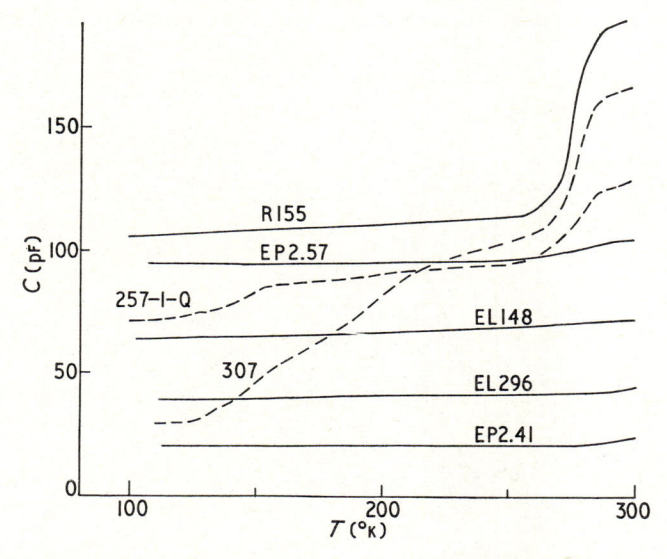

Figure 1. Thermally stimulated variations of capacitance, after saturation of the electron traps. Samples Nos. EL 148 and EL 296 have been grown in gallium solution; samples Nos. EP 2.57 and EP 2.41 have been grown by vapour phase epitaxy; sample R 155 has been obtained by the horizontal Bridgman method and annealed at 750°c for 48 hours. Samples Nos. 257.1.Q and 307, obtained by the H.B. method and not annealed, have been added in order to show the position of the drops ascribed to emptying of defects.

Table 1. Orders of magnitude of the energy depth and capture cross-section values, previously studied in bulk H.B. crystals

Temperature (°K)	Energy depth (ev)	Capture cross-section (cm²)	Nature
120–160	0·11	10^{-23}	lattice defect
120–160	~0·3	~10^{-14}	lattice defect
210–240	~0·4	~10^{-16}	lattice defect
260–300	0·68	10^{-15}	oxygen related

solution-grown layers it appears that the deep centre concentrations are one order of magnitude lower than the electron concentrations. Layers obtained by gaseous epitaxy are comparable to those obtained with a Ga solution. However, they sometimes contain oxygen as in the case of layer EP 2.41 which was grown without purification of the hydrogen. The third column of table 4 gives typical values of the total deep centre concentrations N_T for the different types of material. From this it may be concluded that the deep centre concentration is negligible, compared to the shallow impurity concentration and a two-level compensation model can be used for interpreting the Hall characteristics of the solution grown layers, in the next section.

3. Hall effect analyses

Conductivity and Hall effect measurements have been performed from 4° to 300°K on cloverleaf samples, carefully selected for their symmetry. The method described by Van der Pauw was used, with a magnetic field of 4000 gauss.

3.1. Temperature dependence of the Hall coefficient

The variations of the Hall coefficient R_H with the reciprocal of the temperature are shown in figure 2, for different solution grown layers. A typical heat-treated crystal, obtained by the horizontal Bridgman method, has been added for comparison.

As the temperature is reduced below 50°K, deionization of the shallow donors occurs and the increase of R_H may be used for determination of the donor ionization energy and donor

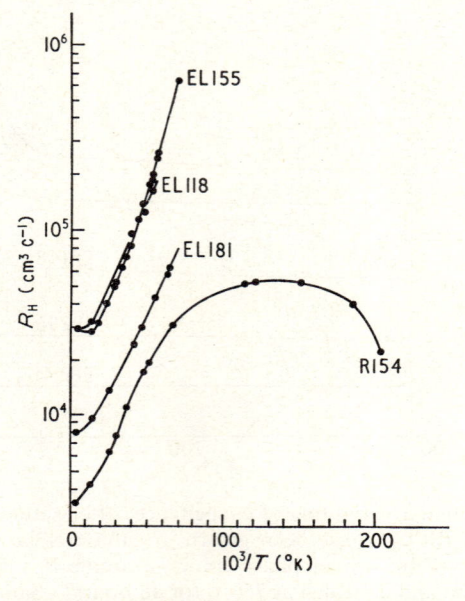

Figure 2. Hall coefficient curves, at low temperature, of solution grown layers Nos. EL 155, EL 118, and EL 181. The annealed H.B. crystal No. R 154 has been added for comparison.

and acceptor concentrations. Using a two-level compensation model, the temperature dependence of the carrier concentration is given by the well-known equation:

$$\frac{n(N_A+n)}{(N_D-N_A-n)\,N_c} = \frac{1}{g}\exp\left(-\frac{E_D}{kT}\right) \qquad (1)$$

where N_c is the effective density of states of the conduction band, N_D and N_A are the donor and acceptor concentrations, E_D the ionization energy of the donors and g is the degeneracy of the donor level.

We have taken $n = r/R_{He}$ with $r = 1$. This approximation is quite valid near 50°K, when $\mu B \gg 1$, the dominant scattering mechanism being the polar optical mode scattering. But at temperatures close to 20°K it no longer holds because of the scattering by the ionized impurities which would give $r \simeq 1.9$. Nevertheless for simplicity, we have used in the calculations $r = 1$.

Taking $n_{300°K}$ as $(N_D - N_A)$ and estimating the total ionized impurity concentration from the mobility values, iteration of the formula (1) leads to reasonable results, except for the value of g which is of the order of 10.

We have taken into account the excited states of the impurity atoms, as was done by Edolls (1966). Equation (1) is modified to:

$$\frac{n(N_A+n)}{(N_D-N_A-n)} = \frac{N_c}{\sum_{r=1}^{q} g_r \exp[(E_D-E_r)/kT]} \qquad (2)$$

which may be written in the form:

$$\frac{n(N_A+n)(1+F)}{(N_D-N_A-n)\,N_c} = \frac{1}{g_1}\exp\left(-\frac{E_D}{kT}\right) \qquad (3)$$

where

$$F = \sum_{r=2}^{p} \frac{g_r}{g_1}\exp\left(-\frac{E_r}{kT}\right) \qquad (4)$$

g_r is the degeneracy factor and E_r the ionization energy of the rth excited state; g_1 is the degeneracy factor of the ground state. Using the hydrogen model, the expression for $(1+F)$ then becomes:

$$1+F = \sum_{r=1}^{p} r^2 \exp\left(\frac{(1-r^2)\,E_D}{kT}\right) \qquad (5)$$

Results computed by taking only the first two terms, are shown in table 2. Values obtained for the sample R 154 are not very significant because of the occurrence of impurity band conduction. The ionization energy depends upon the impurity concentration. It is somewhat lower than the value estimated from the hydrogen model (Hilsum and Rose-Innes, 1961).

Table 2. Results of the analysis of the Hall coefficient curves represented in figure 2.

Sample number	E_D (mev)	g_1	N_D (cm^{-3})	N_A (cm^{-3})	(N_D+N_A) (cm^{-3}) deduced from $\mu_{77°K}$
EL 155	4·5	1·9	8×10^{14}	$5·5 \times 10^{14}$	$1·3 \times 10^{15}$
EL 118	~4·6		~$9·7 \times 10^{14}$	~7×10^{14}	$1·7 \times 10^{15}$
EL 181	2·6	2·8	$1·45 \times 10^{15}$	6×10^{14}	$1·9 \times 10^{15}$
R 154	4·9	1·8	$3·1 \times 10^{15}$	9×10^{14}	$3·8 \times 10^{15}$

3.2. Mobility study

Mobility variations of the samples studied in the previous section are shown in figure 3. Two main scattering mechanisms limit the electron mobility: polar optical mode scattering

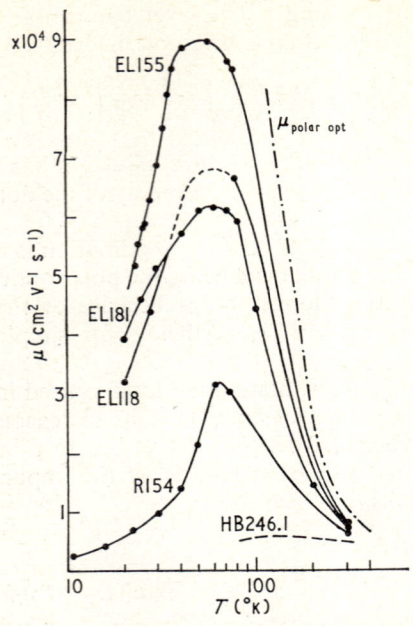

Figure 3. Electron mobility variations with temperature of the same samples as in figure 2. The sample No. 246.1 is a classical unannealed H.B. crystal.

and ionized centre scattering. The phonon contribution μ_0 has been calculated by Howarth and Sondheimer (1953) and the ionized impurity term μ_I by Brooks (1955)

$$\mu_0 \simeq AT^{1/2}\left(\exp\frac{\theta}{T}-1\right) \qquad (6)$$

$$\mu_I \simeq \frac{BT^{3/2}/N_I}{\log bT^2/n_1 - 1} \quad \text{if} \quad b\frac{T^2}{n_1} \gg 1 \qquad (7)$$

where

$$n_1 = n + (n+N_A)\left(1 - \frac{n+N_A}{N_D}\right) \qquad (8)$$

$$N_I = 2N_A + n$$

The optical mode scattering is important above 50°K. Under 50°K, the dominant mechanism is seen to be ionized impurity scattering, μ_I varying as $1/N_I$.

A first possibility is to estimate the ionized impurity concentration at 20°K by considering only the ionized centre contribution to the mobility. Table 3 shows the values obtained, compared to the values deduced from the Hall coefficient curves. The agreement is not satisfactory. Sample R 154 is not to be considered because of impurity band conduction.

On the other hand, using the computations of Bolger et al. (1966) which give a lattice mobility of 200 000 cm^2 v^{-1} s^{-1} at 77°K, it is possible to deduce the ionized impurity

Table 3. Ionized impurity concentrations estimated from the mobility values at 20°K compared to those deduced from the Hall coefficient curves

Sample number	N_I (cm^{-3}) deduced from $\mu_{20°K}$	$N_I = 2N_A + n_{20°K}$ (cm^{-3}) deduced from Hall coefficient curves
EL 155	$0·8 \times 10^{15}$	$1·1 \times 10^{15}$
EL 118	$1·3 \times 10^{15}$	
EL 181	$0·9 \times 10^{15}$	$1·2 \times 10^{15}$
R 154	$> 10^{16}$	$1·9 \times 10^{15}$

PLATE I—R. Solomon: *paper 2*

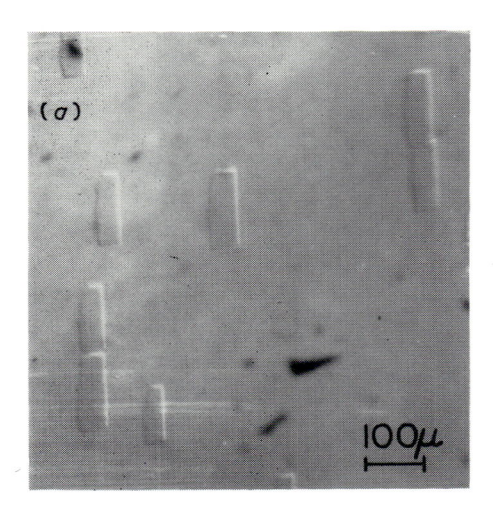

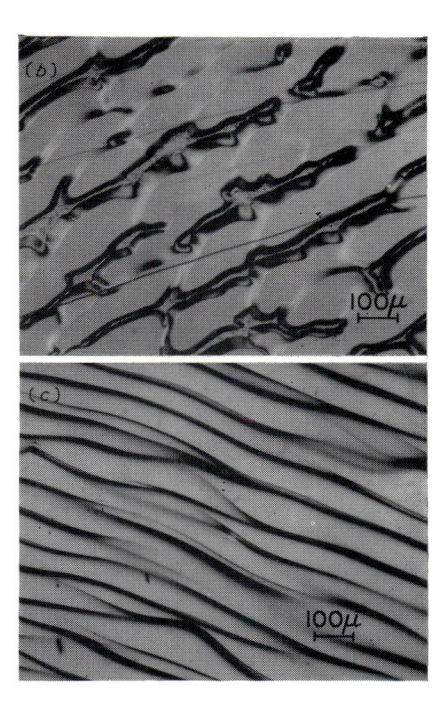

Figure 4. (*a*) Nucleating islands for a (100) orientation. Initial growth temperature 650°C. Tilt furnace with cooling rate of 12°C h^{-1}. (*b*) Intermediate growth stage with layer nearly complete. (*c*) Completed film showing characteristic structure for (100) orientation.

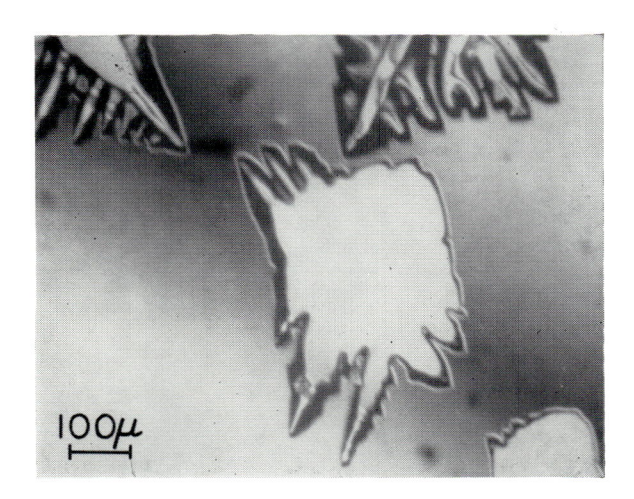

Figure 5. Irregularly shaped islands on (100) orientation. Cooling rate of 100°C h^{-1}.

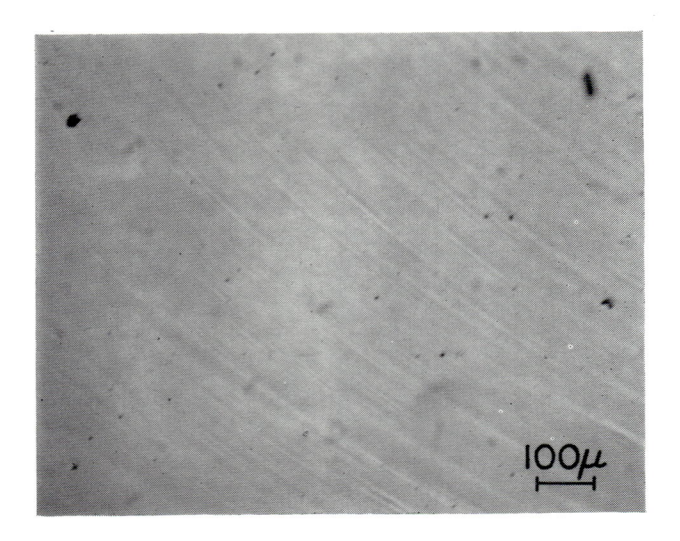

Figure 6. High density of initial nucleating sites results in smooth films. Wafer with (100) orientation, initial growth temperature 700°C.

PLATE III—*A. R. Goodwin et al.: paper 6*

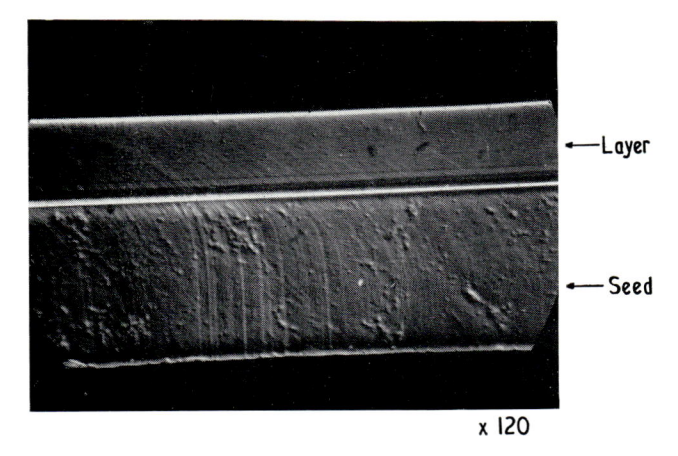

Figure 2. Cleaved and etched TGT grown layer, A–B etch.

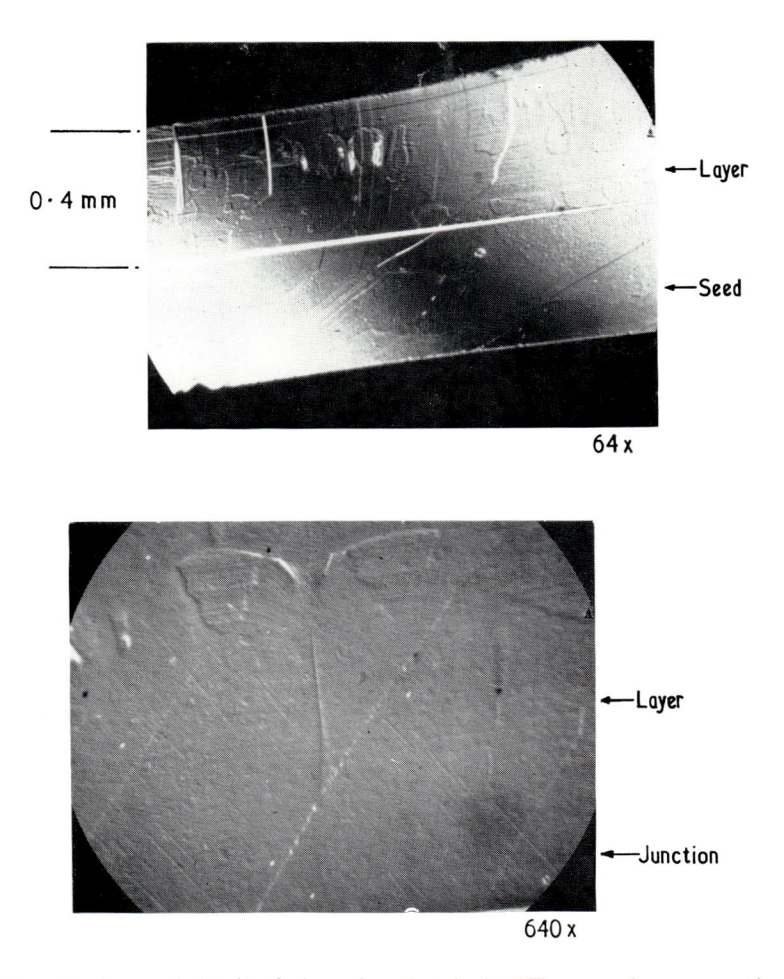

Figure 4. Overall view and detail of cleaved and etched TST grown layer, growth rate approx 150 μm h^{-1}, long A–B etch (2 min).

PLATE IV—A. R. Goodwin et al.: paper 6

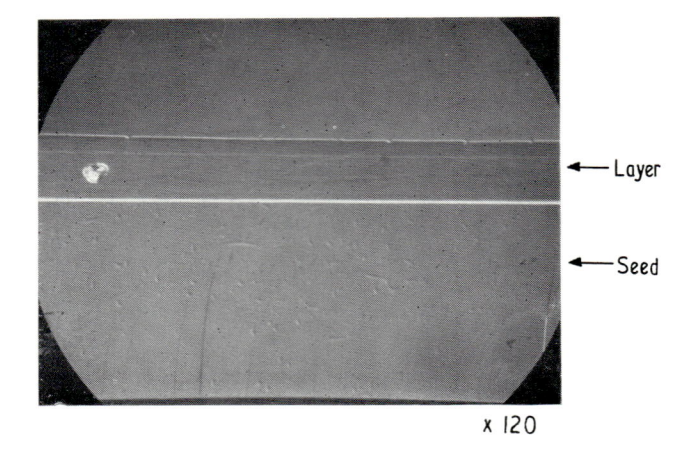

Figure 5. Cleaved and etched TST grown Zn doped layer on n⁺ seed, growth rate approx 60 μm h^{-1}, A–B etch.

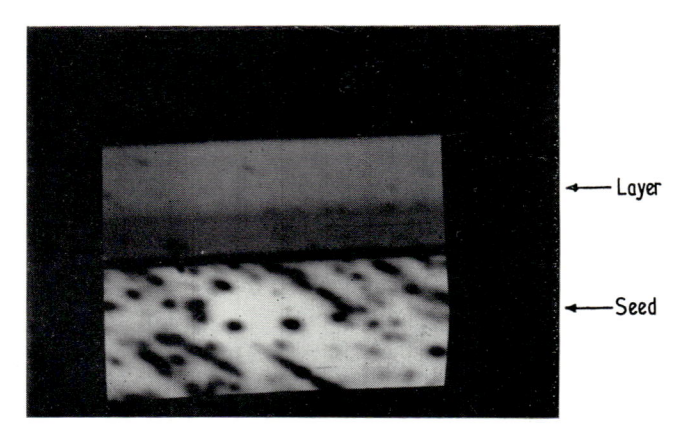

Figure 10. Cathodoluminescence from Zn doped layer on n⁺ seed ({110} cleave).

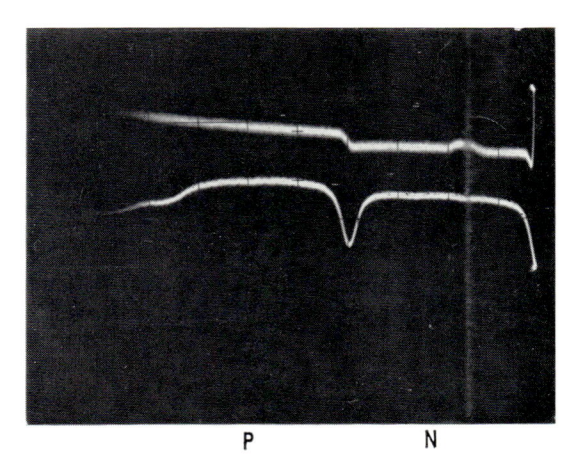

Figure 11. Line scans of secondary electron current (upper trace) and luminescence from p–n junction ({110} cleave) (1 cm = 10 μm).

PLATE V—*Don W. Shaw: paper* 8

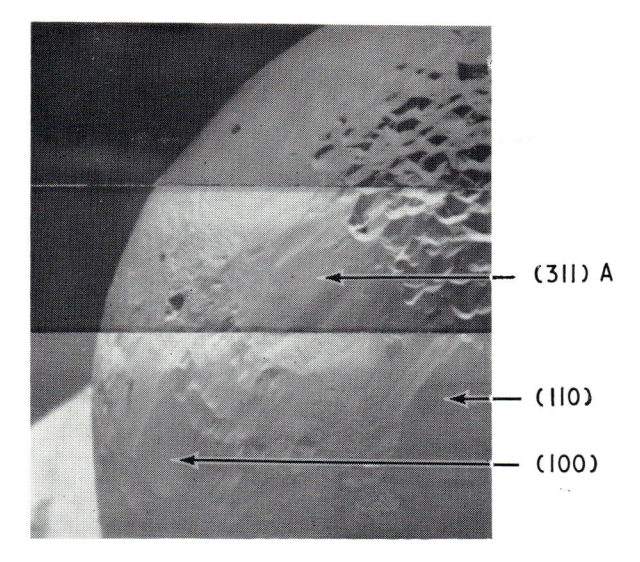

Figure 3. Scanning electron micrograph of epitaxial GaAs deposit on spherical GaAs substrate × 66.

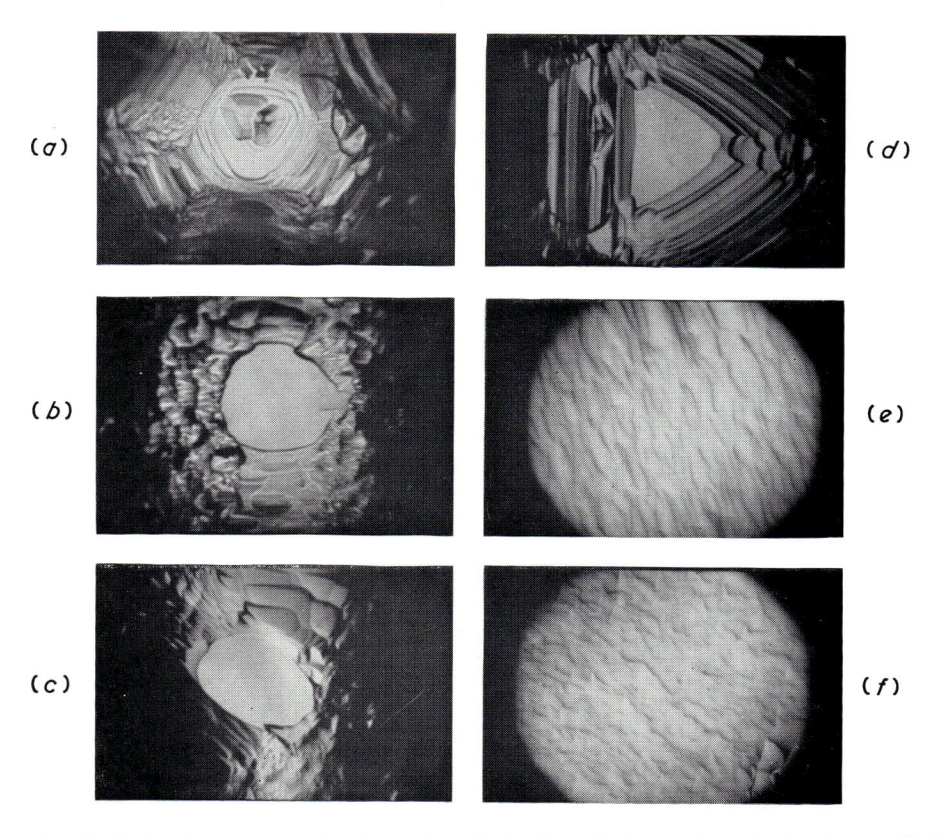

Figure 4. Optical micrographs of sphere deposit (*a–d*) and planar deposits (*e–f*). *a*, {111} B; *b*, {001}; *c*, {113} A; *d*, {110}; *e*, {001}; *f*, {113} A.

PLATE VI—*Don W. Shaw: paper* 8

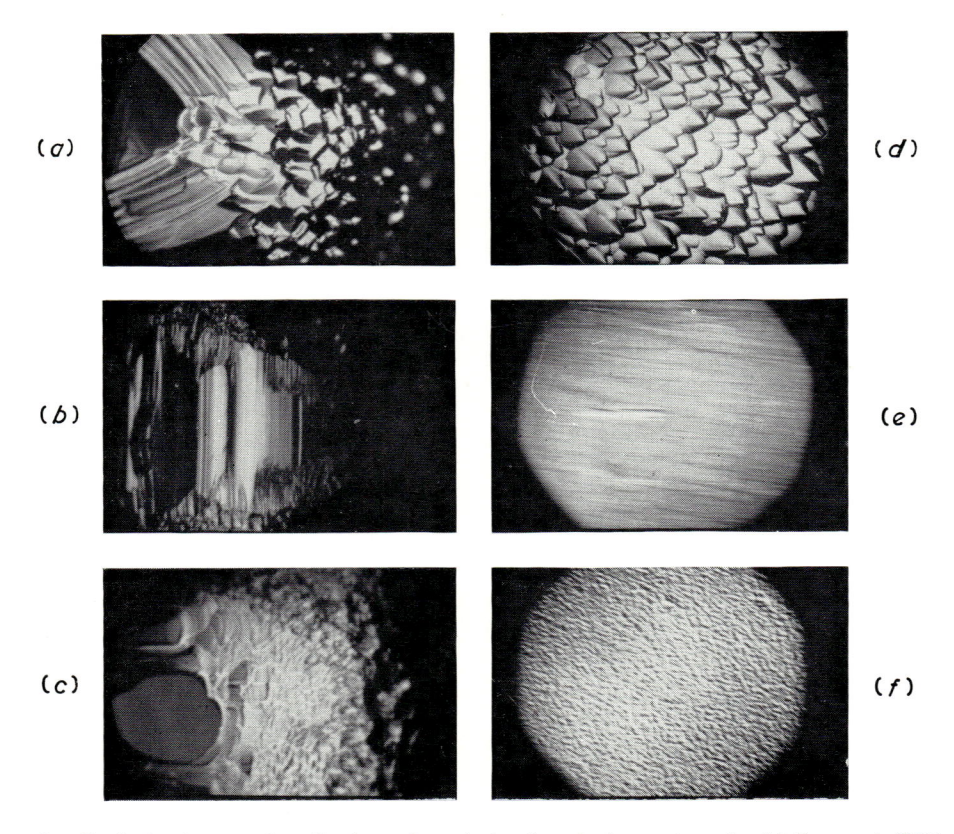

(a) (b) (c) (d) (e) (f)

Figure 5. Optical micrographs of sphere deposit (*a–c*) and planar deposits (*d–f*). *a, d,* {331} A; *b, e,* {331} B; *c, f,* {115}.

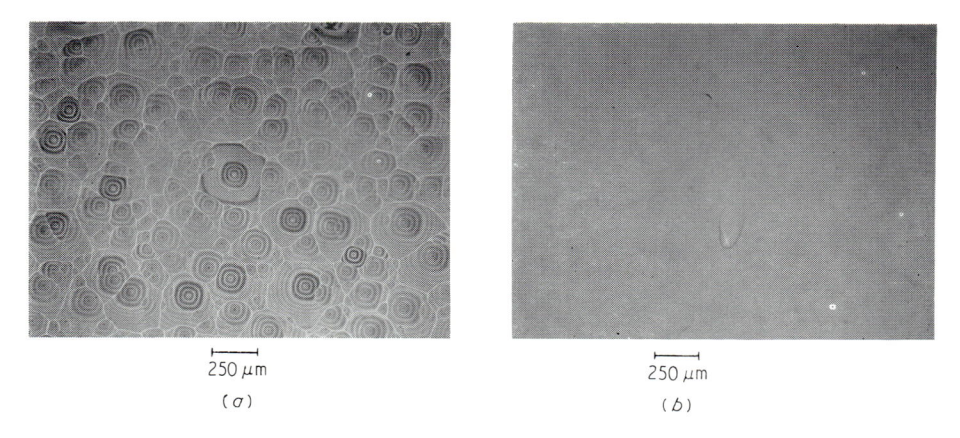

Figure 1. The effect of saw-cutting damage on the surface perfection of epitaxial GaAs.

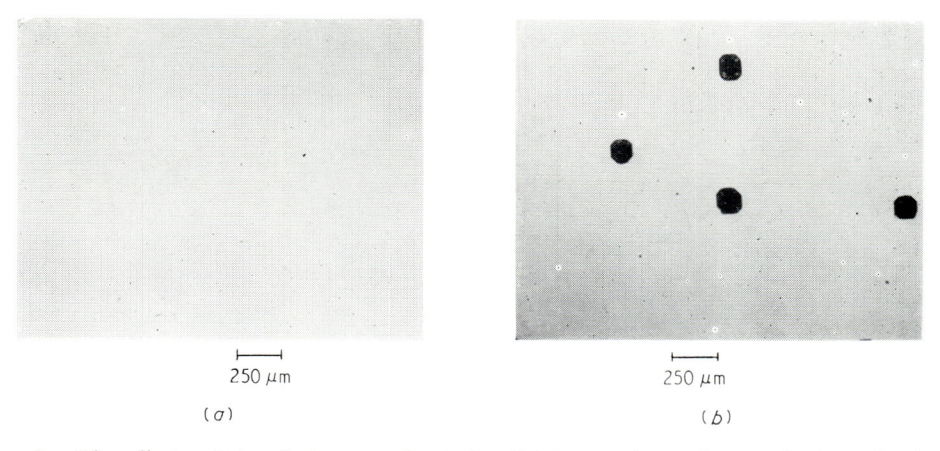

Figure 2. The effects of chemical vs. mechanical polishing on the surface perfection of epitaxial GaAs.

(a)

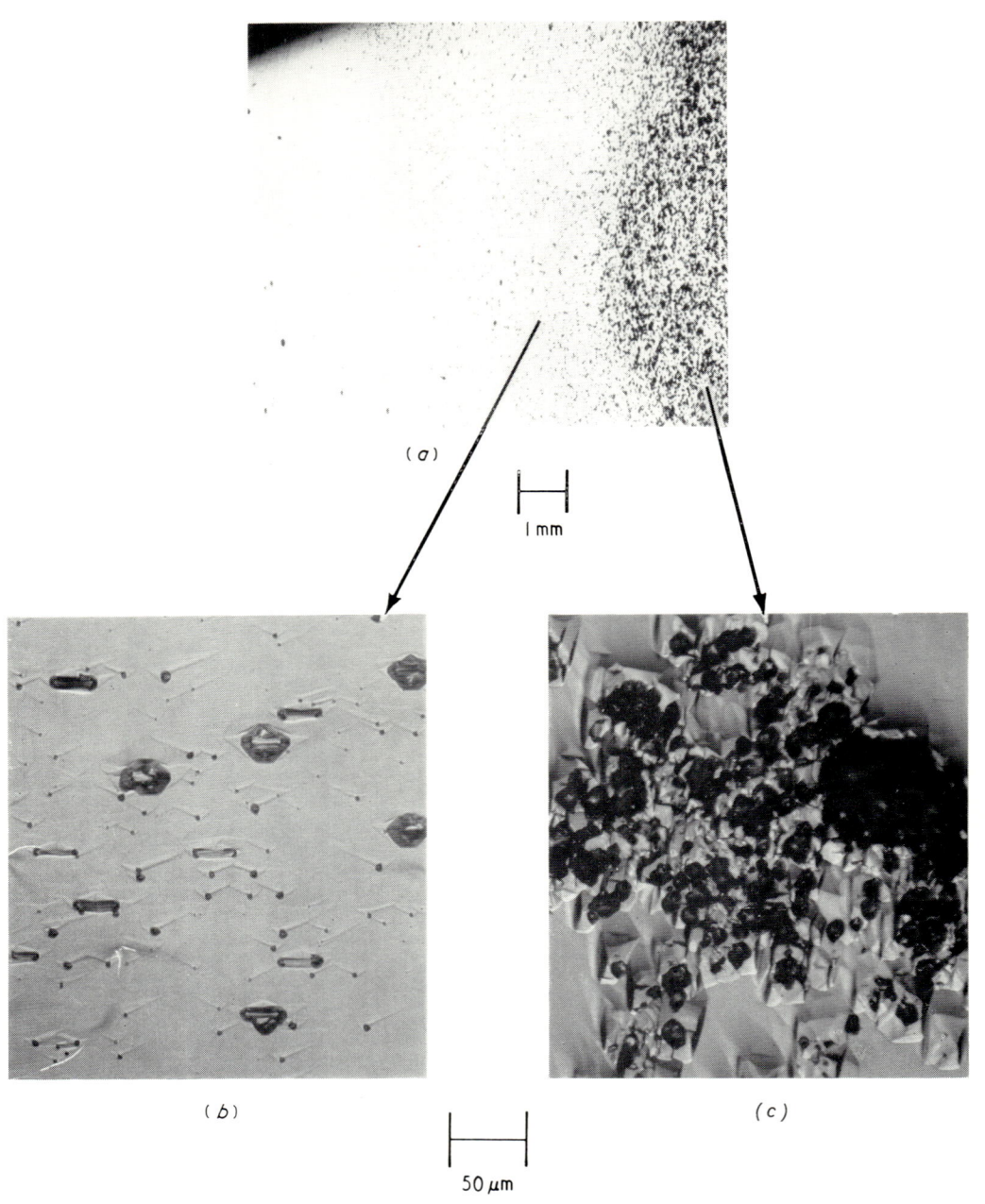

1 mm

(b)   (c)

50 μm

Figure 3. The effect on the surface perfection of an epitaxial GaAs layer of seeding the substrate surface with alumina particles.

concentration at this temperature. These values are shown in the last column of table 2. It is interesting to notice that they are not very different from the values of $(N_D + N_A)$ deduced from the Hall coefficient curves.

These results confirm the possibility of using the mobility values for estimating the impurity concentrations. The approximations used at 20°K do not appear to agree as well as at 77°K. So it can be concluded that the simple determination of the donor and acceptor concentrations from the mobility value at 77°K and the carrier concentration value at 300°K is valid.

It should be noted that, frequently, the room-temperature mobility is lower than expected from the low-temperature mobility values. The more pure the material, the greater is the difference from the theoretical value.

4. Photoionization experiments

It would be interesting to ascertain the nature of the acceptor centres, that is to say shallow substitution impurities or deep lattice defects. In order to reveal the imperfections situated beneath the centre of the forbidden gap, we have performed photoinduced detrapping measurements at liquid-nitrogen temperature, after initial saturation of traps by ultraviolet illumination.

Figure 4 represents the capacitance variations with wavelength. These values are the equilibrium values after irradiation of the samples from longer to shorter wavelengths. A noticeable increase of capacitance occurs near $\lambda \simeq 1\cdot05$ µm, ascribed to electron transition

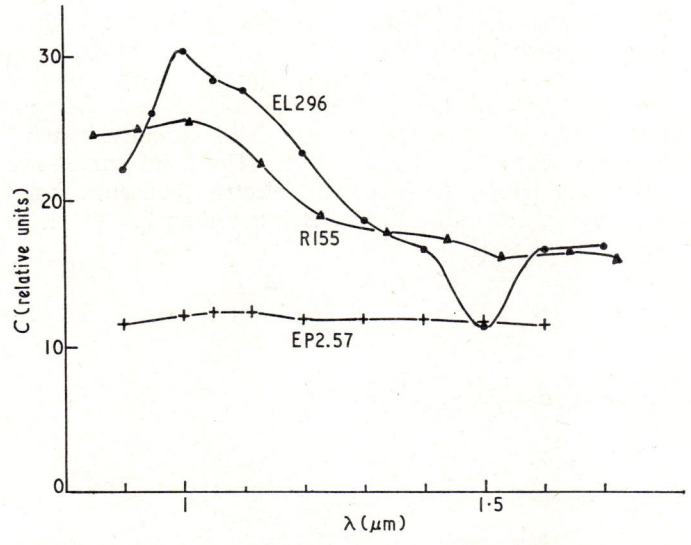

Figure 4. Photoinduced variations of capacitance of three typical samples previously studied by T.S.V.C. measurements.

from an imperfection band up to the conduction band. This level is situated at about 1·15 ev from the conduction band. It appears in H.B. crystals, vapour phase epitaxy, and solution grown layers in variable concentrations. In particular, this concentration of centres seems to be relatively important in the solution grown layers whose electron concentration is low and mobility decreases very sharply above 280°K. Emptying of these centres also appears in T.S.V.C. experiments above room temperature and in Hall coefficient variations near 400°K. The activation energy of the Hall coefficient curves may agree with the value measured by photoionization if we suppose that these deep centres are donor type, that is to say neutral when occupied and positively charged when empty. This interpretation would also partially explain the anomalous values of mobility at and above room temperature. On the other hand it could be possible that the usual scattering model does not apply to the case of such a low concentration material. This point is under further investigation.

5. Discussion

From T.S.V.C. analyses, it appears that the undoped medium conductivity solution grown layers usually contain a fairly low concentration of deep trapping centres. This agrees with the observations of Panish *et al.* (1966) and of Williams and Blacknall (1967) from photoluminescence analyses. So in most cases a compensation model containing only one shallow donor level may be used for interpreting the Hall coefficient variations. On the other hand, it has been shown that the values of the donor and acceptor concentrations may simply be estimated to a good approximation from the electron mobility value at 77°K.

Table 4 summarizes the comparison of purity characteristics of the three kinds of materials studied. Liquid phase growth and vapour phase epitaxy lead to about the same quality of

Table 4. Orders of magnitude of the donor and acceptor concentrations and of the total deep centre concentration for the three kinds of materials

	N_A (cm^{-3})	N_D (cm^{-3})	N_T (cm^{-3})
Solution growth	$3-5 \times 10^{14}$	$\sim 10^{15}$	$< 10^{14}$
Vapour phase epitaxy	$2 \times 10^{14}-10^{15}$	$\sim 10^{15}$	$\sim 10^{14}$
Heat-treated H.B. crystals	$\sim 3 \times 10^{15}$	$5 \times 10^{15}-10^{16}$	$\sim 3 \times 10^{15}$

medium conductivity material ($n \simeq 10^{15}$ cm^{-3}). However, we consider that the characteristics of undoped solution grown layers ($N_A \sim 3$ to 5×10^{14} cm^{-3}) may be a little more reproducible than those obtained with vapour phase epitaxy (N_A varying from 2×10^{14} to 10^{15} cm^{-3}), for the same layer thickness. This appears more clearly on figure 5 which represents the values of the electron concentration and mobility, obtained in successive growing experiments (Boucher and Hollan 1969). The solid curves are calculated for several values of the parameter N_A. When the electron concentration n is lower than 5×10^{14} cm^{-3}, the compensation ratio N_D/N_A is lower than 2. When $n > 5 \times 10^{14}$ cm^{-3}, $N_D/N_A \simeq 3$.

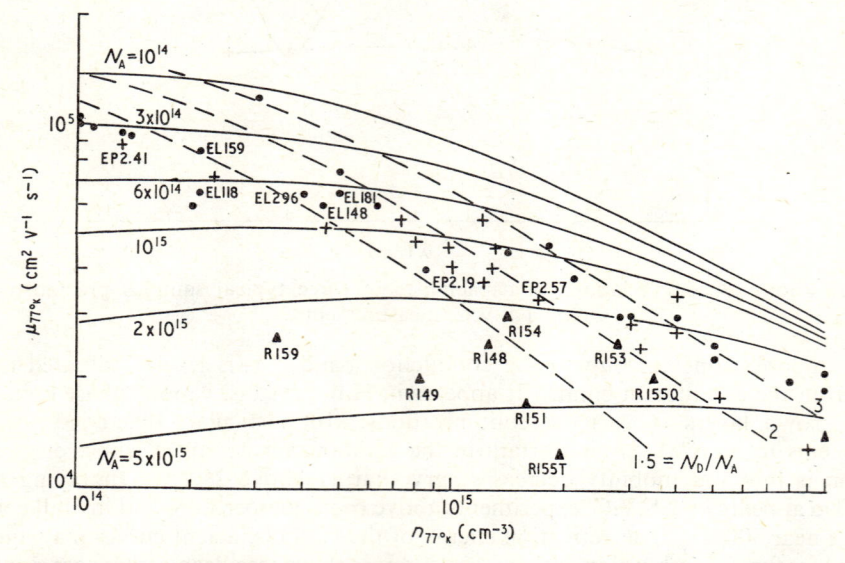

Figure 5. Electron concentration and mobility values of samples obtained in successive experiments. The full lines have been calculated for a given value of the acceptor concentration. The dotted lines correspond to a given value of the compensation ratio N_D/N_A. The labelled points refer to samples selected for detailed analysis.

From trapping measurements it may be concluded that most of the acceptor centres present in the heat-treated H.B. ingots are 'oxygen' centres. On the contrary, most of the centres present in the medium conductivity solution grown layers are shallow centres. In particular, the constancy of the acceptor concentration may be explained by the possibility that these centres are tetravalent impurities, which are substituted for arsenic atoms and which may be mainly silicon contained in the GaAs source.

Table 5. Characteristics of selected samples

Sample number	Nature†	Electron concentration (cm^{-3}) at 77°K	300°K	Electron mobility (cm^2 v^{-1} s^{-1}) at 77°K	300°K
EL 118	sol.	2×10^{14}	$2 \cdot 3 \times 10^{14}$	66 000	
EL 148	sol.	$1 \cdot 5 \times 10^{14}$	$2 \cdot 5 \times 10^{14}$	60 000	6500
EL 155	sol.	$2 \cdot 3 \times 10^{14}$	$2 \cdot 3 \times 10^{14}$	86 000	8000
EL 181	sol.	$6 \cdot 9 \times 10^{14}$	$7 \cdot 8 \times 10^{14}$	60 000	7900
EL 296	sol.		5×10^{14}	65 000	
EP 2.57	V.P.	4×10^{15}	$4 \cdot 9 \times 10^{15}$	33 000	7700
EP 2.41	V.P.	$1 \cdot 3 \times 10^{14}$	$1 \cdot 4 \times 10^{14}$	88 000	8800
R 154	H.T.	$1 \cdot 5 \times 10^{15}$	$1 \cdot 9 \times 10^{15}$	29 000	6500
R 155	H.T.	$3 \cdot 2 \times 10^{15}$	$3 \cdot 8 \times 10^{15}$	15 500	5600
257–1–Q	H.B.		$1 \cdot 6 \times 10^{15}$		5140
307	H.B.		$6 \cdot 4 \times 10^{14}$		5500

† Sol.=grown in a gallium solution; V.P.=vapour phase epitaxy; H.B.=obtained by the horizontal Bridgman method; H.T.=heat-treated crystal after growing by the horizontal Bridgman method.

However, solution grown layers which could be supposed the purest, considering only the mobility value at liquid-nitrogen temperature, sometimes exhibit anomalous mobility decrease and deep centre ionization at and above room temperature. This could explain why the best results in Gunn effect are obtained with layers whose electron concentration is rather high ($n \sim 3 \times 10^{15}$ cm^{-3}).

Acknowledgments

Part of this work was supported by La Délégation Générale à la Recherche Scientifique et Technique. We are grateful to E. André, L. Hollan, and E. Deyris for supplying respectively the liquid and vapour phase epitaxy layers and the bulk H.B. crystals; Professor Fortini, J. M. Le Duc, and E. André for fruitful discussions; J. Fertin for a critical review of the manuscript; M. Boulou, R. Duchesne, and J. Varon for their technical assistance.

References

ANDRE, E., DEYRIS, E., and LE DUC, J. M., 1968, Spring Meeting of the Electrochemical Society, Boston.
BOLGER, D. E., FRANKS, J., GORDON J., and WHITAKER, J., 1966, *Proc. Int. Symp. on GaAs*, I.P.P.S. Conf. Series No. 3, 16–22.
BOUCHER, A., and HOLLAN, L., 1969, to be published.
BROOKS, H., 1955, *Adv. Electronics Electron Phys.*, **7**, 158.
CARBALLÈS, J. C., and LEBAILLY, J., 1968, *Solid State Com.*, **6**, 167–171.
EDOLLS, D. V., 1966, *Phys. Stat. Sol.*, **17**, 67–76.
FERTIN, J. L., LEBAILLY, J., and DEYRIS, E., 1966, *Proc. Int. Symp. on GaAs*, I.P.P.S. Conf. Series no. 3, 46–51.
HILSUM, C., and ROSE-INNES, A. C., 1961, *Semiconductor III–V Compounds*, Pergamon Press, 72.
HOWARTH, D. J., and SONDHEIMER, E. H., 1953, *Proc. Roy. Soc.* **A219**, 53–74.
PANNISH, M. B., QUEISSER, H. J., DERICK, L., and SUMSKI, S., 1966, *Solid State Electron*, **9**, 311–314.
WILLIAMS, E. W., and BLACKNALL, D. M., 1967, *Trans. Met. Soc. A.I.M.E.*, **239**, 387–394.
WOODALL, J. M., and WOODS, J. F., 1966, *Solid State Com.*, **4**, 33.

Liquid phase epitaxial growth of gallium arsenide

A. R. GOODWIN, C. D. DOBSON† and J. FRANKS†

Standard Telecommunication Laboratories, Harlow, Essex, England.

Abstract. Doped layers of GaAs were grown by liquid phase epitaxy from solutions of GaAs and dopant in Ga. Layers were grown either by tipping the solution over the seed near the growth temperature under a temperature gradient of about $10°c\,cm^{-1}$ (TGT), or by a modified travelling solvent technique in which the seed and solution remain in contact throughout the temperature cycle (TST). In both methods GaAs in the solution may be prepared by reacting pure Ga with pure $AsCl_3$. The electrical characteristics of GaAs doped with Se, Te, Si, and Zn in a range of concentrations were measured. With Se and Te doping, good reproducibility was obtained for donor concentrations in the range 10^{16} to $6 \times 10^{18}\,cm^{-3}$. Doping with Zn gave reproducible acceptor concentrations of the order $10^{19}\,cm^{-3}$. Doping with Si produced GaAs varying in both carrier type and concentration.

For a given electron concentration the measured mobilities are somewhat less than those predicted by Ehrenreich for all electron concentrations studied but for high electron concentrations are essentially in agreement with the more recent predictions of Moore.

Junctions were formed by deposition of a Zn doped layer on a Se doped seed and examined in a scanning electron microscope for secondary electron current and cathodoluminescence.

1. Introduction

Liquid epitaxy has made an increasing contribution to GaAs technology in recent years. Since Nelson (1963) first demonstrated the value of the technique for devices requiring high doping levels, applications to various aspects of the fabrication of electroluminescent lasers and lamps have been reported, e.g., Winogradoff and Kessler (1964) and Rupprecht (1966). Undoped liquid epitaxial GaAs has very recently been reported with electron mobilities comparable to the highest mobilities quoted for GaAs produced by other methods (Kang and Green 1967; Goodwin, Gordon and Dobson 1968; Andre and Le Duc 1968) opening up possibilities for the technique for devices requiring low doping levels and high purity. In the present paper, methods are described of growing GaAs with a range of dopants of various concentrations from solution in Ga, and some electrical and luminescent properties are given.

2. Epitaxial techniques

Initially GaAs layers were grown by heating a GaAs seed next to a solution of As and dopant in Ga in a silica boat, and at the appropriate temperature tipping the boat so that the solution covered the seed (Nelson 1963). Growth on the seed occurs as the temperature is dropped. The temperature distribution in the furnace was such that at any time the seed and solution were at approximately the same temperature. These layers, grown for 1–2 hours between 850 and 600°C were typically 60–80 μm thick and generally had rough surfaces. It was found that most of the growth occurred very rapidly (500 μm h^{-1}) near the end of the process. Occasionally growth was so rapid that gallium droplets several μm in diameter were included in the interface between the substrate and the layer.

To obtain more control over the growth of the layer, the system was modified to allow growth to occur under temperature gradient conditions (figure 1b). The furnace was heated from the top with a semicylindrical element, so that the bottom of the seed was coolest. The gradient across solution and seed was estimated to be about $10°c\,cm^{-1}$. As before,

† Now at Electrotech Equipments Ltd, Abercarn, Newport, Mon.

the boat was tipped when the temperature was about 850°C and growth occurred down to about 600°C, the temperature being reduced at a rate of about 1°C min^{-1}.

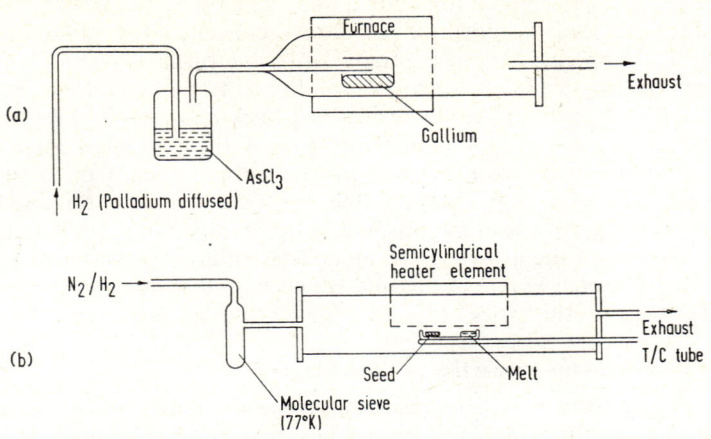

Figure 1. (a) Gallium saturation unit. (b) Solution growth unit for temperature gradient tipping technique (TGT).

To avoid the introduction of unwanted impurities, the solution was prepared by cracking AsCl$_3$ at the growth temperature over Ga in a separate reactor (figure 1a). To achieve good reproducibility in doped epitaxial layers of GaAs, elemental dopants were added to the solution rather than heavily doped GaAs.

With this technique of tipping in the presence of a thermal gradient (TGT), layers between 150 and 200 μm thick were grown from 850°C over periods of about 3 hours. The surfaces were generally good (figure 2), on occasion being mirror bright.

If the Ga solvent is saturated with As above the temperature at which layers are eventually grown, GaAs crystallites remain floating on the surface of the solution throughout the growth cycle. It is then possible to grow the layer at a constant temperature gradient without reducing the temperature. This method is essentially the travelling solvent technique (TST) reported by Mlavsky and Weinstein (1963). The usual requirement of providing a source layer of GaAs above the solvent is made unnecessary, because the floating crystallites act as source, which can therefore be of high purity. With this technique growth and doping should be more uniform than with the TGT since a constant growth temperature is used and growth occurs at useful rates at lower temperatures.

With the TST, the seed is placed on a silica boat under the solution (i.e., the solution is not tipped at any stage of the process). The boat is then loaded into the furnace tube as shown in figure 3 and raised to the growth temperature in a dry hydrogen atmosphere.

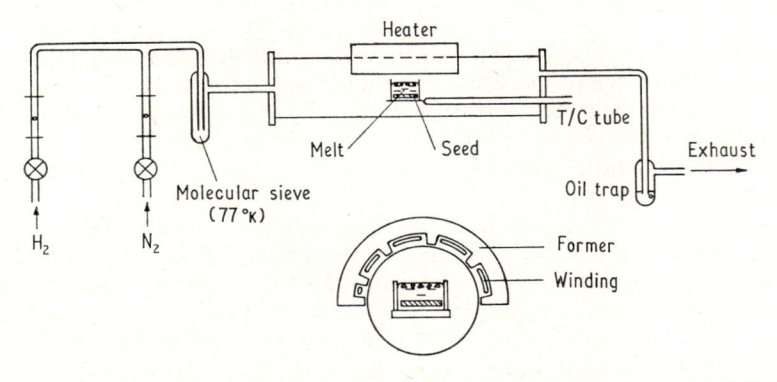

Figure 3. Solution growth unit for modified travelling solvent technique (TST).

Difficulties were encountered due to the melt not wetting the slice, but these were overcome by preliminary baking of the seed in a vacuum at 800°C. In the steady state (i.e., after the system has been at growth temperature for a sufficient time), As is transported from the surface region to the seed by diffusion as an As concentration gradient is established associated with the temperature gradient imposed by the furnace.

Prior to reaching this steady-state growth condition, two opposed fluxes of As may be considered to exist, one from the seed causing etch-back and a second larger flux from the surface region, which in time dominates the first so that the etch-back rate smoothly decreases and growth commences, the growth rate approaching a maximum as the melt reaches the steady state. At 800°C, layers 400 μm thick have been obtained after 3 hours (figure 4). At 700°C, layers about 60 μm thick were obtained in 1 hour (figure 6). As the system cools at the end of the growth time, an additional layer of about 30 μm is deposited (figures 4 and 5). The etch of Abrahams and Buiocchi (1965) was used in obtaining figures 2, 4, and 5. Throughout this work seeds cut on a {100} plane were used.

3. Electrical and luminescent properties of doped layers

Doped layers were grown by adding dopant to the solution. The weight of dopant against carrier concentration for TGT grown layers with 15 g melts is shown in figure 6

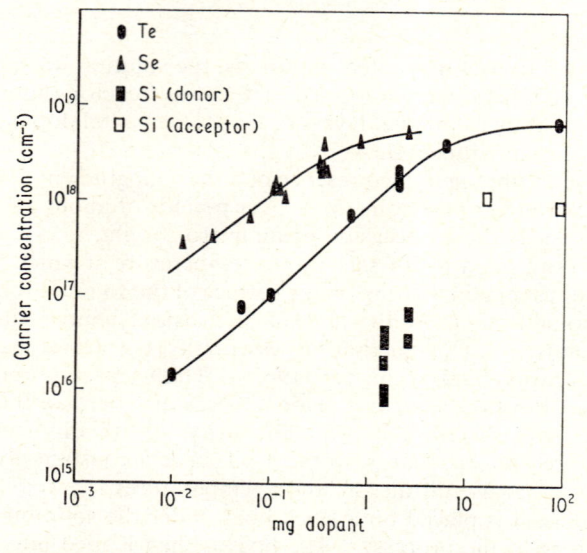

Figure 6. Carrier concentration of TGT grown layers against weight of dopant added to melt for Se, Te, and Si dopants. Melt weights were about 15 g.

for Se, Te, and Si. It is clear that the distribution coefficient for Te is lower than for Se. The behaviour of Si is caused by its amphoteric nature in GaAs: at low concentrations the layers were generally n-type, occasionally n and p layers grew alternately in the same run.

Zinc doped layers, produced by adding 0·3 g of Zn to the solution, had hole concentrations of about 2×10^{19} (mobilities typically about 70 cm^2 V^{-1} s^{-1}) for both the TG and TS techniques.

The electrical properties of the layers were assessed by the method of van der Pauw (1958). Experimental details were as reported by Bolger, Franks, Gordon, and Whitaker (1966). Figure 7 shows the mobility at room temperature plotted against carrier density for Se doped layers grown by the TG and TS techniques. Preliminary measurements indicate that very high doping levels of Se in GaAs can be obtained with TST. Figure 8 shows the mobility of Se doped layers at 77°K. The reproducibility of the characteristics was good, mobilities were generally 15% below the theoretical curve of Ehrenreich (1960),

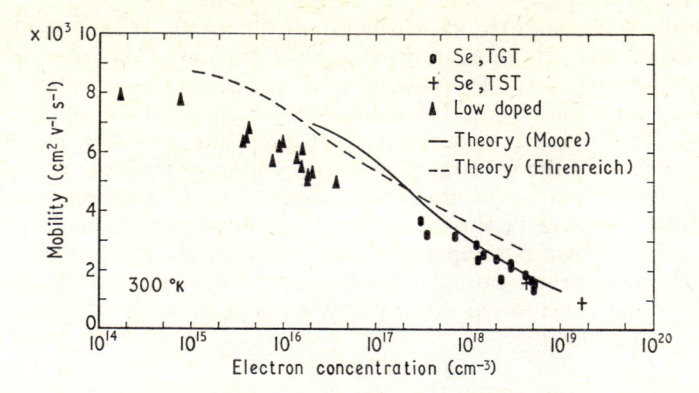

Figure 7. Mobility at room temperature plotted against electron density for Se doped layers and low doped layers (nominally undoped, but in the presence of residual dopant, in most cases Se).

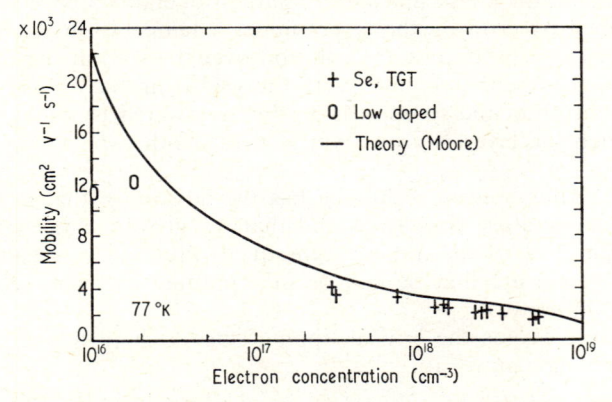

Figure 8. Mobility at liquid-nitrogen temperature plotted against electron density for Se doped layers and low doped layers.

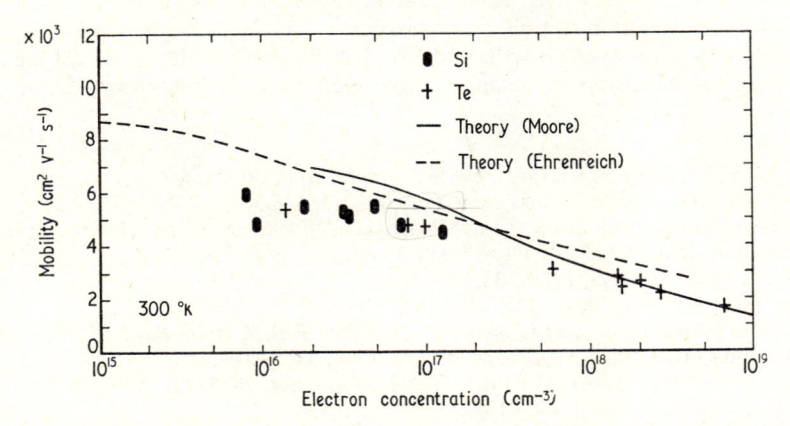

Figure 9. Mobility at room temperature plotted against electron density for Te and Si doped TGT grown layers.

but fitted well to the predicted values of Moore (1967), where his treatment is valid, i.e., at high concentrations.

Tellurium doped layers (figure 9) were grown only by TGT and have characteristics similar to those of Se.

The characteristics of Si doped layers (TGT) are shown only over the low concentration range, in which the layers are predominantly n-type (figure 9). There was little consistency between the carrier concentration and the amount of dopant added, but it is interesting that the mobility values are on a similar curve to those for Se and Te.

N and p doped layers grown by the various liquid epitaxial techniques were examined with a scanning electron microscope for luminescence and secondary electron emission.

The luminescence from the substrates is generally inhomogeneous; the grown layers are much more uniform but may show gradients. Figure 10 is an example of a Zn doped layer on an n+ substrate. Figure 11 shows a scan of the luminescence and secondary electron current across a TST grown p–n junction. Laser efficiencies of this type of junction were found to increase after heat treatment but thresholds were raised. This behaviour is in contrast to that found for diffused junctions by Carlson (1967) for which thresholds were reduced after heat treatment.

4. Conclusion

For many applications in the fabrication of devices, stringent control must be achieved over the doping content, thickness, and surface flatness of epitaxial layers. A good measure of control is obtained by growing the layers under a temperature gradient of 10°c cm^{-1}. The layers may be grown by tipping the solution over the seed at high temperature (800–900°C) or the solution may be in contact with the seed from the outset. With the tipping method, the temperature of the solution is gradually lowered to about 600°C, so that the temperature at which the layer is grown varies with depth, which may affect the doping content.

The modified travelling solvent technique has the advantage that growth occurs under virtually constant temperature conditions and that the growth temperature may be lower than the peak temperature of the tipping technique. Preliminary results at a low growth temperature (700°C) indicate that Se can be incorporated at higher concentrations than have been achieved hitherto.

Since this technique avoids the critical tipping stage at high temperature, it is easy to operate and may be made automatic.

Acknowledgments

The authors wish to thank Mr. S. B. Marsh, Director of Standard Telecommunication Laboratories, for permission to publish this paper. They also thank Messrs. K. Brown and R. Hinton for help with layer growth, and Mr. J. Whitaker for the electrical measurements. The support of the Ministry of Defence (Navy Department) is acknowledged.

References

ABRAHAMS, M. S., and BUIOCCHI, C. J., 1965, *J. Appl. Phys.*, **36**, 2855–63.
ANDRE, E., and LE DUC, J. M., 1968, *Mat. Res. Bull.*, **3**, 1–6.
BOLGER, D. E., FRANKS, J., GORDON, J., and WHITAKER, J., 1967, *Proc. Int. Symp. on GaAs*, Reading 1966 (London, Inst. Phys. and Phys. Soc.), Conf. Ser. No. 3 pp. 16–22.
CARLSON, R. O., 1967, *J. Appl. Phys.*, **38**, 661–68.
EHRENREICH, H., 1960, *Phys. Rev.*, **120**, 1951–63.
GOODWIN, A. R., GORDON, J., and DOBSON, C. D., 1968, *Brit. J. Appl. Phys. (J. Phys. D)*, **1**, 115–6.
KANG, C. S., and GREEN, P. E., 1967, *Appl. Phys. Lett.*, **11**, 171–3.
MLAVSKY, A. I., and WEINSTEIN, M., 1963, *J. Appl. Phys.*, **34**, 2885–92.
MOORE, E. J., 1967, *Phys. Rev.*, **160**, 618–26.
NELSON, H., 1963, *RCA Rev.* **XXIV**, 603–15.
RUPPRECHT, H., 1967, *Proc. Int. Symp. on GaAs*, Reading 1966 (London: Inst. Phys. and Phys. Soc.), Conf. Ser. No. 3 pp. 57–61.
VAN DER PAUW, L. J., 1958, *Philips Res. Rep.*, **13**, 1–9.
WINOGRADOFF, N. N., and KESSLER, H. K., 1964, *Sol. Stat Comm.*, **2**, 119–122.

CHAPTER 2

Vapour phase epitaxial growth and bulk material

Tin doping of epitaxial gallium arsenide

C. M. WOLFE, G. E. STILLMAN and W. T. LINDLEY

Lincoln Laboratory,[†] Massachusetts Institute of Technology, Lexington, Massachusetts, U.S.A.

Abstract. Gallium-tin melts have been used with the $AsCl_3$–Ga flow system to prepare n-type epitaxial GaAs with carrier densities from 10^{15} to 10^{17} cm^{-3}. The transfer ratio of net tin donor concentration in the epitaxial layer to tin concentration in the gallium melt has been examined as a function of seed orientation, growth rate, seed and melt temperatures, and flow rates. Over a wide range of conditions the tin transfer ratio is dominated by its growth-rate dependence. This dependence and the dominant growth-rate behaviour can be qualitatively explained by a time-dependent adsorption of tin at the vapour–solid interface. Tin acceptors have the same growth-rate dependence as the tin donors and are incorporated into the layer in the ratio of about one acceptor for every three donors. Because of the dominant growth-rate dependence, uniformly doped layers can be obtained under conditions where small variations in temperature and flow rate have little effect on the growth rate and therefore the doping. Under these conditions average measured doping variations as low as $\pm 2 \cdot 6 \%$ have been attained.

1. Introduction

This investigation of tin-doped epitaxial GaAs was initiated by a desire to incorporate in the $AsCl_3$–Ga flow system (Knight, Effer, and Evans 1965; Effer 1965) a means of uniformly and controllably doping epitaxial layers n-type for device purposes. Since the simplicity and relative purity of the system as used at the residual impurity level was also desired, it was decided to introduce known impurities into the system by doping one of the starting materials. By adding impurities to the gallium melt, the doping level could conceivably be changed from run to run and uniformly doped layers should be attainable. N-type impurities such as selenium and tellurium were found to deplete from the melt because of their high vapour pressures, so impurities which could be transported at a lower rate by the formation and reduction of their chlorides were examined.

Initially, both germanium and tin were used as dopants to give n-type layers with average doping variations of about $\pm 20 \%$. However, the effective transfer ratio for these impurities and the uniformity of doping varied considerably under different growth conditions and it became necessary to examine the growth parameters which could conceivably affect the transfer of impurities from the gallium melt to the epitaxial layer. In this paper, the effects of crystallographic orientation, growth rate, melt and growth temperatures, and flow rates on the incorporation of tin into epitaxial GaAs are examined.

2. Procedure

The epitaxial layers were grown in a horizontal reactor which was provided with independent control of the gallium melt temperature and the growth temperature. Two input hydrogen lines, one to the $AsCl_3$ bubbler and one bypass line, allowed changes in the $AsCl_3$-to-H_2 ratio, while controlling the $AsCl_3$ temperature at 20°c. The seeds were set horizontally on the seed holder and the layers were grown in a temperature gradient which could be varied from 0 to 30°c in^{-1}. In each run the seeds were left out of the furnace until the gallium melt was saturated with arsenic. The seeds were then inserted into the furnace while keeping the reactor airtight and the growth cycle was initiated.

Doping was obtained by introducing known amounts of tin into the gallium melt. Sized tin spheres from 0·005 in. to 0·02 in. in diameter were used for doping levels from 10^{15} to

[†] Operated with support from the U.S. Air Force.

10^{16} cm^{-3}, while the tin was weighed for heavier doping. In each run the residual impurity level was determined before tin was introduced into the system. This residual doping level was typically in the high 10^{14} to low 10^{15} cm^{-3} range. Once this level was determined, an amount of tin was introduced to ensure that all observable effects were due to tin doping and not to residual impurities or tin solubility limits. Thus, the range of tin doping was between 5×10^{15} and 2×10^{17} cm^{-3}; that is, about a factor of five above the residual impurity level and a factor of five below possible solubility limits.

In the range of conditions examined no evidence was obtained which could be attributed to an accumulation or depletion of tin in the gallium melt. Therefore, the tin is probably transported from the melt by the formation of a chloride at about the same rate as the gallium.

In the following experiments one growth parameter was varied while the others were held constant. The gallium melt was controlled at 850°C, the growth temperature at 740°C, the growth temperature gradient at 25°C in^{-1}, the AsCl$_3$ flux at $4 \cdot 5 \times 10^{-5}$ mole min^{-1}, and the total hydrogen flow at 250 ml min^{-1}, while one parameter or another was varied. The tin transfer ratio has been defined as the ratio of the net concentration of tin donors in the epitaxial layer to the concentration of tin introduced into the gallium melt.

3. Growth-rate dependence

Figure 1 shows the variation of transfer ratio with growth rate for the four low-index planes. The growth rate was changed by varying the seed temperature gradient. The melt and seed temperatures and the AsCl$_3$ and hydrogen flow rates were held constant at the

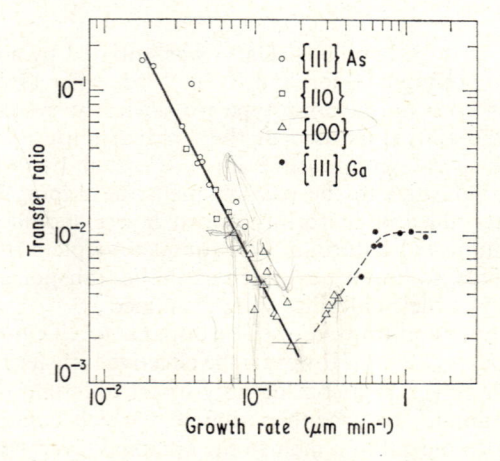

Figure 1. Growth-rate dependence of the tin transfer ratio for the {111} As, {110}, {100}, and {111} Ga planes. The growth rate was varied by changing the temperature gradient in the growth region. Other growth parameters were: melt temperature, 850°C; growth temperature, 740°C; AsCl$_3$ flux, $4 \cdot 5 \times 10^{-5}$ mole min^{-1}; and hydrogen flow, 250 ml min^{-1}.

values previously indicated. These data were plotted as a function of growth rate since this, rather than temperature gradient, appears to be the controlling factor. As shown, the transfer ratio decreases with increasing growth rate for the {111} As, {110}, and {100} surfaces up to a growth rate of about 0·15 μm min^{-1}. A solid line has been drawn through these points. Beyond this value, corresponding to the dashed line in the figure, the transfer ratio begins to increase with growth rate for the {100} and the {111} Ga surfaces, and above about 0·5 μm min^{-1} the {111} Ga growth becomes noticeably rough. Under the conditions required to obtain reasonable growth rates for the other low-index planes, the {111} Ga growth rate was high and the layers were rough.

Several features of these data should be emphasized. First, the growth rates and the tin transfer ratios were markedly different for layers of different orientation which were grown under the same conditions with the same temperature gradient. However, the transfer ratios were similar for layers such as the {111} As and the {100} which were grown under the same conditions with different temperature gradients to give comparable growth rates. Also, the doping dependence on growth rate for {100} layers was changed from negative to positive by increasing the growth rate. This positive growth-rate dependence appears to be the same as that observed for {111} Ga layers in the same growth-rate range. Thus, the data indicate that all low-index planes have the same general growth-rate dependence.

We believe this growth-rate behaviour can be explained qualitatively in the following manner. At the lower growth rates the incorporation of tin into the epitaxial layer is controlled by a time-dependent adsorption. That is, the tin is readily adsorbed on the growing surface and the amount desorbed is very small. Because of this, the longer a growing surface is exposed to the flux of tin, or the slower the growth rate, the larger the amount of tin incorporated into the epitaxial layer. This behaviour holds up to a growth rate which is so large that the crystal surface becomes non-singular and a larger surface area per unit volume is available for adsorption. With the larger surface area for adsorption, the tin incorporated into the epitaxial layer begins to increase with growth rate and, as the growth rate increases further, the growth becomes dendritic corresponding to a noticeably rougher surface.

Adsorption has been previously used to account for variations in impurity segregation with crystallographic orientation and growth rate by Hall (1953) for melt-grown Ge and Si, by Banus and Gatos (1962) for melt-grown InSb, and by Williams (1964) for vapour-grown GaAs.

Figure 2 illustrates the variation of transfer ratio for {100} seeds with melt temperatures from 840 to 940°C. Other growth parameters were held constant. The incorporation of tin into the epitaxial layers (solid line) decreases as the temperature increases from 840 to

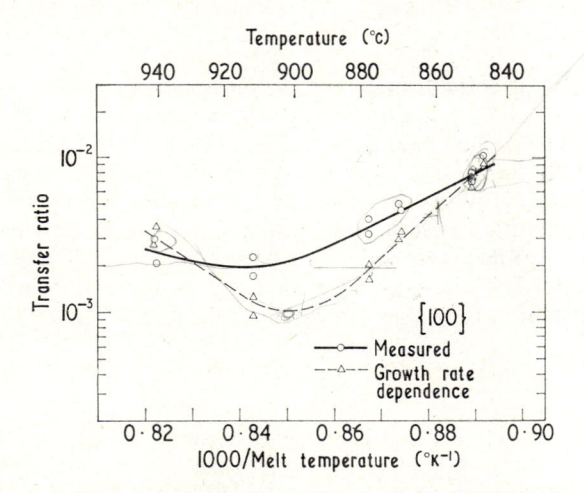

Figure 2. Dependence of the transfer ratio on gallium melt temperature for {100} epitaxial layers. The dashed line shows the growth rate variation to be expected from the data in figure 1. Other growth variables were held constant at the following values: growth temperature, 740°C; growth temperature gradient, 25°C in^{-1}; AsCl$_3$ flux, 4·5 × 10^{-5} mole min^{-1}; and hydrogen flow, 250 ml/min^{-1}.

about 900°C where it levels off and begins to increase. However, since the GaAs transported to the layer increases with temperature, the growth rate continually increases from 840 to 940°C. The dashed line in figure 2 indicates the variation in transfer ratio with growth rate expected from the data in figure 1 where the GaAs transport was changed by varying the temperature gradient in the immediate vicinity of the seeds. Both curves show the same general trend with the transfer ratio first decreasing with increasing growth rate and then increasing. The minimum in the measured melt temperature dependence curve occurs at about the same growth rate as the minimum or break in the data of figure 1. Evidently, increasing the melt temperature and changing the seed temperature gradient both affect the incorporation of tin into the epitaxial layer by increasing the growth rate. The difference between the curves indicates that the transfer ratio is also influenced to a lesser extent by other factors.

The growth-temperature dependence of the transfer ratio between 760 and 680°C is shown in figure 3 for the {111} As, {110} and {100} orientations. Melt temperature, temperature gradient, and flow rates were held constant. For the {110} and {100} seeds the transfer

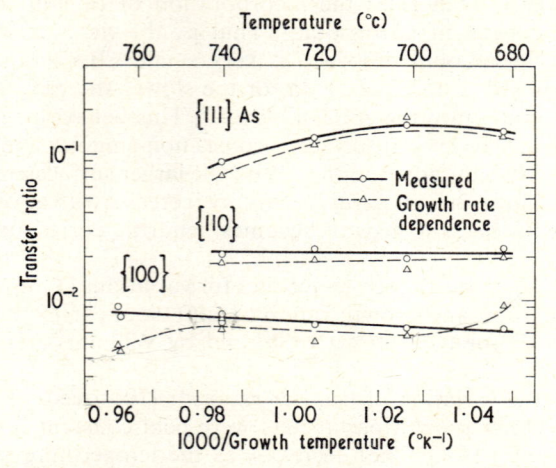

Figure 3. Growth-temperature dependence of the transfer ratio for {111} As, {110}, and {100} surfaces. Other growth variables were held constant at the same values as previously indicated.

ratios (solid lines) and the growth rates remain fairly constant. For the {111} As surface the doping increases (solid line) and the growth rate decreases as the temperature decreases from 740 to 700°C. The dashed lines show the expected growth rate dependence of transfer ratio from figure 1. As shown, the growth temperature behaviour for the incorporation of tin into the epitaxial layers follows the growth-rate dependence rather closely. The growth temperature itself apparently has little or no effect on the tin transfer ratio.

Doping variations with flow rate have also been examined. The dependence on $AsCl_3$ flow for {100} substrates is indicated by the solid line in figure 4. The other growth parameters were held constant at the values previously indicated. The tin incorporated into the

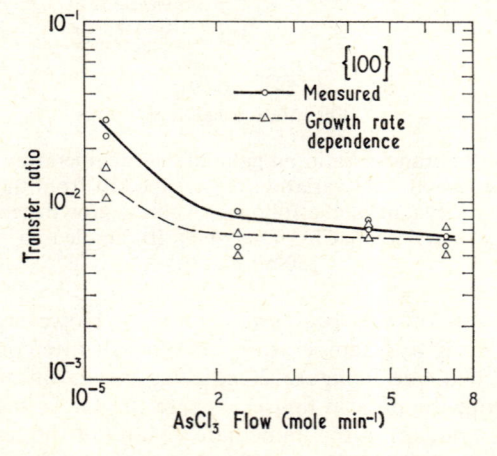

Figure 4. Dependence of transfer ratio on the $AsCl_3$ flux for {100} layers. Other variables were held constant as previously indicated.

epitaxial layer decreases with increasing AsCl₃ flow up to about $2 \cdot 5 \times 10^{-5}$ mole min⁻¹ and then remains relatively constant. However, the growth rate increases with AsCl₃ flux up to about the same point and remains constant. The dashed line shows the growth-rate variation of transfer ratio expected from figure 1. Here also, the measured dependence follows the growth-rate variation reasonably well, so that changes in AsCl₃ flow over this range apparently have little effect on the tin concentration in the epitaxial layer other than by changing the growth rate.

The effects of varying the total hydrogen flow from 100 to 400 ml min⁻¹ have also been examined. No appreciable variation in either growth rate or transfer ratio was observed over this range.

To summarize this series of experiments, it has been found that over a wide range of conditions the transfer of tin from the gallium melt to the epitaxial layer is dominated by a dependence on growth rate. For layers grown under the same conditions the most obvious difference in tin doping is due to the crystallographic orientation of the substrate. This orientation dependence in turn appears to be largely caused by a variation of growth rate with orientation.

4. Tin acceptor concentration

To determine the tin acceptor concentration, layers were doped to levels high enough to ensure that residual acceptors would have little effect on the electrical properties. At these levels it was difficult to analyse the temperature dependence of the Hall constant because of the effects of impurity band or degenerate conduction. The tin acceptor concentration was therefore determined by analysing the Hall mobility with the Brooks–Herring equation (Brooks 1955) for ionized impurity scattering. Since lattice scattering decreases with decreasing temperature while the amount of impurity band or degenerate conduction increases, there exists an optimum temperature for this mobility analysis. These two competing effects produce a minimum in the compensation ratio calculated from Hall mobility and carrier concentration data as a function of temperature using the Brooks–Herring equation. The minimum compensation ratio determined in this manner was used to define an upper limit on the concentration of tin acceptors.

Results of this analysis for two orientations from the same run are shown in table 1. Here the growth rates, the calculated concentration of tin donors and tin acceptors, the compensation ratio, and the optimum measurement temperature are listed for each orienta-

Table 1. Tin acceptor concentration

Plane	{111} As	{100}
Growth rate (μm min⁻¹)	0·043	0·093
N_D (cm⁻³)	$5 \cdot 2 \times 10^{17}$	$2 \cdot 8 \times 10^{16}$
N_A (cm⁻³)	$1 \cdot 8 \times 10^{17}$	$9 \cdot 7 \times 10^{15}$
N_A/N_D	0·35	0·35
T(°K)	130	60

tion. These results indicate that tin is incorporated into the growing crystal in the ratio of about one tin acceptor for every three tin donors independent of growth rate and crystallographic orientation. The tin acceptor concentration must therefore change with growth rate in much the same way as the donor concentration. This compensation ratio of one-to-three is probably determined by the As-to-Ga ratio in this system.

5. Doping uniformity

For uniformly doped layers two effects must be considered: impurity tailing from the seed dopant due to autodoping and/or outdiffusion and perturbations in the system doping. We have previously found that impurity tails from the seed dopant can be minimized by using Se-doped substrates and by not initiating growth until the gallium is completely saturated with arsenic (Wolfe, Foyt, and Lindley 1968). To avoid variations in the system doping, the observed growth-rate dependence of the tin transfer ratio suggests that layers should be grown under conditions where small random changes of the growth parameters have little effect on the growth rate and thus on the tin doping.

Figure 5 shows doping variations parallel to the growth axis for three {100} layers grown under different conditions with the same degree of control over the growth parameters. In each case none of the parameters were intentionally changed during growth and observed variations are due to small fluctuations in the system parameters which unavoidably occur.

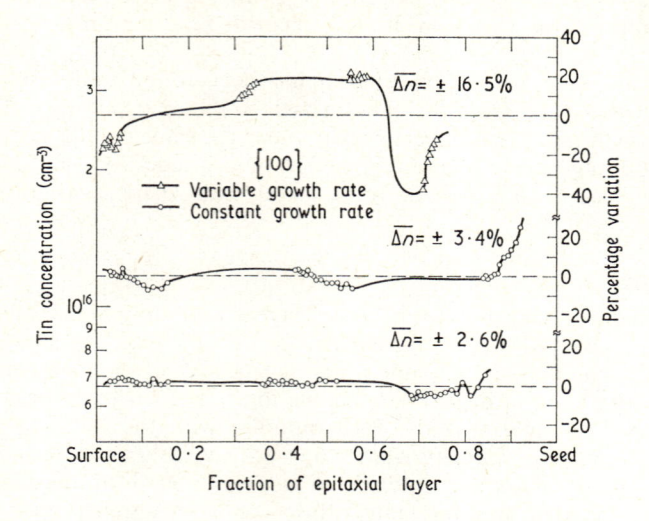

Figure 5. Doping variations parallel to the growth axis for {100} epitaxial layers grown on low-resistance substrates. The vertical scale has been amplified to indicate small variations in doping. The upper profile corresponds to a layer grown under conditions where small variations of temperature and flow will vary the growth rate and thus the doping. The lower two profiles were grown under conditions where the growth rate is insensitive to variations in temperature and flow.

These profiles were obtained by differential capacitance measurements (Thomas, Kahng, and Manz 1962; Amron 1967) on Schottky barrier diodes. The vertical scale has been greatly enlarged to amplify small variations in the doping level. The upper profile is for a layer grown under conditions where small perturbations of the growth parameters would be expected from the curves in figures 1–4 to result in appreciable changes in growth rate. The lower two profiles are for layers which were grown under conditions where the growth rate would be expected from figures 1–4 to vary little with growth parameter changes. The upper profile has an average measured deviation of $\pm 16.5\%$ with a maximum variation of about 40%. The lower profiles have average variations of $\pm 3.4\%$ and $\pm 2.6\%$ with maximum measured deviations of about 6%. Considering the accuracy of the measurement, much of the variation in the lower two profiles could easily be due to measurement error. Thus, to obtain uniformly doped layers it is clearly desirable to grow under conditions where the growth rate and thus the transfer ratio is insensitive to small variations of the growth parameters.

Table 2 lists the results of measurements to determine doping variations perpendicular to the growth axis for both low- and high-resistance substrates grown under constant growth rate conditions. For a layer grown on a low-resistance substrate a series of zero bias capacitance measurements on different Schottky barrier diodes yielded a concentration as shown with an average variation of $\pm 4.4\%$. However, at the edges of the epitaxial layer, where surface preparation prior to growth produced rounding of the substrate, the layer has a faster growth rate. Therefore, tin is incorporated into the edges of the layer at a lower concentration than in the centre. The lower concentration at the edges and its estimated growth rate agree with the growth-rate dependence of the transfer ratio shown in

Table 2. Doping variations perpendicular to growth axis

	Low-resistance substrate		High-resistance substrate
	Centre	Rounded edge	
$n(\text{cm}^{-3})$ Schottky barrier zero bias capacitance	$1 \cdot 13 \times 10^{16}$ $\pm 4 \cdot 4\%$	$5 \cdot 47 \times 10^{15}$	—
Growth rate (μm min^{-1})	0·063	0·089	
$n(\text{cm}^{-3})$ van der Pauw measurement	—	—	$1 \cdot 85 \times 10^{16}$ $\pm 5 \cdot 2\%$
Resistance ratio van der Pauw measurement	—	—	$1 \cdot 02$ $\pm 0 \cdot 02$

figure 1. The last rows in table 2 show the deviation in carrier concentration, $\pm 5 \cdot 2\%$, from a series of four van der Pauw measurements on a high-resistance substrate from the same run and the corresponding variation in resistance ratio, $\pm 0 \cdot 02$.

6. Conclusions

We derive the following conclusions from this investigation of tin-doped epitaxial GaAs.

(a) Over a wide range of experimental conditions the incorporation of tin into the epitaxial layer is dominated by its growth-rate dependence.

(b) This growth-rate behaviour can be explained qualitatively by a time-dependent adsorption of tin at the vapour-solid interface.

(c) Tin is incorporated in the growing crystal in the ratio of approximately one tin acceptor for every three tin donors with the tin acceptors having apparently the same growth-rate dependence as the donors.

(d) For uniform doping, growth conditions should be selected where small random variations in temperature and flow rate have little effect on the growth rate and thus on the doping.

References

AMRON, I., 1967, *Electrochem. Tech.*, **5**, 94.
BANUS, M. D., and GATOS, H. C., 1962, *J. Electrochem. Soc.*, **109**, 829.
BROOKS, H., 1955, *Advan. Electronics and Electron Phys.*, **7**, 158.
EFFER, D., 1965, *J. Electrochem. Soc.*, **112**, 1020.
HALL, R. N., 1953, *J. Phys. Chem.*, **57**, 836.
KNIGHT, J. R., EFFER, D., and EVANS, P. R., 1965, *Solid-State Electronics*, **8**, 178.
THOMAS, C. O., KAHNG, O., and MANZ, R. C., 1962, *J. Electrochem. Soc.*, **109**, 1055.
WILLIAMS, F. V., 1964, *J. Electrochem. Soc.*, **111**, 886.
WOLFE, C. M., FOYT, A. G., and LINDLEY, W. T., 1968, *Electrochem. Tech.*, **6**, 208.

Influence of substrate orientation on GaAs epitaxial growth rates

DON W. SHAW

Texas Instruments Incorporated, Dallas, Texas, U.S.A.

Abstract. Gallium arsenide epitaxial deposition rates were studied as a function of substrate orientation using an HCl/Ga/As$_4$/H$_2$ vapour deposition system. The vapour composition, gas stream velocity, and deposition temperature were held constant while the deposition rates were determined for the following substrate orientations lying in a $\langle 1\bar{1}0 \rangle$ zone: {111} A, {112} A, {113} A, {115}, {001}, {113} B, {112} B, {111} B, {331} B, {221} B, {110}, {551} A, {331} A, and {221} A. Most of the high index orientations were found to exhibit polarity effects. In order to check the rate data, GaAs was deposited on a spherical single crystal substrate under the same conditions. Such a substrate simultaneously exposes all orientations. As expected, facets were found at certain orientation regions on the sphere. These facets corresponded to minima on a polar plot of deposition rate versus orientation. The surface morphologies on the sphere were very similar to those obtained on planar substrates of the same orientations.

1. Introduction

Only a limited number of substrate orientations are commonly employed for epitaxial growth of gallium arsenide. Although a single orientation is not universally accepted, the great majority of investigators confine themselves to {001}, {111} A, {111} B, and {110}. Even though Sangster (1962) predicted theoretically that {113} should be a favourable growth orientation for III–V compound semiconductors, very few experimental studies have considered the high index orientations. In practice, substrate crystals are often cut a few degrees off the principal direction, presumably to improve the surface characteristics of the epitaxial layers. This, in effect, results in high index orientations, and as Blakeslee (1968) has recently shown, can result in sizeable differences in the growth rates and electrical properties of the layers. At least two references (Ing and Minden 1962; Pizzarello 1963) are available for heteroepitaxial growth of GaAs on {113} Ge substrates. However, the high index orientations have not been systematically investigated and related to the low index orientations. This is the goal of the present investigation.

2. Experimental

An elemental reactor system, previously described (Shaw 1968a), was employed throughout the study. Gallium was transported as GaCl by reaction of HCl with the liquid element at 900°C. Arsenic was transported as the element with a hydrogen carrier gas. The calculated reactant partial pressures were $P_{\text{GaCl}}{}^0 = 7\cdot9 \times 10^{-3}$ atm and $P_{\text{As}_4}{}^0 = 3\cdot7 \times 10^{-3}$ atm. A small amount of free HCl was added to the reactant gas stream. This was found to minimize deposition on the tube walls which might deplete the reactant gas stream before it reached the substrate region. The deposition temperature was 750°C.

The substrates were cut from Czochralski grown, Sn doped crystals. These crystals were grown in $\langle 001 \rangle$, $\langle 111 \rangle$ B, or $\langle 110 \rangle$ directions. In order to specify crystal polarity, the high index slices were always cut from $\langle 111 \rangle$ B grown crystals. The slices were checked after sawing to determine that their surfaces were within $\pm 0\cdot5°$ of the desired orientation. After sawing, the substrates were polished with sodium hypochlorite on a rotating Pellon cloth in the manner described by Reisman and Rohr (1964).

Usually, four different substrate orientations were simultaneously subjected to deposition in a given run. Preliminary experiments established that the deposition rates were independent of the relative positions on the substrate holder. In every run one of the substrate orientations was either {001} or {110}. This served as a control and permitted assessment of the experimental reproducibility. The deposition rates were found to be reproducible to within approximately 10%. Just prior to deposition, the substrates were briefly etched with HCl in the deposition apparatus. The duration of the deposition was always one hour and the deposit thickness was identified with the deposition rate in microns per hour. However, preliminary experiments established that the deposition rates were independent of time for times up to 90 minutes. Thickness measurements were obtained by measuring cleaved or angle lapped cross-sections, with the substrate-deposit interface being revealed by anodic oxidation.

In all, 14 orientations lying in a ⟨110⟩ zone were studied. These were the following: {111} A, {112} A, {113} A, {115}, {001}, {113} B, {112} B, {111} B, {331} B, {221} B, {110}, {551} A, {331} A, and {221} A. The deposition rates for the first eight orientations were described by Pizzarello (1963). The choice of the high index orientations was based on the concepts of surface steps such as described by Sangster (1960, 1962). In this case the high index surfaces are represented as stepped structures built from fundamental units of the three basic orientations, {001}, {110}, and {111}. For example, the {113} surface may be considered as composed of equal numbers of {001} and {111} unit steps. Figure 1 illustrates

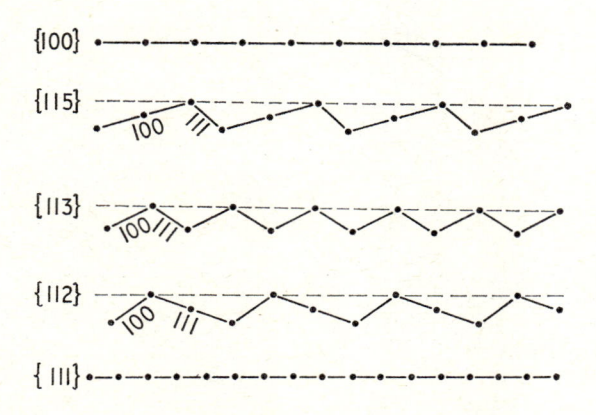

Figure 1. Theoretical surface arrangements.

this concept for the high index surfaces lying between {001} and {111}. Of course, these are idealized surfaces, and in reality greater surface roughness would be expected. Nevertheless, the relative proportions of {001} and {111} character would not change.

3. Results and discussion

3.1. *Growth on planar substrates*

The actual deposition rates obtained for all 14 orientations are listed in table 1. However, the results are best summarized in the polar diagram given in figure 2. In this plot the orientation is represented by the azimuthal position and the deposition rate by the distance from the origin. The upper half of the curve represents the 'A' polar orientations or those surfaces terminating in an excess of gallium atoms, while the 'B' orientations with excess arsenic lie in the lower half. The generally higher deposition rates for the 'A' polar orientations are obvious in figure 2.

The lowest deposition rate occurs with {111} B substrates, and the deposition rate increases as the orientation moves from the {111} B toward either the nearest {001} or {110}. The results are consistent if one assumes that the rate-limiting process for growth on {111} B is nucleation of a surface layer which then spreads laterally across the surface. Thus, if the orientation deviates slightly from {111} B, steps are introduced which can then spread across

Table 1

Orientation	Deposition rate (microns h^{-1})
{001}	29
{115}	32
{113} A	34
{112} A	67
{111} A	90
{221} A	92
{331} A	70
{551} A	51
{110}	14
{331} B	15
{221} B	20
{111} B	6
{112} B	24
{113} B	37

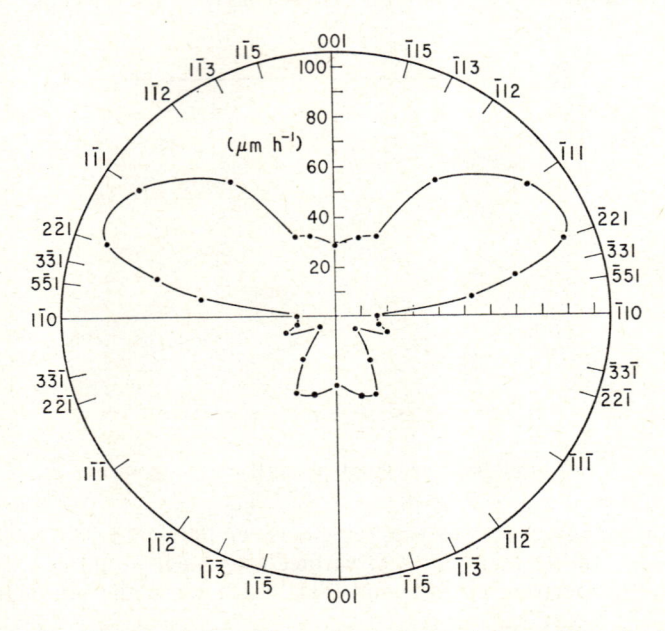

Figure 2. Polar diagram of deposition rate versus crystallographic orientation.

the surface, and the deposition rate should increase relative to the {111} B. Between {111} B and {001} the greatest theoretical density of surface steps occurs at {113} B, and as can be seen from figure 2, the deposition rate passes through a maximum at this orientation.

In contrast to {111} B, the {111} A deposition rate is probably not limited by the rate of step nucleation since the deposition rate actually decreases as the orientation deviates from {111} A toward {001}. As the orientation is shifted from {111} A toward {110}, the rate increases very slightly, and beyond the {221} A, decreases rapidly as the influence of the slow growing {110} becomes significant. Note that as the orientation changes 8° from {110} to {551} A, the deposition rate increases by a factor of three. However, if the orientation deviates from the {110} in the opposite direction by as much as 13° ({331} B), the deposition rate remains essentially unchanged. This illustrates the significance of specifying the direction as well as the degree of misorientation when a crystal is intentionally cut off the major plane.

The small dip in the curve at {113} A is of interest. All other minima occur at low-index orientations ({001}, {110}, and {111} B). The experiments were repeated several times and in every case the minimum at {113} A was observed. Sangster (1962) described the {113} orientation as particularly favourable for the growth of III–V compound semiconductor crystals; however, he did not consider the differences between the {113} A and the {113} B. Significant differences are apparent in figure 2, since {113} A occurs at a minimum while the {113} B occurs at a maximum even though the differences in their absolute rates are very small. No significant differences in deposition rate between the {115} A and {115} B were observed. The difference in deposition rates for the {331} polarities is much greater than for the {113}.

3.2. *Growth on a spherical substrate*

Growth on a single crystal sphere exposes all orientations to simultaneous deposition (see for example, Buckley 1951). In addition, minima in the polar diagram would be expected to result in facets on a sphere at the orientations corresponding to the minima. Thus, deposition on a spherical substrate provides an opportunity to verify to some extent the results described earlier, since facets should appear at the {113} A, {111} B, {110}, and {001} regions of the sphere.

A single crystal of GaAs was mechanically ground into a spherical shape. The surface damage was then removed by polishing the sphere in 40 : 4 : 1 $HCl : H_2O_2 : H_2O$. This etch was found to be non-preferential and did not result in formation of facets during etching. The polished sphere was then subjected to epitaxial growth under the same conditions as for the planar growth experiments. A scanning electron micrograph showing the sphere after deposition is given in figure 3. The micrograph shows regions of different texture with certain relatively flat, faceted regions. An optical microgoniometer was used to determine the orientations of these facets. The facets were found only at the orientations corresponding to minima on figure 2, i.e., {111} B, {110}, {001}, and {113} A. In addition, at each position on the sphere corresponding to one of these orientations a facet was present.

Optical micrographs representing examples of these facets are presented in figure 4a–d. Two small growth defects are apparent on the {111} B facet (figure 4a). Other {111} B facets were present on the sphere without apparent growth defects. Comparison of figure 4b and c reveals a morphological similarity between the {001} and {113} A facets. Both facets have roughly circular, irregular peripheries without crystallographic symmetry. In addition, unlike the {111} B and {110} facets, neither the {001} nor the {113} A facet has a terraced structure leading up to it. This similarity is also apparent from micrographs of surfaces grown on planar {001} and {113} A substrates (figure 4e and f). These figures are interference contrast micrographs, and both reveal wavy, irregular surfaces (both surfaces appear smooth and brilliant under normal microscopy).

Definite similarities were present between the morphologies of the various high index regions on the sphere and the surfaces of layers grown on planar substrates of the corresponding orientations. This is illustrated by figure 5 where optical micrographs (left) of the {331} A, {331} B, and {115} regions of the sphere are compared with those of the corresponding planar growth surfaces (right). In figure 5a the {331} A region corresponds to the bright area in the centre of the micrograph. In the far right of figure 5a may be seen the adjacent {110} facet. Likewise, the {115} region is the bright pebble-like area in figure 5c. In the dark areas to the left the near {001} facet may be observed. Note also the pronounced differences in surface morphologies between the {331} A and {331} B polar orientations.

The similarities in surface morphologies between the sphere and the planar deposits of the same orientation demonstrate the usefulness of growth on a spherical substrate. This is particularly true for materials or growth processes where the most promising orientation is to be selected for further study. It should be stressed that the results described herein were obtained for a particular set of growth conditions and would be somewhat different if the vapour composition or some other important parameter were altered. Indeed, it has been shown that the relative growth rates of {111} A and {111} B facets are drastically altered when the GaAs ratio in the vapour phase is changed (Shaw, 1968b).

Acknowledgments

The author wishes to express his appreciation to Dr. R. D. Dobrott for the scanning electron microscopy. In addition, appreciation is extended to Drs. L. G. Bailey, L. D. Dyer, and G. O. Krause for their helpful suggestions and discussions.

References

BLAKESLEE, A. E., 1968, Presented August, 1968 meeting of the Met. Soc. of A.I.M.E., Chicago Paper 32.
BOBB, L. C., HOLLOWAY, H., MAXWELL, K. H., and ZIMMERMAN, E., 1966, *J. Appl. Phys.*, **37**, 4687.
BUCKLEY, H. E., 1951, *Crystal Growth* (New York: John Wiley), p. 130.
ING, S. W., Jr., and MINDEN, H. T., 1962, *J. Electrochem. Soc.*, **109**, 995.
PIZZARELLO, F. A., 1963, *J. Electrochem. Soc.*, **110**, 1059.
REISMAN, A., and ROHR, R., 1964, *J. Electrochem. Soc.*, **111**, 1425.
SANGSTER, R. C., 1962, *Compound Semiconductors*, Vol. 1 (New York: Reinhold), pp. 241–253.
SANGSTER, R. C., 1960, *Electrochem. Soc. Extended Abstracts*, **9**, 172.
SHAW, D. W., 1968a, *J. Electrochem. Soc.*, **115**, 405.
SHAW, D. W., 1968b, *J. Electrochem. Soc.*, **115**, 777.

The origin of macroscopic surface imperfections in vapour-grown GaAs

J. J. TIETJEN, M. S. ABRAHAMS, A. B. DREEBEN and
H. F. GOSSENBERGER

RCA Laboratories, Princeton, New Jersey 08540, U.S.A.

Abstract. The origin of surface imperfections on vapour-grown epitaxial layers was studied using the 'arsine' growth technique. To delineate the role of the substrate, the effect of saw-cutting damage, mechanical polishing, impurity striations in the substrate, surface particles, and exposure to laboratory atmosphere were examined. With respect to growth conditions, the effect of oxygen contamination and deposition of GaAs on the walls of the growth apparatus were investigated. The major defects observed were macroscopic surface pits, and their origin was shown to be localized non-uniformities, especially particles adherent to the growth surface. Special precautions are indicated to reduce the density of surface imperfections.

1. Introduction

Although great care is taken in the preparation of epitaxial GaAs, both with regard to substrate preparation and growth conditions, macroscopic surface imperfections are a common occurrence in these layers. Such imperfections can be the direct cause of, or be related to, p–n junction nonplanarity, microplasmas, and electrical inhomogeneities in devices such as injection lasers or Gunn oscillators. Nevertheless, little discussion of these defects appears in the literature, indicating that the cause of these imperfections is not well understood.

Most of the work reported to date concerning surface defects in vapour-grown GaAs has involved the role of contaminants such as oxides or gallium droplets (Goldsmith and Oshinsky 1963; Joyce and Mullin 1966, Holloway and Bobb 1967). In the present study, it is shown that localized surface non-uniformities, such as adherent particles or impurities segregated in the substrate surface, provide an additional source of macroscopic defects.

This study has utilized the 'arsine' method of vapour-phase growth (Tietjen and Amick 1966), and the dominant defects observed were pits. This is in contrast to the hillock-like protuberances seen in the above-mentioned studies. Nevertheless, it might be expected that the specific polishing and handling procedures suggested here can be of value for improving the surface quality of GaAs layers prepared by other techniques.

2. Experimental

Homo-epitaxial GaAs layers were prepared by a vapour-phase growth technique (Tietjen and Amick 1966) in a flow system. In this technique arsine serves as the source of arsenic vapour, and gallium is transported as its subchloride via the reaction of gallium and hydrogen chloride.

The substrates used for this investigation were obtained from Czochralski-grown single-crystalline ingots purchased from Bell and Howell. The substrate surfaces were misoriented 3° from the {100} plane.

For chemical polishing, a solution of 0·5% by volume of bromine in methanol was used. Mechanical polishing was accomplished in a series of steps involving alumina particles having diameters in the range of 0·05 to 5 μm.

Both before and after growth, selected surfaces were examined by either or both optical and carbon-replication electron microscopy.

3. Results and discussions

The effects of substrate preparation were examined independently of the condition of the vapour-growth system, and, accordingly, these experiments are discussed separately in the following two sections.

3.1. *Substrate preparation and selection*

Figure 1 shows the effect of damage introduced by saw-cutting on the surface perfection. In both cases, 15-μm-thick GaAs layers were deposited on substrates which were chemically polished after having been cut to the proper orientation. However, the substrate of figure 1*a* was polished to remove 25 μm of material, while 100 μm were removed from the substrate of figure 2*b*. The defects seen in figure 1*a* are terraced hillocks containing one or more pits at the apex of the structure. However, as seen in figure 1*b*, the removal of 100 μm of material by an all-chemical polishing procedure provides a nearly defect-free surface. It is evident, therefore, that the work damage introduced in the cutting process extends at least 25 μm into the substrate. It should also be noted that even when relatively small amounts of the surface are removed, as in figure 1*a*, there are no macroscopic pits, indicating that pits are not caused by cutting damage.

Figure 2 shows a comparison between the effects of mechanically polishing and chemically polishing a substrate. In both cases, 50-μm-thick GaAs epitaxial layers were deposited on substrates which were mechanically polished to remove approximately 100 μm of material to eliminate the above discussed work-damage introduced in the crystal sawing process. However, the substrate of figure 2*a* was subsequently chemically polished to remove an additional 100 μm of material while the substrate of figure 2*b* had only 25 μm removed by this means. Macroscopic pits are evident in figure 2*b* but absent in figure 2*a*. Since this defect was not observed in the presence of work damage when an all-chemical polishing procedure was used (figure 1*a*), it must be concluded that these pits are introduced by the mechanical polishing procedure. In particular, they are ascribed to the presence of the abrasive particles which are imbedded in the surface of the substrate, and act as nucleating centres for pit formation. These particles are not easily removed from the surface of the substrate and, therefore, it requires prolonged chemical polishing to minimize their presence.

To test the hypothesis of pit nucleation at surface-particles, seeding experiments were performed in which 0·3-μm-diameter alumina particles were placed on the substrate surface prior to growth. These particles were distributed in three bands across the surface corresponding to low, medium, and high densities. The surface of a 15-μm-thick layer grown on this seeded substrate is shown in figure 3*a*, with the medium and high density bands displayed at higher magnification in figures 3*b* and 3*c*. It is obvious that the formation of pits can result from the presence of particles on the crystal surface, with the pit-density being proportional to the number of particles. The faceted octagonal-shaped pits observed are very similar in appearance to those seen in the mechanical polishing experiment.

When melt-grown GaAs is doped to high concentrations with impurities such as selenium or tellurium, thermal fluctuations during growth can result in a striated segregation of these impurities in the crystal. It is likely that localized regions of enhanced segregation will occur in such crystals, which can serve as nucleation centres for defects when these crystals are used as substrate material.

Figure 4 shows the effect of impurity striation on the perfection of a GaAs epitaxial layer. Figure 4*a* shows the appearance of a substrate surface after chemical polishing to remove more than 100 microns of material from a wafer having a high (greater than 2×10^{18} cm^{-3}) tellurium concentration. Figure 4*b* shows the nature of the surface of a 15-μm-thick GaAs layer deposited on this substrate. In addition to the stria which propagate into the grown layer, a high density of pits is observed.

An additional cause of localized surface non-uniformity can be contamination resulting from exposure to the laboratory atmosphere. Figure 5 is a comparison between layers having thicknesses of 15 μm (*a*) and 50 μm (*b*) which were grown on substrates which, after chemical polishing, were exposed to the laboratory atmosphere for periods as short as 2 and 5 minutes, respectively. Two observations are readily apparent. First, the uniform size of the pits indicates that they are undoubtedly nucleated by defects present on the substrate surface, and that additional pits are not formed during the growth process. Second, even slightly longer exposure to the laboratory atmosphere results in more pitting. These two results are both explicable in terms of the substrate surface becoming increasingly contaminated in the laboratory atmosphere, as, for example, by dust particles, which could again serve as nucleation centres for pits.

3.2. *Effects related to the growth system*

As has been described by previous workers, (Goldsmith and Oshinsky 1963; Holloway and Bobb 1967) contamination of the growth system, especially by oxygen or water vapour, can play an important role in the formation of macroscopic surface imperfections. Accordingly, layers were grown with and without the addition of water vapour to the growth apparatus. Further insights were gained in these experiments, because the manner in which the water was introduced to the system affected the perfection of the growing layer.

The water vapour contamination experiments were divided into two parts. For one growth, the water vapour was added to the hydrogen chloride flow and, therefore, came in contact with the gallium source. For a second growth, the water was added to the arsine flow, and because of the design of the growth apparatus, did not contact the gallium. In both cases the concentration of water in the vapour phase was estimated to be about 10 p.p.m.

The results of these experiments are presented in figure 6. Figures 6a and 6b show the surface of a 15-μm-thick layer prepared without the addition of water vapour, as seen by a light and electron microscope, respectively. For this experiment, care was taken in the selection and preparation of the substrate surface in order to avoid the presence of localized surface non-uniformities. The uniformity and smoothness of this deposit is apparent.

A virtually identical surface appearance to that in 6a and 6b was obtained (not shown here) when water vapour was introduced to the growth system along with the arsine and, as previously stated, did not react with the gallium source. However, when the water vapour is permitted to pass over the gallium with the hydrogen chloride, the surface of the deposited layer appears rough to the unaided eye. Typical surfaces are shown in the light and electron micrographs of figures 6c and 6d, respectively. A density of about 10^4 pits cm^{-2} are present, which have a variety of sizes, implying that they form continually during the growth process. In addition, a background of smaller pits is present at a density of about 10^7 cm^{-2}.

A plausible explanation for these results is that the pits arise from Ga_2O_3 particles present on the surface of the growing crystal. At the low gallium-source temperature (about 775°C) employed in this experiment, the primary reaction product of gallium and water would be Ga_2O_3 which is relatively non-volatile but which could be swept through the system in the form of fine particles. Then, when the water vapour is added so that it does not contact the gallium, Ga_2O_3 particles do not form and smooth surfaces result.

Another source of particles that can fall onto the substrates can be any loose deposit that forms in the system. In particular, GaAs deposits on the walls during growth, and the position of the deposit relative to the substrate is a function of the temperature of the gallium source and the total growth time. An experiment was carried out in which the conditions were selected such that the wall-deposits of GaAs would slowly grow back into the substrate region during the growth of the epitaxial layer. Two layers were grown sequentially in the system, the first deposition occurring prior to the formation of the deposit near the substrate, while the second occurred as the deposit formed in the vicinity of the substrate. In both cases, the substrate was prepared to avoid contamination with particles, and about 15 μm of GaAs was grown. In the first deposition, virtually no pits were seen (figure 7a), while a very high density of pits was observed (figure 7b) for the second deposition.

4. Conclusions

The results of the experiments discussed above show that localized surface non-uniformities are an important cause of macroscopic surface imperfections. Therefore, the ability to prepare GaAs homo-epitaxial films free of such defects relies on paying strict attention to a number of details. These include choosing proper substrate material, particularly material free of macroscopic imperfections such as impurity stria; polishing the substrate by an all-chemical method so as to remove the work damage introduced by cutting without introducing particles to the crystal surface; and adjusting the growth conditions to prevent deposition from occurring on the walls of the reaction tube in the vicinity of the substrate. In addition, it is necessary to avoid surface contamination either during substrate handling or in the growth system. When these precautions are followed, it is frequently possible to prepare layers having a high degree of surface smoothness, especially when the layer thickness is less than 50 μm.

Acknowledgments

The authors are grateful to D. Richman, L. R. Weisberg, and R. E. Enstrom for many valuable discussions, and to C. J. Buiocchi and A. J. Tocci for important technical assistance.

References

GOLDSMITH, N., and OSHINSKY, W., 1963, *RCA Review*, **24**, 546–54.
HOLLOWAY, H., and BOBB, L. C., 1967, *J. Appl. Phys.*, **38**, 2893–6.
JOYCE, B. D., and MULLIN, J. B., 1966, *Inter. Symp. on GaAs*, Inst. of Physics and the Physical Society Conf. Series No. 3, pp. 23–6.
TIETJEN, J. J., and AMICK, J. A., 1966, *J. Electrochem. Soc.*, **113**, 724–8.

Preparation of epitaxial gallium arsenide for microwave applications

F. J. REID and L. B. ROBINSON

The Bayside Laboratory, Research Center of General Telephone & Electronics Laboratories Incorporated, Bayside, New York, U.S.A.

Abstract. The preparation and properties of epitaxial layers of GaAs grown by a vapour-phase deposition involving a reaction of $AsCl_3$ with high-purity gallium are described. Special treatment of substrates and starting materials is required to obtain high-quality layers free from interface barrier layers. Several wafers have been prepared with electron mobilities at 77°K in the range 90 000 to 130 000 $cm^2\ v^{-1}\ s^{-1}$. The quality of the $AsCl_3$ is shown to be the most important consideration in determining the purity of the GaAs. Hall coefficient and resistivity values measured on layers deposited on semi-insulating GaAs are found to be in error if tellurium-doped substrates are present during the deposition. This has not been the case when silicon doping is used in the n^+ GaAs substrate.

1. Introduction

Semiconductor microwave devices have recently taken a variety of forms, including avalanche diodes, field-effect transistors, Schottky-barrier mixers, and transferred-electron devices, such as Gunn-type diodes and various non-transit-time diodes. Since the geometry of the device plays an important role in determining characteristics of the device, particularly the frequency of operation, epitaxially grown structures are used almost exclusively. These structures are preferred whether the material is silicon, germanium or GaAs. For transferred-electron devices GaAs is the only material of the three that is suitable.

Two types of epitaxially grown structures are used as the basic structure for each of the devices. These structures are n/n^+ and n/i where the n region is a layer of n-type material grown on a heavily doped n^+ substrate or on an insulating i substrate. In the latter case, the insulating substrate is used as a non-conducting support, the device being made in a planar or co-planar configuration. The n/n^+ structures, or in some cases p/p^+ structures, are building blocks in the fabrication of sandwich-type devices in which the n^+ semiconductor serves as one of the electrical contacts.

In addition to achieving the desired properties in the n layer, one must produce a barrier-free interface in the n/n^+ structures. This has proved to be an important consideration with GaAs particularly when the n layer must contain a low carrier concentration. In this paper, procedures are presented for the preparation of epitaxial layers of n-type GaAs with carrier concentrations in the low $10^{14}\ cm^{-3}$ range and n/n^+ structures free of undesirable barriers. Experience gained in working with n^+ substrates doped with silicon or tellurium and with the determination of the properties of n layers is discussed. A paper by Cohen, et al. (1969) discussing the properties of Gunn-effect oscillators made using these GaAs n/n^+ structures is included in the section on 'Microwave Devices'.

2. Preparation

The vapour-phase growth of GaAs layers was conducted using the $AsCl_3$–Ga–H_2 open flow system. A schematic diagram of the reactor system and furnace-temperature profile is shown in figure 1. The system, from the flowmeters to and including the furnace tube, is constructed of Pyrex and natural quartz. Where indicated, teflon stopcocks are used; however, no stopcock is used directly in-line with the $AsCl_3$ stream. Between the flowmeters and the hydrogen purifier, stainless steel tubing and valves are used. Helium is used to flush the system.

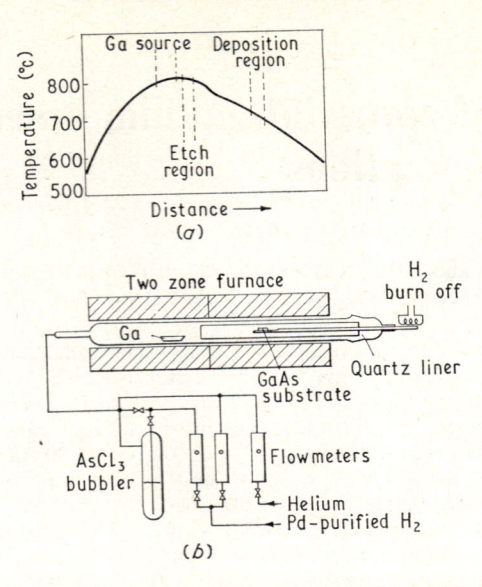

Figure 1. Reactor for epitaxial deposition of GaAs.

Approximately 50 grams of gallium are placed in the boat initially and 100 to 150 grams of AsCl$_3$ in the bubbler so that many runs can be made with a minimum of disturbance to the system. After each run the right-hand zone of the furnace is raised to about 800°C and the substrate region is cleaned by passing AsCl$_3$ through it. The gallium is saturated with arsenic at 800°C using flow rates of 250 cm^3 min^{-1} of hydrogen through the AsCl$_3$ bubbler at about 25°C and 100 cm^3 min^{-1} of hydrogen by-passing the bubbler. These flow rates are also employed during the vapour growth of GaAs. The velocity of the gas stream in the reactor proper is approximately 1 cm sec^{-1}.

According to the data of Effer (1965) on the material balance during the AsCl$_3$–GaAs–H$_2$ transport reaction, the maximum GaCl-to-As ratio formed at the gallium source at 800°C is approximately as shown in the following reaction.

$$4\text{Ga} + \tfrac{10}{3}\text{AsCl}_3 \rightarrow \text{GaCl} + 3\text{GaCl}_3 + \tfrac{5}{6}\text{As}_4. \tag{1}$$

The reason the term maximum ratio is used is that the reaction occurs as shown only if all the AsCl$_3$ entering the system reacts. Then at the substrate the complete reaction of the available GaCl and arsenic yields the following.

$$3\text{GaCl} + \tfrac{5}{2}\text{As}_4 \rightarrow 2\text{GaAs} + \text{GaCl}_3 + 2\text{As}_4. \tag{2}$$

Hence, with the gallium source at 800°C, the by-products in the formation of GaAs contain a considerable quantity of free arsenic. This is indeed observed experimentally. The GaAs deposition rate under these conditions is about 0·2 micron min^{-1}. If the gallium-source temperature is raised high enough to produce free gallium in the by-products, the quality of the epitaxial-layer surface is degraded.

Boat-grown GaAs is used exclusively for n$^+$ substrates, because with this technique moderately low dislocation density material can be prepared. Both silicon and tellurium have been used as dopants; however, more emphasis has been placed on silicon-doped material. The substrates contain dislocation densities in the range 1000 to 3000 cm^{-2}. Chromium doping is used to prepare GaAs for insulating substrates, and this material contains larger dislocation densities (5000 to 10 000 cm^{-2}). The epitaxial layers are grown on surfaces oriented 2 degrees from a {100} plane.

The substrates are given a chemo-mechanical polish using a Lustrox 1550 solution on a rotating Supreme-K cloth. Although this procedure gives a fine optical finish to the GaAs, an examination of the first batches of samples by x-ray diffraction topography showed that latent scratches were often present. Improvements were made to eliminate the occurrence of what may be termed deep latent scratches; however, the chemo-mechanical polish generally leaves a network of shallow defects. These defects are removed by a chemical polish in 3 : 1 : 1 $H_2SO_4 : H_2O_2 : H_2O$. Approximately 10 microns of material are removed from each surface during the chemical etching. The substrates are given a final rinse in ultra-pure water prior to being loaded (wet) into the reactor under a flowing H_2–He mixture. More is said concerning the water rinse in section 3. The substrates are given a chemical vapour etch in the reactor system at about 800°C and another 7 microns or so of material is removed. This step is critical with respect to the resulting surface finish of the epitaxially grown layer. If too much etching occurs, the layer can be pitted. If too little etching occurs, the layer can have a dull appearance. The optimum conditions for etching are dependent on the type of dopant in the substrate.

3. Interface barriers due to copper

In many of the microwave applications sandwich structures, such as $n^+/n/n^+$ and $p^+/p/n/n^+$, are used, where the basic n/n^+ structure is grown epitaxially. An important consideration with GaAs is to produce a low-resistance, ohmic interface between the lightly doped n region and the heavily doped n^+ region. It has been the experience of many workers in the field that barriers can be formed at this interface. The barriers appear to be thin layers of insulating material, or in some cases p-type material. Such a barrier gives rise to a unipolar negative resistance, or in other words, the diode behaves as a switch for one direction of bias (see for example, figure 2 of Weiser 1967 and figure 5 of Wolfe et al. 1968). Structures of this type are useless for microwave devices.

Structures free from interface barriers can be prepared by careful cleaning of the substrate. One cause of such interface barriers has been isolated and involves the contamination of the surface of the GaAs substrate with copper. If such surface contamination is present, the chemical-vapour etch does not remove all the copper before some of it diffuses into the substrate at the temperature (800°C) used during the etching. The solubility of copper in heavily doped n-type GaAs at 800°C is nearly 1×10^{18} cm^{-3}. This copper then can diffuse into the growing n-type layer to produce serious effects near the n–n^+ interface. In the very lightly doped n-layer, sufficient copper is incorporated to convert part or all of the layer to p-type material or to convert part of the layer to semi-insulating GaAs by compensating out the shallow donors.

One particular source of copper is water that is less than ultra-high purity, i.e., less than 12 to 18 megohm-cm water. When 1 to 5 megohm-cm water was used for rinsing the substrates, copper was positively identified by three techniques as follows: emission spectrographic analysis on the substrates before heating; photoluminescence, and electrical measurements both after heating and epitaxial deposition. For the emission spectrographic analysis a surface area of about 0·1 cm^2 and a surface area-to-volume ratio of about 50 was used. Copper concentrations equivalent to about a monolayer of copper on the surface of the GaAs substrate were detected when the impure water was used. The concentration of copper increased as the amount of water used to rinse the substrate was increased. No copper was detected when the substrates were rinsed in the ultra-pure water. With the detection limit prevailing, this makes the concentration less than 0·01 of that detected when the impure water was used.

Substrates rinsed with the impure water were placed in the reactor system, vapour etched, and coated with a layer of GaAs. In every case, the sandwich structures prepared instead of being n/n^+ were $n/i/n^+$, $n/p/n^+$, or p/n^+. Therefore they were not suitable for microwave devices. When p-type layers were formed, the companion p/i structure was used for Hall-effect measurements as a function of temperature. A thermal activation energy associated with the major acceptor impurity was found to be in agreement with the activation energy obtained with samples purposely doped with copper.

In addition, examples of all three undesirable structures were subjected to photoluminescence measurements using a laser beam for excitation (6328 Å emission from He–Ne). The laser beam sampled only to depths of the order of 1 micron into the sample. Regardless of whether the surface of a p-layer, an n-layer, or an n^+ substrate was examined, the photoluminescence spectra showed an emission peak at about 1·36 ev at 77°K. Such an emission has been correlated with the presence of copper. When ultra-pure water was used in rinsing the substrate, this emission did not appear, and no evidence of interface barrier or p-type layers was observed. For water of an intermediate purity of say 5 to 10 megohm-cm, an intermittent occurrence of barrier layers may be observed and the analyses for copper are not as positive.

Various investigators have reported on vapour-grown GaAs structures which perform well in microwave devices. In the following references the substrate-cleaning procedures used suggest that special techniques for avoiding copper contamination were employed. For example, Eddolls et al. (1967) used a final washing in isopropyl alcohol in a Soxhlet apparatus; Bolger et al. (1967) used water doubly distilled in quartz and a final rinse in electronic grade isopropanol; Effer (1965) used a KCN rinse with a final rinse in pure, water-free methanol; Wolfe et al. (1968) used an EDTA rinse with a final rinse in a transistor washer (14 megohm-cm water). A. E. Blakeslee in a private communication indicated that he had recently observed the effects of copper contamination from water with respect to interface barriers.

4. Electrical evaluation of epitaxial layers

Electrical properties were determined on layers grown on high-resistivity substrates, i.e., n/i structures, using the van der Pauw (1958) method and a symmetric geometry. The resistance ratios in all the measurements reported was in the range 1 to 2. The Hall mobility and electron concentration for each layer were calculated from $\mu_H = R_H/\rho$ and $n = r/eR_H$ where ρ is resistivity, R_H is Hall coefficient and the scattering factor, r, was taken as 1. For the lowest-carrier-concentration samples a magnetic field of 10 to 15 kG was used for the Hall measurements. Where data were obtained at 3 kG a notation is made.

To determine if the Hall measurements on the n/i structures give a true indication of the properties of the n layer throughout its thickness and an indication of the properties of the n layer on a companion n/n+ structure, several complementary measurements were made. Carrier concentration profiles of the n layer were made on n/i structure by Schottky-barrier measurements. Details of the measurement technique are given by Cohen et al. When a companion n+ substrate doped with silicon or when no n+ substrate was present, the carrier-concentration profile of the n/i structure was reasonably flat, exhibiting a sharp decrease in carrier concentration near the n–i interface as seen in figure 2. The carrier concentration

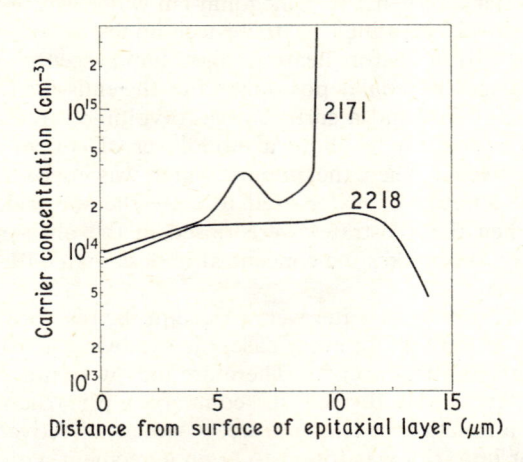

Figure 2. Schottky-Barrier profiles of GaAs n/i structures. Grown in the presence of n+ substrates.

Sample	Doping in n+ substrate	$n = 1/R_H e$ cm^{-3}	Thickness of n layer μm
2171	Te	$7\cdot 8 \times 10^{15}$	9
2218	Si	$1\cdot 9 \times 10^{14}$	13

calculated from a Hall-effect measurement was in good agreement with the Schottky-barrier measurement.

When the n+ substrate was doped with tellurium, meaningful Hall-effect measurements were not possible because of an autodoping effect on the n/i structures. For example, in figure 2 a profile of an n/i structure grown in the presence of a tellurium-doped substrate is shown. Near the n–i interface a high-conductivity region was formed. Hence, Hall measurements on such a structure give erroneous results. The degree to which autodoping occurs is dependent on the relative positions of the i and n+ substrates in the reactor system. The smallest effect is observed if the i substrate is upstream from the n+ substrate. However, the effect is never eliminated completely. All data reported in this paper are therefore restricted to measurements on n/i structures grown in the presence of silicon-doped n+ substrates. Of course, n/n+ structures, where tellurium is the n+ dopant, are useful for devices; but carrier concentrations in the n layer must be determined by Schottky barrier measurements.

Table 1 is a compilation of electrical properties on selected samples illustrative of preparations yielding exceptionally high-quality material. More is said in the next section regarding the conditions required to achieve this level of purity reproducibly.

Table 1. Electrical properties of selected epitaxial layers of GaAs

Layer number	Thickness, μm	ρ at 300°K, ohm-cm	n at 300°K, cm^{-3}	μ_H at 300°K, cm^2 v^{-1} s^{-1}	μ_H at 77°K, cm^2 v^{-1} s^{-1}
2239	11	8·6	$1·1 \times 10^{14}$	6600	120 000
2168	10	4·8	$1·8 \times 10^{14}$	7200	100 000
2218	13	4·7	$1·9 \times 10^{14}$	7000	130 000
2170	8	3·5	$2·6 \times 10^{14}$	6800	86 000
2182A	8	3·3	$3·1 \times 10^{14}$	6100	95 000
2225	10	2·7	$3·4 \times 10^{14}$	6700	83 000
2181	15	2·0	$4·5 \times 10^{14}$	7000	120 000
2182B	8	1·67	$5·2 \times 10^{14}$	7200	100 000
2219	16	1·55	$5·9 \times 10^{14}$	6800	92 000

5. Purity of epitaxial layers

For microwave applications which require low-carrier-concentration GaAs, i.e., transferred-electron devices, one must achieve nearly state-of-the-art purities for GaAs. In most other microwave device applications the addition of dopants is required to prepare material with higher carrier concentrations. For all of the device applications, carrier mobility should be as large as possible; however, the requirements for high-purity GaAs are best stated in terms of the resistivity, ρ, and carrier concentration, n, of the material. Domains will develop in the material only if the product nl is above a limit of about 4×10^{11} cm^{-2} (Hilsum and Morgan 1967), where l is the separation between anode and cathode. This is equivalent to an l/ρ greater than about 5 for l in microns and ρ in ohm-cm. In addition, to maintain cw operation a good design figure to use is l/ρ less than 20 (Hilsum 1968).

Table 2 is a compilation of the materials requirements for transferred-electron devices. Also included are the approximate requirements for other microwave device materials. For cw operation of Gunn devices, material with carrier concentrations in the range 1×10^{14} to 3×10^{15} cm^{-3} is required. To achieve the lowest carrier concentrations, the vapour-phase-growth system is operated in an undoped mode keeping to a minimum the residual impurities emanating from the materials in the system.

It has been found that successful preparation of GaAs with $n < 1 \times 10^{15}$ cm^{-3} depends on the quality of the AsCl$_3$. When the purity of AsCl$_3$ is such as to make it possible to deposit n-type GaAs with carrier concentrations of 1 to 2×10^{15} cm^{-3}, the available analytical techniques are not able to resolve differences in various AsCl$_3$ lots. Therefore, in the selection of AsCl$_3$ lots to generate low 10^{14} cm^{-3} GaAs, the only reliable test of the quality of the AsCl$_3$ is to prepare and evaluate GaAs. In doing this, the AsCl$_3$ can be separated

Table 2. Materials requirements

f (GHz)	l (μm)	ρ(Ω-cm) for l/ρ of 5–20	n (cm^{-3})	nl (cm^{-2})
15	7	1·4–0·35	8×10^{14}–3×10^{15}	
10	10	2·0–0·5	5×10^{14}–2×10^{15}	5×10^{11}–2×10^{12}
6	17	3·4–0·85	3×10^{14}–$1·2 \times 10^{15}$	
4	25	5·0–1·25	2×10^{14}–8×10^{14}	
2	50	10·0–2·5	1×10^{14}–4×10^{14}	

	l(μm)	n(cm^{-3})
Planar Gunn oscillator	nominal	1×10^{14}–1×10^{15}
Avalanche diode	6–10	5×10^{16}–1×10^{17}
Varactor diode	1–2	1×10^{16}–2×10^{16}
Field effect transistor	2–3	1×10^{15}–4×10^{15}
Grown n$^+$ contact	1–5	1×10^{18}–1×10^{19}

into 'good' and 'moderately-good' lots. The moderately-good lots are characterized by the compilation in table 3. Typically, for this grade of AsCl$_3$, the first GaAs run of a series produces a carrier concentration of a few times 10^{15} cm^{-3}. A minimum n of about 1×10^{15} cm^{-3} is achieved in subsequent runs. As the source materials are depleted the carrier concentration increases for each run.

Table 3. Properties of GaAs prepared using a 'moderately-good' lot of AsCl$_3$

Run number	at 300°K cm^{-3}	μ_H at 300°K cm^2 v^{-1} s^{-1}	μ_H at 77°K cm^2 v^{-1} s^{-1}
1	$4·2 \times 10^{15}$	6000	
3	$2·5 \times 10^{15}$	6600	50 000
4	$2·2 \times 10^{15}$	6400	56 000
5	$1·0 \times 10^{15}$	5000	34 000
7	$2·2 \times 10^{15}$	6600	36 000
8	$1·4 \times 10^{15}$	3300	
9	$9·5 \times 10^{14}$	6900	60 000
10	$1·5 \times 10^{15}$	6600	57 000
11	$2·0 \times 10^{15}$	6900	50 000
12	$3·2 \times 10^{15}$	6100	38 000
13	$6·1 \times 10^{15}$	5500	33 000

Note: The Hall-effect measurements for calculating these data were taken at a magnetic field of 3 kG.

Table 4. Properties of GaAs prepared using a 'good' lot of AsCl$_3$

AsCl$_3$ lot and run number	n at 300°K, cm^{-3}	μ_H at 300°K, cm^2 v^{-1} s^{-1}	μ_H at 77°K, cm^2 v^{-1} s^{-1}
A 1[a]	$3·5 \times 10^{15}$	5300	
A 2[a]	$2·0 \times 10^{15}$	4600	48 000
A 3	$6·6 \times 10^{14}$	6300	70 000
A 4[a]	$6·2 \times 10^{14}$	5400	
A 5	$4·5 \times 10^{14}$	7000	120 000
A 6	$5·2 \times 10^{14}$	7200	100 000
A 10	$1·6 \times 10^{14}$	8000	100 000
A 12	$2·6 \times 10^{14}$	7000	86 000

[a] Hall-effect data obtained at 3 kG.

Table 4 is a compilation of the results achieved with a good lot of AsCl$_3$. With this grade of material the carrier concentration is consistently less than 1×10^{15} cm^{-3} and does not show significant increases as the gallium source is depleted. After going through a specific good lot of AsCl₃ followed by two moderately-good lots, an additional quantity of the same good lot was ordered from the supplier. This lot behaved as expected, yielding superior GaAs. There has not been an instance where a lot has deviated from its classification, suggesting that there is uniformity within a given lot.

6. Conclusions

An open-tube flow system utilizing AsCl₃ and high-purity gallium has been developed to vapour-deposit epitaxial layers of GaAs suitable for use in microwave devices. A substrate cleaning procedure is described which is sufficient to remove any damage remaining after a chemo-mechanical polish and to insure the absence of barrier layers at or near the epitaxial layer–substrate interface. The quality of undoped GaAs layers has been found to be dependent on the particular lot of AsCl₃ used. Only selected lots of AsCl₃ are capable of producing n-type GaAs with carrier concentrations in the low 10^{14} cm^{-3} range. It was determined that if tellurium is the n⁺ dopant, sufficient tellurium gets into the vapour phase during the initial growth to affect the evaluation using n/i structures. When silicon-doped substrates are used, the carrier concentration measured using an n/i structure is in agreement with the carrier concentration measured by Schottky barrier techniques on either the n/n⁺ or n/i structures. Several wafers have been prepared with carrier concentrations in the 1×10^{14} to 5×10^{14} cm^{-3} range with electron mobilities at 77°K in the range 90 000 to 130 000 cm² V^{-1} s^{-1}.

Acknowledgments

The authors wish to acknowledge the invaluable assistance of R. B. Alfano, J. F. Black, D. H. Baird, E. D. Jungbluth, S. Perkowitz, B. A. Shortt, M. J. Urban, and S. Weisberger for their contributions to the experimental work.

References

BOLGER, D. E., FRANKS, J., GORDON, J., and WHITAKER, J., 1967, *Proc. Int. Symp. on GaAs*, Reading, 1966, Institute of Phys. and Phys. Soc., London, pp. 10–15.
COHEN, L., DRAGO, F., SHORTT, B., SOCCI, R., and URBAN, M., 1969, 1968 International Conference on GaAs, paper 24.
EDDOLLS, D. V., KNIGHT, J. R., and WILSON, B. L. H., 1967, *Proc. Int. Symp. on GaAs*, Reading, 1966, Institute of Phys. and Phys. Soc., London, 3–9.
EFFER, D., 1965, *J. Electrochem. Soc.*, **112**, 1020–5.
HILSUM, C., and MORGAN, J. R., 1967, *IEEE Trans. Electron Devices*, **ED–14**, 532–4.
HILSUM, C., 1968, *Brit. J. Appl. Phys. (J. Phys.* D), **1**, 265–81.
VAN DER PAUW, L. J., 1958, *Philips Res. Rep.*, **13**, 1–9.
WEISER, K., 1967, *Solid State Electronics*, **10**, 109–11.
WOLFE, C. M., FOYT, A. G., and LINDLEY, W. T., 1968, *Electrochem. Tech.*, **6**, 208–14.

Site distribution of silicon in silicon-doped gallium arsenide[†]

W. P. ALLRED, G. CUMMING, J. KUNG, and W. G. SPITZER

Departments of Materials Science and Electrical Engineering, University of Southern California, Los Angeles, California, U.S.A.

1. Introduction

Gallium arsenide which is grown from a stochiometric melt and doped with silicon is always an n-type semiconductor. However, it is known (Whelan *et al.* 1960; Kolm *et al.* 1957) that at silicon concentrations of $[Si] \geqslant 10^{19}$ cm^{-3}, the carrier concentration n, in the exhaustion range, is substantially less than [Si]. The difference between the values of n and [Si] is related to the amphoteric nature of silicon in GaAs. Silicon on a gallium site, Si_{Ga}, is a donor while Si_{As} is expected to be an acceptor. There is also some evidence (Queisser 1966) from photoluminescence studies for the existence of a substantial concentration of (Si_{Ga}–Si_{As}) nearest-neighbour pairs when the $[Si] \geqslant 10^{19}$ cm^{-3}.

There have been several recent studies of the infrared absorption associated with localized vibrational modes of impurities in semiconductors. In particular, three of these studies (Lorimor and Spitzer 1966; Spitzer and Allred 1968a; Spitzer and Allred 1968b) have dealt with silicon impurities in GaAs. It was found that one could correlate many of the observed absorption bands with the presence of different point defects involving silicon. The present work uses this identification to investigate the influence of the other impurities, specifically tellurium and zinc, on the site distribution of silicon.

2. Review

There has been considerable progress in the understanding of the localized vibrational mode absorption in silicon-doped GaAs. Because of the importance of previous studies for the interpretation of the present results, a brief review will be presented here.

Silicon, as an isolated substitutional impurity on either the gallium or arsenic sublattice, has tetrahedral point group symmetry. To terms quadratic in the displacment of the silicon, the potential energy is that of a spherically symmetric, harmonic oscillator. Because the silicon is lighter than either the gallium or arsenic, one can expect spatially localized vibrational modes associated primarily with the motion of the silicon impurity at frequencies above the highest phonon frequency of the unperturbed lattice (Haas 1967). A single silicon defect will give three vibrational modes, and because of the symmetry of the potential, the modes will be degenerate. Thus a single infrared absorption band is expected for Si_{Ga} and another one for Si_{As}. Substitutional silicon on each of the two sites can have different frequencies because of a slightly different mass defect, i.e., the mass defect is given by $\epsilon = (M_{Si} - M)/M$, where M = mass of gallium or arsenic. Also the different nearest neighbours in the two cases should influence the effective force constants. In table 1 it is seen that the observed local mode frequencies are $\omega = 384$ cm^{-1} for Si_{Ga} and 399 cm^{-1} for Si_{As}.

In some cases the silicon impurity is no longer isolated from other defects but is paired with another silicon or another impurity on a neighbouring site. There can be different reasons for the existence of these ion pairs. There is evidence already mentioned (Queisser 1966; Spitzer and Allred 1968b) that (Si_{Ga}–Si_{As}) pairs tend to form at high [Si]. Moreover,

---

[†] The research reported in this paper was sponsored in part by the Air Force Cambridge Research Laboratories, Office of Aerospace Research under Contract No. F 19628–68–C–0169, but the report does not necessarily reflect endorsement by the sponsor.

Table 1

Defect	Impurity	Mode frequency (cm^{-1})		
		^{7}Li	^{6}Li	Cu
Li-lattice defects	Li	365 379	352 389 406	
Si$_{Ga}$–Li$_{Ga}$	Li$_{Ga}$	438 447 454	470 480 487	
	Si$_{Ga}$	374 379 405	374 379 405	
Si$_{Ga}$–Cu$_{Ga}$	Si$_{Ga}$			374 376 399
Si$_{Ga}$ Si$_{As}$	Si$_{Ga}$ Si$_{As}$	384 399	384 399	384 399
Si$_{Ga}$–Si$_{As}$	Si$_{Ga}$–Si$_{As}$	367 393 464	367 393 464	367 393 464

the samples are strongly n-type and hence the resulting free carrier absorption must be reduced by the introduction of an electrically compensating impurity. Lithium or copper is diffused at elevated temperature to produce the compensation. Since the material is initially n-type, the compensating diffusant must produce acceptors, which are generally assumed to be Li$_{Ga}$ or Cu$_{Ga}$. There is also a tendency for the Si$_{Ga}$ donors and the Li$_{Ga}$ or Cu$_{Ga}$ acceptors to interact, resulting in (Si$_{Ga}$–Li$_{Ga}$) or (Si$_{Ga}$–Cu$_{Ga}$) pairs. Clearly, in these cases, the silicon is no longer in a site with tetrahedral symmetry.

For the (Si$_{Ga}$–Li$_{Ga}$) pairs where the defect involves impurities on second neighbour sites, a weak coupling model appears to give reasonable agreement with experiment (Spitzer and Allred 1968b). The effect of one atom is to alter the potential at the site of the other impurity atom. Since the nearest-neighbour gallium sites are along a $\langle 110 \rangle$ direction, which is not a symmetry rotation axis of the tetrahedral point group, the triply degenerate Si$_{Ga}$ band will be split into three bands with frequencies nearly independent of the lithium isotope. These bands will be located near the Si$_{Ga}$ frequency of 384 cm^{-1}. A simple perturbation argument (Lorimor and Spitzer 1967b) gives

$$\omega^2 = \tfrac{1}{3} \sum_{i=1}^{3} \omega_i^2,$$

where ω is the frequency of the triply degenerate mode and the ω_i values are the perturbed frequencies. In the (Si$_{Ga}$–Li$_{Ga}$) defect case there will also be three bands involving primarily lithium motion. The frequencies of both the predominantly silicon modes and those due to lithium are given in table 1.

The (Si$_{Ga}$–Si$_{As}$) defect is more complicated (Spitzer and Allred 1968b; Elliott and Pfeuty 1967) in that the two impurities are strongly coupled, and both are light compared to the atoms they replace. The pair axis is in a $\langle 111 \rangle$ direction which is a three-fold rotation axis, and the six vibrational modes are expected to contain two sets of doubly degenerate modes. Thus, there can be four frequencies, of which three are identified in table 1. The fourth mode is either too small in absorption to be observed or, as calculation indicates, falls in a region of large absorption from the GaAs host lattice (Spitzer and Allred 1968b). The location of all bands observed in lithium or copper diffused, silicon-doped GaAs is listed in table 1. For completeness, the location of some bands related to lithium paired with lattice defects in undoped GaAs is also given (Levy and Spitzer 1968). These latter bands are not related to the silicon impurity but can occur in all samples which are lithium diffused at sufficiently high temperature (Lorimor and Spitzer 1966).

There have been a number of experiments which have suggested that the distribution of silicon among the various defects mentioned will be influenced by the presence of other impurities. Some electrical and radio-tracer measurements of selenium-plus-silicon-doped GaAs indicated that the Se$_{As}$ donors could change $\eta = [Si_{As}]/[Si_{Ga}]$ from <1 to >1 (Whelan et al. 1960). Local mode measurements of a sulphur-plus-silicon-doped sample (Lorimor and Spitzer 1966) showed the Si$_{As}$ band at 399 cm^{-1} to be larger than the Si$_{Ga}$ band at 384 cm^{-1}. In similarly silicon-doped samples but without sulphur (Spitzer and Allred 1968b), the 384 cm^{-1} band is ~ 4–5 times as strong as the 399 cm^{-1} band. However, the bands in the sulphur-doped sample were small and close to some very large Li-lattice defect bands, thus making the results uncertain. Another study (Spitzer and Allred 1968a) showed that silicon site transfer could take place as a result of lithium diffusion at temperatures $\geqslant 800\,°$C. The sulphur plus silicon-doped sample (Lorimor and Spitzer 1966) was diffused at $\sim 800\,°$C again making this result uncertain.

3. Experimental results and discussion

Table 2 lists several ingots of GaAs with estimates of the doping levels anticipated near the seed end on the basis of the impurities added and published (Willardson and Allred 1966) distribution coefficients. Samples were taken from each ingot and diffused with lithium. In some cases samples were taken from both the seed end and near the back end of the ingot. The temperature of diffusion was $T_d = 700\,°$C in order to avoid the site redistribution observed for $T_D \geqslant 800\,°$C (Spitzer and Allred 1968a). The diffusion was done from a surface alloy phase while in an argon atmosphere. The method of diffusion has been described previously (Lorimor and Spitzer 1967).

Table 2

Ingot No.	Dopants	Doping concentrations
1	Si	$\sim 2 \times 10^{18}$ cm^{-3}
2	Si	$\sim 2 \times 10^{19}$ cm^{-3}
3	Si, Te	[Si]$\sim 1 \times 10^{18}$ cm^{-3}, [Te]$\sim 3 \times 10^{18}$ cm^{-3}
4	Si, Te	[Si]$\sim 7 \times 10^{18}$ cm^{-3}, [Te]$\sim 5 \times 10^{18}$ cm^{-3}
5	Si, Zn	[Si]$\sim 6 \times 10^{18}$ cm^{-3}, [Zn]$\sim 1 \times 10^{18}$ cm^{-3}
6	Si, Zn	[Si]$\sim 6 \times 10^{18}$ cm^{-3}, [Zn]$\sim 3 \times 10^{18}$ cm^{-3}
7	Si, Zn	[Si]$\sim 6 \times 10^{19}$ cm^{-3}, [Zn]$\sim 3 \times 10^{19}$ cm^{-3}

Samples from the silicon-doped ingots 1 and 2 were lithium diffused and the infrared absorption measured at liquid-nitrogen temperature is given in figure 1. Also shown for comparison is the measured absorption curve for pure GaAs. The curves show the spectral region containing most of the local mode bands which are labelled with the related defect. Although, as indicated in table 1, some bands occur at $\omega > 430$ cm^{-1}, they have been shown in previous measurements (Lorimor and Spitzer 1966; Spitzer and Allred 1968b) and will not be of major importance for the present work.

The crosses of figure 2 show, on a log-log graph, the peak absorption coefficient α_P for the Si$_{Ga}$ band at 384 cm^{-1} plotted against α_P for the Si$_{As}$ band at 399 cm^{-1}. The data are taken from a number of silicon-doped samples which are all diffused with lithium at $T_D \leqslant 800\,°$C. Most of these samples were measured as part of an earlier study (Spitzer and Allred 1968b). It is observed that the points are close to a line of slope ~ 0.9. It is to be emphasized that silicon is the only major dopant in these samples.

The absorption of samples from ingots 3 and 4, which are doped with silicon plus tellurium and diffused with ^{7}Li is shown in figure 3. Comparison with figure 1 shows several new features. The bands near 475 cm^{-1} and 391 cm^{-1} are due to (Li$_{Ga}$–Te$_{As}$) and have been previously studied (Lorimor and Spitzer 1967a). The band at 379 cm^{-1} is largely a

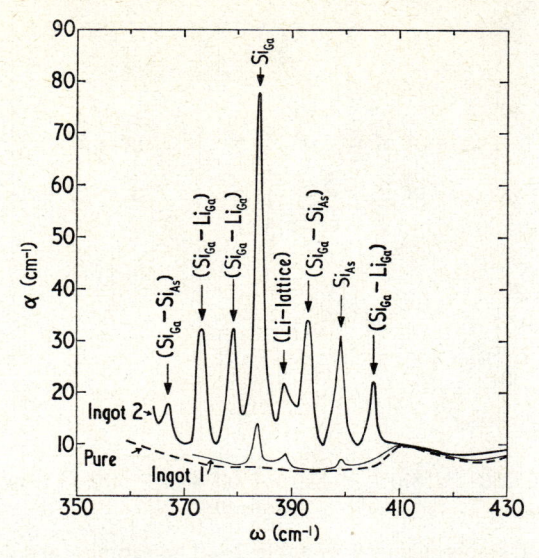

Figure 1. Absorption, at liquid-nitrogen temperature, of samples from ingots 1 and 2 of table 2. The samples were ^{7}Li diffused at 700°C for 24 hours.

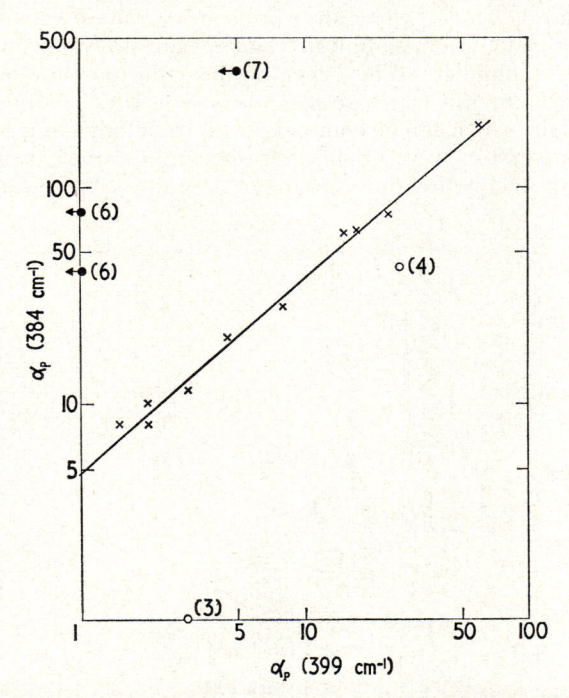

Figure 2. A plot of the peak absorption coefficient of the Si$_{Ga}$ band at 384 cm^{-1} vs that for the Si$_{As}$ band at 399 cm^{-1}. ×: silicon doped samples; ○: silicon and tellurium doped samples; ●: silicon and zinc doped samples. The numbers indicate the ingot of table 2. All samples are lithium diffused.

Li-lattice defect band and shows a large lithium isotope shift, i.e., 379 cm^{-1} → 406 cm^{-1} when ^{7}Li → ^{6}Li. The bands at 464, 393, and 367 cm^{-1} in the sample from ingot 4 are the (Si$_{Ga}$–Si$_{As}$) bands mentioned earlier and listed in table 1. A principal difference between the samples of figure 3 and figure 1 is the change in the relative strengths of the Si$_{Ga}$ and

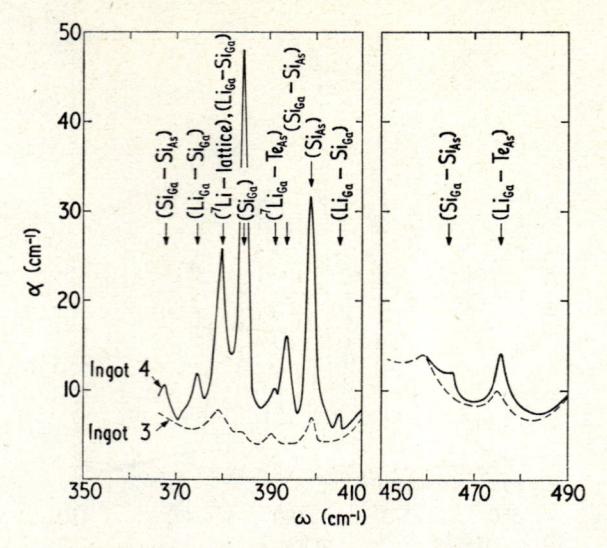

Figure 3. Absorption of ingots 3 and 4 of table 2. The samples are doped with silicon and tellurium and diffused with lithium.

Si$_{As}$ bands at 384 and 399 cm^{-1}. The absorption peak values less background for these bands of the silicon plus tellurium samples of figure 3 are shown as circles on figure 2 and labelled with the ingot number. The effect of the tellurium has been to substantially enhance the fraction of the [Si] which resides on arsenic sites. More extensive measurements should give results which can be compared with thermodynamic predictions.

The most interesting results were obtained from samples taken from ingots 5, 6 and 7, which were doped with silicon plus zinc. Figure 4 shows the absorption for several samples

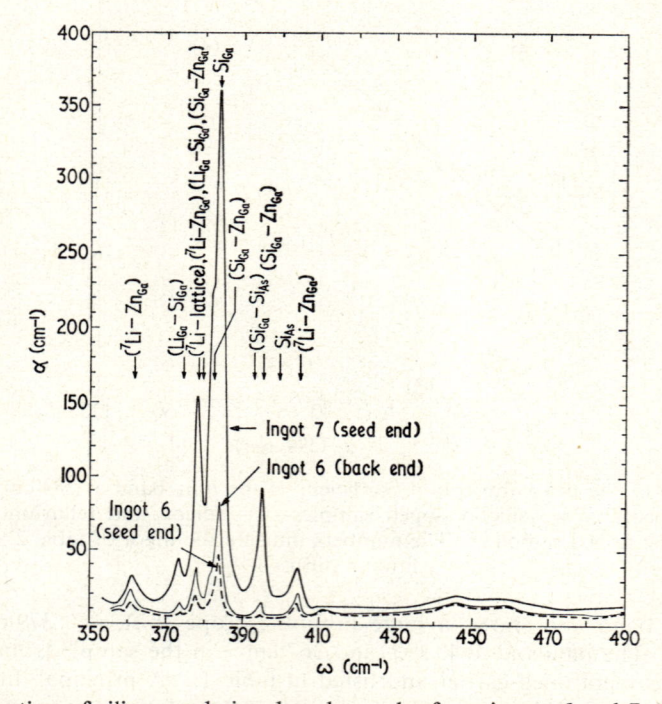

Figure 4. Absorption of silicon and zinc doped samples from ingots 6 and 7 of table 2.

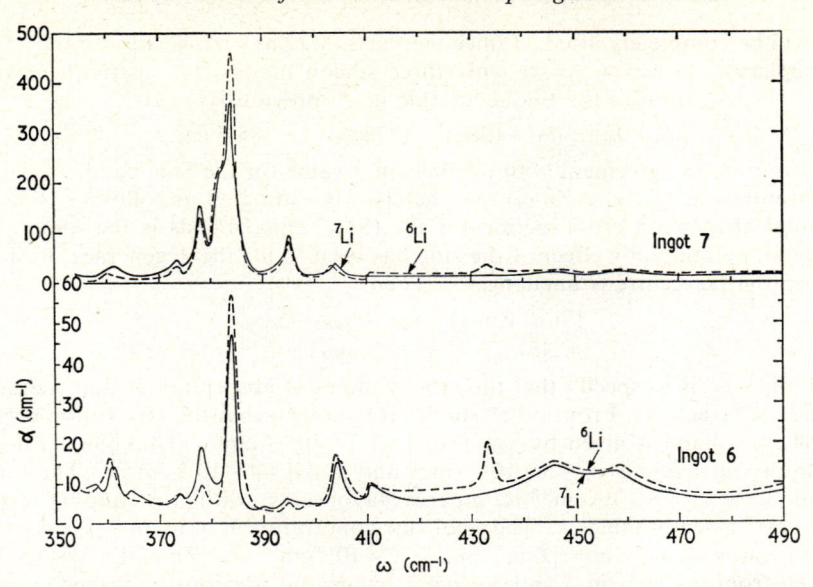

Figure 5. Comparison of the absorption of ^{6}Li and ^{7}Li diffused samples from ingots 6 and 7 of table 2.

diffused with ^{7}Li. Figure 5 shows, for each of two ingots, a comparison of two samples taken from adjacent wafers and diffused with ^{7}Li and ^{6}Li respectively. From previous studies (Lorimor and Spitzer 1967b) of zinc-doped and lithium-compensated samples it is known that there are four lithium bands, presumably associated with (Li–Zn) pairs. These bands are nearly equal for a given [Zn] and are located at 405, 378, 361, and 340 cm^{-1} for ^{7}Li and at 433, 404, 385, and 361 cm^{-1} for ^{6}Li. Many of these bands are observed in figures 4 and 5 and are taken into account in some of the quantitative discussion which will follow.

The Si$_{As}$ acceptor band at 399 cm^{-1} is observed only in the most lightly zinc-doped ingot (5). Within experimental accuracy this band is totally absent in figure 4 from the samples of ingots 6 and 7. We also note from figures 4 and 5 that the (Si$_{Ga}$–Si$_{As}$) bands are totally absent (at 464, 393, and 367 cm^{-1}) in the heavily zinc-doped ingots. Moreover, several of the samples show the appearance of three new bands of approximately equal strength. These bands are at 395, ~382, and 378 cm^{-1}.

The 378 cm^{-1} band requires some discussion. In figure 5, the band at 378 cm^{-1} could involve contributions from three absorption bands seen in previous work. In GaAs with (^{7}Li–Zn), there is a band at 378 cm^{-1}, but it should be equal in strength to the (^{7}Li–Zn) band at 405 cm^{-1}. There is also a (^{7}Li–lattice) defect band at 379 cm^{-1} which could be contributing. However, both the (^{7}Li–lattice) and the (^{7}Li–Zn) bands shift to near 405 cm^{-1} if ^{6}Li is used. Comparison of the ^{7}Li and ^{6}Li samples of ingot 7 in figure 5 shows little change in the 378 cm^{-1} band. The remaining previously known contribution to the absorption near 378 cm^{-1} comes from a 379 cm^{-1} silicon mode for (Si$_{Ga}$–Li$_{Ga}$) defect (see table 1). However, this contribution should only be as large as the much weaker band at 374 cm^{-1} (see figure 5, ingot 7). Therefore, we have the result that most of the strength of the 378 cm^{-1} band of ingot 7, figure 5 cannot be accounted for in terms of any of the previously known defects.

The enhanced strength of the 378 cm^{-1} band occurs only in those cases where the 395 and 382 cm^{-1} bands are also observed. The three new bands do not show any lithium isotope shift, and they occur only in samples with large [Si$_{Ga}$] and [Zn$_{Ga}$]. It is reasonable to assume that these new bands arise from (Si$_{Ga}$–Zn$_{Ga}$) defects where the silicon and zinc are on nearest-neighbour gallium sites (second neighbour positions). For such a pair, the axis is $\langle 110 \rangle$ and, as discussed for (Si$_{Ga}$–Li$_{Ga}$), the three-fold degeneracy of the tetrahedral potential will be completely lifted. Since the mass, $M(\mathrm{Zn})=65$, is close to the $M(\mathrm{Ga})=70$ which it replaces, we expect to see only three silicon modes split in frequency about the $\mathrm{Si_{Ga}}$ band. Application of the frequency rule given previously yields

$$\omega = [\tfrac{1}{3}\{(395)^2 + (382)^2 + (378)^2\}]^{1/2} = 384 \cdot 5 \text{ cm}^{-1}.$$

This result is in good agreement with the 384 cm^{-1} value for the $\mathrm{Si_{Ga}}$ band.

The concentration of ($\mathrm{Si_{Ga}}$–$\mathrm{Zn_{Ga}}$) may be crudely estimated as follows: It is assumed that the total absorption cross section for the ($\mathrm{Si_{Ga}}$–$\mathrm{Zn_{Ga}}$) bands is the same as that for the $\mathrm{Si_{Ga}}$ band, i.e., the only effect of the zinc has been to lift the degeneracy of $\mathrm{Si_{Ga}}$ but the total absorption per centre is unchanged. Then

$$\frac{[\mathrm{Si_{Ga}\text{–}Zn_{Ga}}]}{[\mathrm{Si_{Ga}}]} = \frac{\alpha_{395} + \alpha_{382} + \alpha_{378}'}{\alpha_{384}}.$$

The prime on α_{378}' is to specify that the other sources of absorption at that frequency have been already subtracted. From other studies it is known that the absorption cross section for the 384 cm^{-1} band is given by $\alpha_{384}/[\mathrm{Si_{Ga}}] \simeq 7 \cdot 3 \times 10^{-18}$ cm^2. Thus $[\mathrm{Si_{Ga}\text{–}Zn_{Ga}}]$ for the ingot 7 samples of figure 5 is $\sim 3 \times 10^{19}$ cm^{-3} and $[\mathrm{Si_{Ga}}] \sim 4 \times 10^{19}$ cm^{-3}. The total silicon concentration of $\sim 7 \times 10^{19}$ cm^{-3} compares favourably with the doping estimate of $\sim 6 \times 10^{19}$ cm^{-3} given in table 2. The total zinc concentration of ingot 7 is $\sim 4 \times 10^{19}$ cm^{-3}. This concentration includes both $[\mathrm{Zn_{Ga}\text{–}Si_{Ga}}] \sim 3 \times 10^{19}$ cm^{-3} and $[\mathrm{Zn_{Ga}\text{–}Li_{Ga}}] \sim 1 \times 10^{19}$ cm^{-3} as estimated from the 405 cm^{-1} and 361 cm^{-1} bands and previous published (Lorimor and Spitzer 1967) α vs. $[\mathrm{Zn_{Ga}\text{–}Li_{Ga}}]$ data. Again the total [Zn] is regarded as in reasonable agreement with the estimate of 3×10^{19} cm^{-3} given in table 2. It may also be noted that if the $\mathrm{Zn_{Ga}}$ and $\mathrm{Si_{Ga}}$ were randomly distributed on the gallium sublattice then the $[\mathrm{Zn_{Ga}\text{–}Si_{Ga}}]$ would be 10^{17} to 10^{18} cm^{-3} for these samples. This concentration is substantially smaller than the estimate given above which indicates that the pairing energy plays a significant role in determining the $\mathrm{Zn_{Ga}}$–$\mathrm{Si_{Ga}}$ pair concentrations.

The measurements of samples taken from ingot 6 and 7 have been used to obtain the points given by solid dots on figure 2. Recall that in these samples no detectable band was observed at 399 cm^{-1}, and thus the arrows indicate that the points have been placed at estimates for the maximum possible value for α_P (399 cm^{-1}). The correct value is probably substantially smaller. The points indicate that for ingots 6 and 7, the $[\mathrm{Si_{As}}]$ has been decreased by at least one order of magnitude, and a more realistic estimate gives approximately two orders of magnitude for a given $[\mathrm{Si_{Ga}}]$. Although not plotted in figure 2, a comparison of figure 1 with figures 4 and 5 gives a similar conclusion for $[\mathrm{Si_{Ga}\text{–}Si_{As}}]$.

It is reasonably clear that studies of silicon site distribution as functions of [Si], [Te], and [Zn] should lead to quantitative estimates of distribution coefficients and the dependence of specific defects on impurity concentrations. Moreover, correlations with electrical measurements should yield information on the electrical nature of each of the defects. This work is now being carried out.

References

ELLIOTT, R. J., and PFEUTY, P., 1967, *J. Phys. Chem. Sol.*, **28**, 1627.
HASS, M., 1967, *Semiconductors and Semimetals*, ed. Willardson, R. K. and Beer, A. C., Academic Press, New York, Vol. 3, p. 3.
KOLM, C., KULIN, S. A., and AVERBACH, B. L., 1957, *Phys. Rev.*, **108**, 965.
LEVY, M., and SPITZER, W. G., 1968, *J. Appl. Phys.*, **39**, 1914.
LORIMOR, O. G., and SPITZER, W. G., 1966, *J. Appl. Phys.*, **37**, 3687.
LORIMOR, O. G., and SPITZER, W. G., 1967a, *J. Appl. Phys.*, **38**, 2713.
LORIMOR, O. G., and SPITZER, W. G., 1967b, *J. Appl. Phys.*, **38**, 3008.
QUEISSER, H. J., 1966, *J. Appl. Phys.*, **37**, 2909.
SPITZER, W. G., and ALLRED, W., 1968a, *Appl. Phys. Letters*, **12**, 5.
SPITZER, W. G., and ALLRED, W., 1968b, *J. Appl. Phys.*, **39**, 4999.
WHELAN, J. M., STRUTHERS, J. D., and DITZENBERGER, J. A., 1960, *Proc. Intern. Conf. on Semiconductors*, Czechoslovak Academy of Science, Prague, p. 943.
WILLARDSON, R. K., and ALLRED, W. P., 1966, *Proc. Intern. Symp. on GaAs, Reading*, Institute of Physics and the Physical Society, p. 35.

Correlation between diffusion and precipitation of impurities in dislocation-free GaAs[†]

H. R. WINTELER and A. STEINEMANN

Battelle Institute, Geneva, Switzerland

Abstract. In an ideal diffusion process, the host crystal lattice is not altered during the diffusion process and the position of the diffusing atoms is restricted to interstitial and substitutional lattice sites. In this case, the effective diffusion coefficient has a minimum value. For low surface concentrations, this type of diffusion is possible in dislocation-free GaAs for all the impurities investigated (Be, Zn, Cd, Mn, Si, S). Even with high surface concentrations, the diffusion remains ideal for some cases (e.g., Mn, Zn) so long as the mean penetration depth is limited to about 10 to 15 μm. The numerical value of this limit is a function of the diffusion conditions and of the local crystal quality. The difference of the mean penetration depths in samples with perfect and with imperfect diffusion is (a) large for the elements Zn and Mn, and (b) of minor importance for Cd, Si and S. During the diffusion of Zn or Mn, the locally-varying concentration of impurities and/or precipitations of Czochralsky grown dislocation-free crystals leads to drastic variations of the penetration depth within the same sample.

1. Introduction

The use of dislocation-free GaAs (Zimmerli 1968; Steinemann and Zimmerli 1967; Zimmerli and Steinemann 1965) does not directly affect electron transport phenomena, but it is of special interest for diffusion experiments. The practical importance for device applications is demonstrated by the improvement of diffused laser junctions (Hatz, Deutsch and Mohn 1967), where homogeneous light emission instead of emission spots is very direct evidence for this improvement. The use of dislocation-free bulk GaAs is a necessary but not sufficient condition to achieve ideal (flat) diffused p–n junctions over large areas. It has been pointed out (Winteler and Steinemann 1966[†]; Black and Jungbluth 1967; Iizuka 1968; Maruyama 1968) that the diffusion of Zn into GaAs can induce a large density of precipitations and dislocations in the originally perfect crystal lattice. The formation of these defects always causes an enhancement of the diffusion rate. The determination of diffusion mechanisms and the calculation of diffusion coefficients based on penetration profiles of such non-ideal experiments may lead to erroneous results, i.e., too high values of the diffusion coefficient.

The ideal diffusion process—i.e., without creation of crystalline defects (precipitations and dislocations)—can be realized experimentally. However, serious deviations are likely to occur, not only as a function of (avoidable) inappropriate external diffusion conditions, but also as a function of the characteristics of the raw material. The main parameters are the initial distribution of precipitation centres and their modification as a function of thermal treatment. The inhomogeneous incorporation of Te in highly-doped GaAs is a typical example (Schwuttke and Rupprecht 1966; Casey 1967).

This paper will specify phenomenologically the conditions for perfect diffusion (with device applications in view); as well as the transition from perfect to imperfect diffusion.

[†] Most of the work on preparation of dislocation-free GaAs single crystals as well as the early work on diffusion has been sponsored by the Research Laboratories, N.V. Philips Gloeilampen-fabrieken, Eindhoven, Holland.

2. Raw material

Czochralsky pulling of GaAs single crystals is possible without the creation of any dislocation (Steinemann and Zimmerli 1967; Zimmerli 1968). However, impurity atoms picked up from the melt are inhomogeneously distributed in the crystal. At a local doping level, even as low as $n=10^{17}$ cm^{-3} or $p=10^{18}$ cm^{-3}, the impurities are no longer homogeneously dissolved as point defects in the crystal but they are partly concentrated in precipitations. Figure 1 exhibits the pattern of the radial distribution of precipitations near the crystal surface as revealed by appropriate etching of a circular (111) cross-section of a Te-doped dislocation-free crystal. By prolonged heat treatment at elevated temperatures, stacking faults surrounded by polygonized partial dislocation loops are created in the bulk of such crystal samples. They are mostly concentrated in the neighbourhood of the very sharp boundary between regions of high and low initial precipitation density. The effect is demonstrated in figure 2. It shows the same region of large impurity—and precipitation—density gradient as figure 1, but after annealing at 1100°C for 4 hours. Closed dislocation loops are formed in the critical region.

The largest variation of the concentration of point defects is observed along the growth axis. This considerably influences the thermal conversion effect. Figure 3a shows this influence (lower, interrupted line) at an early stage of a regular Be-diffusion (upper continuous line). The phenomenon starts on some preferential {111} plane which during the growth of the crystal coincided with the local solid–liquid interface. It then extends further, but only on one side of the given plane. This is a consequence of the variation of the local growth rate during pulling and of the asymmetric profile of the point defect concentration (supercooling, etc.). Figure 3b shows the thermal conversion at a later stage.

3. Experimental diffusion methods

The diffusion process is assumed to be perfect as long as the collective migration of foreign atoms into the crystal sample does not produce any precipitation centres or new dislocations. As an experimental rule, this is possible (for all investigated elements) under excess arsenic pressure (p_{As} in the vapour phase equal to or higher than the corresponding partial equilibrium As pressure over the solid GaAs sample at the given temperature) and for surface concentrations of the diffusing atoms sufficiently below the solid solubility limit (at the given temperature). For higher surface concentration, the depth of the perfectly diffused zone is, in general, limited to 5–15 μ. The lower value is typical for Si diffusion in p-GaAs, the upper limit for Mn diffusion.

The methods used for diffusion with low surface concentration depend on the type of diffusing atoms:

(a) Small quantities of acceptors of group II with high vapour pressure (Zn, Cd) are added to the diffusion ampoule in elemental form. The diffusion in the sealed ampoule proceeds via the vapour phase (under arsenic pressure);

(b) Doped GaAs powder is used as a diffusion source for elements with low vapour pressure (Mn, Si, Sn);

(c) Special precautions are necessary with group VI donors (S, Se) with high vapour pressure and high reactivity at the GaAs surface for ordinary diffusion temperatures ($T>900°$C). As an example, for S, even at a vapour pressure below the equilibrium pressure of S over Ga_2S_3, a transport reaction of GaAs takes place even for extremely small temperature variations across the sample. These difficulties are overcome by a two-step process. First, a small amount of elemental sulphur is incorporated via the vapour phase into a very thin surface layer of the GaAs sample during a low temperature ($250<T<300°$C) run. Afterwards, the sample is covered with a pyrolytic SiO_2 layer and annealed at temperatures between 900 and 1150°C.

4. Experimental results

The difference in the penetration depth (and the impurity concentration profile) between perfect and imperfect diffusion—i.e., with simultaneous formation of new precipitations and dislocations—is most evident for the rapidly-diffusing acceptors Mn and Zn. The diffusion coefficients of Cd, Si (concentration dependent) and S (concentration independent) are only very slightly affected by precipitations and dislocations.

4.1. Perfect diffusion

Figure 4 demonstrates the perfect diffusion of Zn, (a) in dislocation-free GaAs and (b) in GaAs with a dislocation density of about 10^4 cm^{-2}. The diffusion fronts are planar everywhere. There is no disturbance in figure 4a, but in figure 4b a spike due to the lattice distortions and open bonds along a dislocation line is visible. The impurity profile resulting from perfect diffusion is step-like but less pronounced than for imperfect diffusion: The diffusion coefficient $D(c)$ depends on the concentration c, though its numerical value is much lower than that for imperfect diffusion. Diffusion of Mn, Cd, and Si gives similar pictures as long as the conditions for perfect diffusion are fulfilled.

4.2. Imperfect diffusion

The simultaneous formation of crystalline defects during diffusion leads to an enhanced diffusion rate for Zn and Mn diffusion. In the case of Zn, the phenomena differ for high and low As pressure in the vapour phase of the diffusion ampoule.

4.2.1. Zn diffusion under excess As pressure.
Even for high Zn vapour pressure, the diffusion is perfect at temperatures $T \leqslant 850°$C and for a penetration depth up to 10–15 μm. If the diffusion experiment is further prolonged, (under the same operating conditions), precipitations are formed near the p–n junction as soon as it reaches a depth of 15–20 μm below the surface. This is shown in figure 5a. Moreover, the junction is no longer planar. The further movement of the junction is accompanied by a steadily growing zone full of precipitations. Its upper limit is practically independent of time, i.e., it is localized at about 15 μm below the surface (where the first precipitations appear). Only for very long diffusion times, can a movement towards the surface be observed. The lower limit of the perturbed zone coincides with the junction. If the diffusion process is continued further until the junction depth exceeds ca. 100 μ, dislocations are formed throughout the bulk of the originally dislocation-free crystal, as shown in figure 5b. In the diffused zone, it is no longer possible to distinguish dislocations and precipitations by the etching technique.

4.2.2. Zn diffusion without As addition.
Under the same diffusion conditions, but without any additional elementary As in the diffusion ampoule, the depth of perfect diffusion is smaller (10 μm instead of 15 μm) and it depends strongly on the type and level of doping of the bulk material.

Beyond the limit of perfect diffusion, a zone of precipitations is found, which starts right at the surface of the sample and which proceeds into the bulk of the sample much faster than does the p–n junction until its lower limit reaches the p–n junction. From that moment on, the p–n junction is pushed forwards by the lower limit of the precipitation zone. Consequently, the effective diffusion rate is much higher than for perfect diffusion. For deep diffusion into homogeneous parts of the bulk material, the p–n junction becomes again nearly planar, as shown in figure 6.

4.2.3. Other elements.
For the diffusion of Mn under excess As pressure, the same mechanism of formation of precipitations is operative as in the case of Zn diffusion without As addition. Figure 7 gives an example, the diffused zone of 100 μm being full of precipitations (and some dislocations).

The diffusion of Si (with elemental Si as a diffusion source), at a moderate penetration depth of ca 10 μm, is also accompanied by formation of precipitations, and dislocations may appear below the p–n junction. However, the position and the form of the p–n junction are practically unaffected by this additional phenomenon, as shown in figure 8. The diffusion profile is step-like indicating a concentration-dependent diffusion coefficient.

The investigation of the diffusion of sulphur in GaAs leads to the following two results: (a) whenever the surface state of the sample is not affected during the diffusion, practically no new precipitations are formed in the diffusion zone, (b) elsewhere, a zone of freshly created precipitations moves more slowly into the crystal than does the p–n junction. The diffusion profile (corresponding to a constant D) and the junction characteristics are practically independent of crystalline defects.

4.3. *Rôle of the bulk material*

For given diffusion conditions, the bulk properties influence the transition between perfect and imperfect diffusion and the rate of extension of the zone of precipitations—which then affects the rate of diffusion. Again, the effects are most easily shown for Mn diffusion with excess As and for Zn diffusion with and without excess As. We compared (*a*) samples with different bulk properties in the same ampoule and (*b*) the diffusion patterns in one sample for zones of different bulk homogeneity.

Figure 9 refers to Zn diffusion. Figure 9*a* shows that the p–n junction (together with the zone of freshly-created precipitations) progresses more rapidly and in a more homogeneous way in Te-doped crystal regions where the as-grown density of precipitations is high (right-hand side). Figure 9*b* shows that the threshold depth between perfect and imperfect diffusion is influenced by the local crystal quality. At the left-hand side, the as-grown density of precipitations is high, and the imperfect diffusion begins earlier than at the right-hand side of the figure, where the as-grown density of precipitations is zero and where at this stage, only a few dislocations have been formed in the central part of the diffusion zone.

Figure 10 demonstrates the variation of the penetration depth of Mn over several hundred microns due to the inhomogeneous formation of precipitation zones.

5. Conclusion

Ideal diffusion in GaAs is possible, but irreversible deviations are very likely to occur during the experiments. It has been shown that the penetration depth of the Zn and Mn diffused layers is drastically altered if precipitations are formed during the diffusion process. In Czochralsky-grown crystals (where the as-grown inhomogeneity of the impurity concentration and of the precipitation distribution is large), this effect causes large variations of the penetration depth within the same sample. The diffusion of Cd, Si, S may also be accompanied by the formation of some new precipitations (in the diffusion zone) and of new dislocations, but for these elements, the diffusion profile is not seriously affected.

All the elements investigated can be diffused without creation of any lattice defects (precipitations, dislocations) as long as sufficiently low surface concentrations of the diffusing constituents are maintained. Perturbations of the lattice structure and composition near the surface (transport reactions) are less important for experiments with excess As in the vapour phase. The Mn and Zn diffusions remain ideal, even with high surface concentrations, up to an average penetration depth of about 10 to 15 μ. For many device applications, these conditions can be fulfilled. It can be anticipated that this should permit the reproducible realization of stable devices and avoid their premature deterioration under extreme operating conditions.

References

BLACK, J. F., and JUNGBLUTH, E. D., 1967, *J. Electrochem. Soc.*, **114**, 181.
CASEY, H. C. Jr., 1967, *J. Electrochem. Soc.*, **114**, 153.
HATZ, J., DEUTSCH, CH., and MOHN, E., 1967, *IEEE J. of Quantum Electronics*, **3**, 643.
IIZUKA, T., 1968, *Japan J. Appl. Phys.*, **7**, 485 and 490.
MARUYAMA, M., 1968, *Japan J. Appl. Phys.*, **7**, 476.
SCHWUTTKE, G. W., and RUPPRECHT, H., 1966, *J. Appl. Phys.*, **37**, 167.
STEINEMANN, A., and ZIMMERLI, U., 1967, *Crystal Growth*, Pergamon Press, p. 81.
WINTELER, H. R., and STEINEMANN, A., 1966, *Helv. Phys. Acta*, **39**, 182.
ZIMMERLI, U., and STEINEMANN, A., 1965, *Z. f. angew. Math. und Phys.*, **16**, 555.
ZIMMERLI, U., 1968, *Thesis No. 4076*, Federal Inst. of Technology, Zürich.

Diffusion through and from solid layers into gallium arsenide

W. von MÜNCH

Technische Hochschule, Aachen, Germany

Impurity diffusion from the gas phase has become a powerful tool in germanium and silicon device technology. It has failed to acquire a similar importance in the field of gallium arsenide devices. This is mainly due to surface erosion problems and, probably, to the introduction of deep level impurities. It is the purpose of this contribution to demonstrate that diffusion of impurities through and from solid layers into gallium arsenide not only eliminates surface erosion problems but can also facilitate the generation of specific concentration profiles in gallium arsenide.

The principle of the two methods is shown in figure 1a, b. The method of impurity diffusion from the gas phase (figure 1a) through a solid layer into gallium arsenide has been used by Yeh (1964) for group VI donor impurities and by Shortes, Kanz and Wurst (1964)

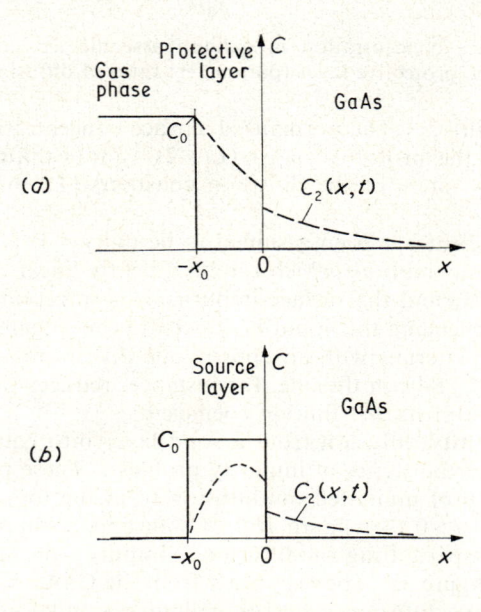

Figure 1. Impurity diffusion into gallium arsenide (a) through protective layer (b) from solid source layer.

for zinc. Silicon monoxide served as the protective layer in the former case whereas reactively sputtered silica layers were used in the latter case. The technique of zinc and tin diffusion from pyrolytic silica layers into gallium arsenide (figure 1b) has been studied by von Münch (1966).

An important feature of the diffusion through protective layers is the effective control of the surface impurity concentration in gallium arsenide. Some quantitative data for this

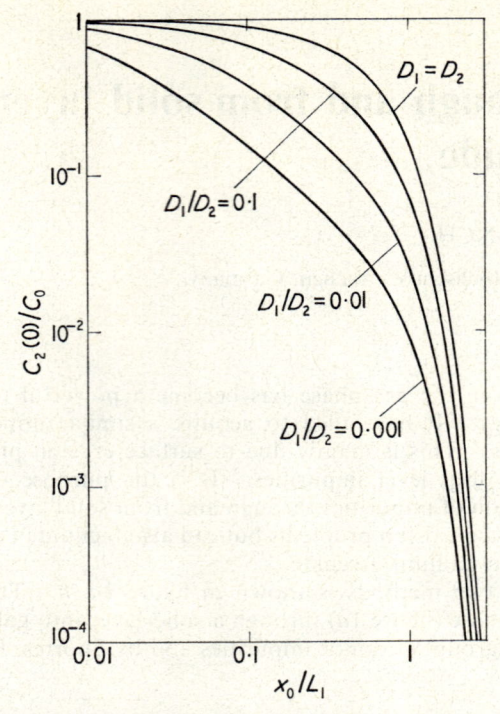

Figure 2. Impurity surface concentration in gallium arsenide as function of the normalized thickness of the protective layer (parameter: ratio of diffusion constants).

method are shown in figure 2. The normalized surface concentration is plotted versus the normalized thickness of the protective layer ($L_1 = 2\sqrt{(D_1 t)}$ = diffusion length in the protective layer) for various ratios of the diffusion constants (D_1 in protective layer, D_2 in gallium arsenide).

The distribution coefficient has been assumed to be unity. It is seen from figure 2 that there are certain diffusion conditions which exhibit a nearly linear relationship between the inverse layer thickness $1/x_0$ and the surface impurity concentration.

A further means of influencing the impurity concentration in gallium arsenide is given by the choice of protective materials with an appropriate distribution coefficient with respect to the diffusing impurity. Silicon dioxide, for instance, reduces the zinc concentration in gallium arsenide by virtue of its distribution coefficient.

The technique of impurity diffusion from a solid layer into gallium arsenide offers the possibility of achieving certain types of impurity profiles. These profiles have been calculated with the assumption of unlimited outdiffusion from the top of the source layer, i.e., $C_1 = 0$ at $x = -x_0$ for all $t > 0$ (see figure 1b). Owen and Schmidt (1968) have treated a similar case, but without permitting outdiffusion. Impurity profiles for various diffusion conditions are shown in figure 3. The distance x from the GaAs-surface is normalized with respect to L_2 ($= 2\sqrt{(D_2 t)}$ = diffusion length in gallium arsenide). For the sake of clarity, a factor $\sqrt{(D_1/D_2)}/[1 + \sqrt{(D_1/D_2)}]$ has been omitted from the concentration scale, i.e., the impurity concentration $C_2(x)$ in gallium arsenide is obtained by multiplying with this factor:

$$C_2(x) = C_2^*(x) \cdot \sqrt{(D_1/D_2)}/[(1 + \sqrt{(D_1/D_2)})].$$

All curves are drawn assuming the distribution coefficient to be unity.

As seen from figure 3, the solid-to-solid diffusion method is particularly suited to obtain an impurity profile with a flat portion, or with a limited region of positive concentration gradient. The limiting case of a simple complementary error-function distribution is also shown in figure 3 ($x_0 \to \infty$).

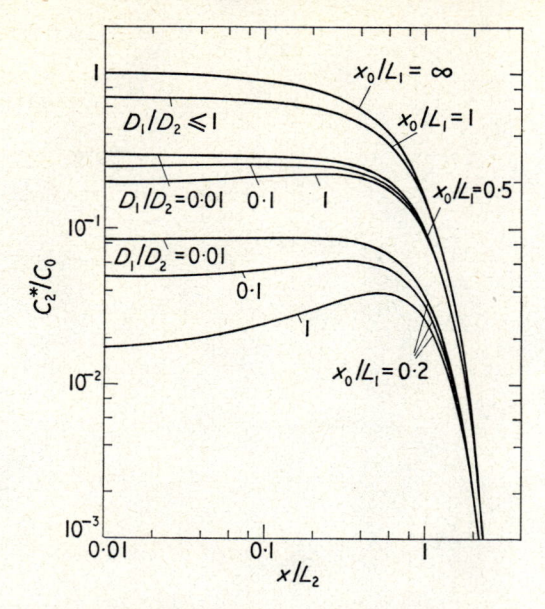

Figure 3. Impurity concentration profiles in gallium arsenide obtained by solid-to-solid diffusion (parameters: normalized thickness of source layer, ratio of diffusion constants).

References

Owen A. E. and Schmidt P. F. *J. Electrochem. Soc.* **115** 548–53.
Shortes S. R. Kanz, J. A., and Wurst, E. C., 1964, *Trans. Met. Soc. AIME*, **230**, 300–6.
von Münch, W., 1966, *Solid-State Electron.*, **9**, 619–24.
Yeh, T. H., 1964, *J. Electrochem. Soc.*, **111**, 253–5.

CHAPTER 3

Stimulated emission

Correlation of GaAs junction laser thresholds with photoluminescence measurements

C. J. HWANG and J. C. DYMENT

Bell Telephone Laboratories, Incorporated, Murray Hill, New Jersey, U.S.A.

Abstract. Threshold currents of diffused GaAs junction lasers are measured for several slices all cut from the same ingot. These laser thresholds are found to be strongly dependent on the donor concentration of the slice and the diffusion time. Three quantities which determine the laser thresholds are examined for each slice after the diffusion: (1) ρ_0, the state density at the quasi-Fermi level in the conduction band tail at threshold; (2) E_0, the band tailing constant; and (3) τ, the total lifetime of the injected electrons. It is found the ρ_0 and E_0 are nearly the same for all diodes and only decreases in τ can account for the increases in the thresholds. Comparison of photoluminescence intensities of the n-substrate before and after the diffusion indicates the formation of a considerable concentration of defects in the crystal during the diffusion. Our model assumes that these defects capture electrons in the p-region of the junction and thus increase laser thresholds. An expression relating τ and hence the threshold current with the near edge photoluminescence intensity of the n-substrate *after* Zn diffusion is then derived and is shown to agree well with the experimental data. It is proposed that the defects produced during the diffusion probably involve complexes containing Te atoms.

1. Introduction

A continuing problem for GaAs lasers made by diffusion of Zn has been the great variability of the threshold currents for different melt-grown substrate materials and fabrication conditions. In extreme cases, diodes fabricated from different slices of the same ingot have exhibited variations in threshold by as much as a factor of 10. These variations are not due to changes in cavity loss since all these lasers have nearly identical cavity structures. In an effort to determine which factors cause these large threshold changes, we examined parameters of several slices from the same ingot both before and after diffusion. No correlation was found with the prediffusion parameters since these were all about the same for each slice. However, after the diffusion, we discovered a definite correlation existing between the threshold current and the near edge photoluminescence intensity of the bulk n-type substrate material. The first purpose of this paper is to demonstrate this correlation both experimentally and theoretically. These results are both given here for the first time. The second purpose of the paper is to discuss the nature of the defects responsible for the increase in threshold.

Before proceeding with our experimental results, we first give an expression for the laser threshold which will allow us to define some quantities to be examined in this study. Assume an exponential density of states in the conduction band tail $\rho_c = \rho_0 \exp(E/E_0)$ where E_0 is the band tailing constant and ρ_0 is defined as the state density at the electron quasi-Fermi level at threshold. The threshold current I_{th} at temperature T can be expressed approximately as (Dousmanis and Staebler 1966)

$$I_{th} = \frac{qdAE_0\rho_0 \exp(kT/E_0)}{\tau}, \tag{1}$$

where q, d, and A are respectively, the electron charge, width of the active region, and cross-sectional area of the laser. For the lasers in this study, both d and A are assumed approximately constant. The quantity τ is the total lifetime of the electrons on the p-side of the junction and includes both radiative and non-radiative processes. E_0 and ρ_0 will depend on the active donor concentration and since, ρ_0 is measured at the electron quasi-Fermi level at threshold, its magnitude will also depend on cavity losses (Dousmanis and Staebler 1966).

In subsequent sections we will show that both ρ_0 and E_0 are nearly constant for each slice and that only decreases in τ can account for increases in I_{th}. Photoluminescence measurements in the n-substrate after diffusion (section 3) indicate that considerable concentrations of defects are formed during diffusion. In section 4 we propose a model which assumes that these thermally generated defects capture electrons on the p-side of the junction. An expression relating τ and hence I_{th} to the photoluminescence intensity of the n-substrate is then derived and is shown to be in good agreement with experimental results. Finally, in section 5, the nature of the defect centre is discussed.

2. Experimental

An ingot of GaAs doped with Te and grown by the Czochralski method was used. Slices of about 0·38 mm thickness were cut from the ingot. Initial measurements of carrier concentration and photoluminescence were made on these slices. The carrier concentration was determined from the infrared reflectance minimum near the plasma frequency (Spitzer and Whelan 1959) and the photoluminescence spectrum was measured both at 20°K and 77°K with the arrangement described in a previous paper (Hwang 1967). Diffusions were carried out at 800°C with the same composition of the diffusion source (2% Zn by weight in a solution of Ga saturated with GaAs (Dyment and D'Asaro 1967)) so that the surface concentrations of Zn were the same for all slices. The junction depths varied from 1·5 μ to 2·7 μ depending on the diffusion time and the substrate doping. The accuracy of the junction depth measurement is about 0·3 μ. Part of each diffused slice was fabricated into laser structures with stripe geometry contacts (Dyment 1967) of dimensions 13 $\mu \times$ 380 μ and 6 or more diodes were tested to determine I_{th} and the laser wavelength. The rest of the slice was used to measure the carrier density and photoluminescence of the substrate after the diffusion.

In order to ensure identical surfaces so that the infrared reflectance spectrum and the photoluminescence intensity of each slice could be compared, the slices were all lapped at one time with wet 1800 carborundum to remove about 3 mils. The lapping damage was then removed by etching off another ~ 10 μ in a solution composed of H_2SO_4, H_2O_2, H_2O (3 : 1 : 1). Since the junction depth was much smaller than the layer removed, the carrier concentration and photoluminescence thus measured can be considered as appropriate for the bulk n-type substrate after the diffusion.

As discussed in the next section, the values of E_0 were obtained from electroluminescence measurements. The detection system was the same one used for the photoluminescence experiments.

3. Results

Table 1 shows the I_{th} range, the laser wavelength, and fabrication conditions for each slice in which the 2% zinc-gallium diffusion source and 800°C temperature were used. The slices are numbered in sequence from one end of the ingot to the other. We see that I_{th} values are generally higher for longer diffusion times and higher numbered slices. The best and worst thresholds differ by more than a factor of 10. No marked changes in laser wavelength are observed.

Figure 1 plots the electron concentration before and after diffusion. We can equate the electron concentration to the active donor concentration (N_D) since, at these doping levels, the donor band overlaps the conduction band and all active donors are ionized. After the diffusion, N_D decreases about 4% to 2.6×10^{18} cm^{-3} at the low doped end and about 9% to 2.8×10^{18} cm^{-3} at the more heavily doped end. Since the donor concentration variation after diffusion is less than 10%, the band tail introduced by these donors will be approximately the same for each slice.

Table 1. Junction depth, diffusion time, laser threshold current range at 77°K for each slice and laser wavelength for best diode

Slice number	Junction depth (microns)	Diffusion time (hours)	I_{th} range at 77°K (mA)	Laser wavelength (Å) at 77°K
9	2·1	3	145–290	8485
10	2·1	3	180–220	8474
11	2·1	3	160–240	8487
12	2·1	3	180–195	8488
13	2·1	3	180–185	8493
14	1·8	3	180–225	8477
15	1·8	3	220–260	8489
16	2·1	6	240–540	8497
17	1·5	3	280–640	8462
18	1·5	3	155–240	8491
24	1·5	3	300–520	8484
28	1·8	4	310–500	8473
29	1·8	4	600–760	8480
30	2·7	6	1350–1800	8469
31	2·7	6	1600–2000	8475
33	2·4	5	2400–2700	8469
36	2·4	6	1280–1700	8485
37	2·4	6	1260–1440	8490
38	1·5	3	700–910	8481

Electroluminescence measurements were made at 77°K in a current region (10 mA to 120 mA) where the injection can be described as a band tail filling process (Nelson *et al.* 1963). The shift in peak energy indicates that the band tail is closely exponential. The slopes of semi-logarithmic plots of peak energy versus current determine the values of E_0 which are given in figure 1 for several slices. For all the diodes tested, E_0 ranges from 9 mev to 10 mev.

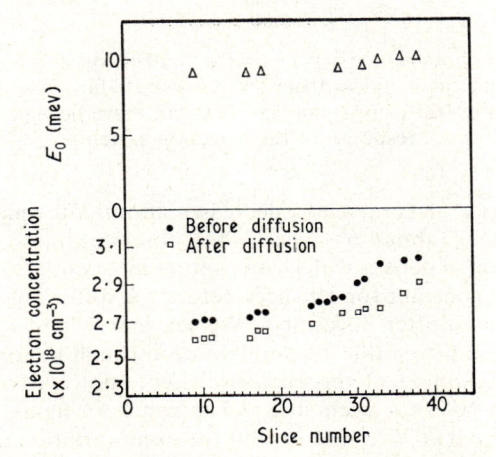

Figure 1. E_0 and the electron concentration before and after the diffusion for several slices from the same ingot.

Our observations that the laser wavelengths in table 1 are essentially constant indicate that the cavity losses are almost the same for all lasers tested (Dousmanis and Staebler 1966). Therefore, increases in I_{th} cannot be explained by increases in ρ_0 (Dousmanis and Staebler 1966). The maximum wavelength difference in table 1 is only 35 Å (or 2·8 mev) which should be approximately equal to the maximum difference in the positions of the electron quasi-Fermi levels assuming a constant hole quasi-Fermi energy for all diodes. This latter assumption will be discussed in section 5. The values of ρ_0 for all our diodes thus vary no more than a factor of $\exp(2\cdot8/E_0)$ or $\simeq 1\cdot3$ for $E_0 \simeq 10$ mev. Since $E_0 \exp(kT/E_0)$ changes less than $3\cdot5\%$ for all the values of E_0 shown in figure 1, the quantity $qdAE_0\rho_0 \exp(kT/E_0)$ in equation (1) can be considered approximately constant for all diodes. Changes in I_{th} should therefore be determined by changes in τ. We will subsequently show that the defects formed during diffusion influence both of these quantities. In order to study these defects, we show in figure 2 the 20°K photoluminescence spectra of a typical n-type substrate before (dashed curve) and after (solid curve) the diffusion. Band 1 at 1·540 ev is associated with transitions from the conduction band tail to the valence band with an indication of a Burstein-Moss shift (Burstein 1954; Moss 1954). Band 2 always appears in donor-doped GaAs crystals grown from the melt (Hill 1964). After diffusion,

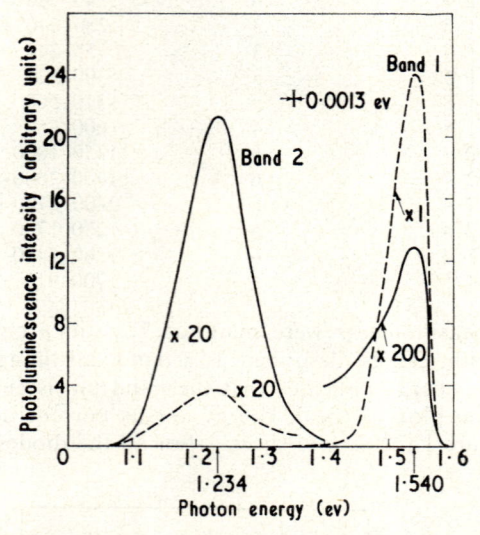

Figure 2. Typical n-type photoluminescence spectra at 20°K for slice number 36. Dashed curve: before the diffusion. Solid curve: taken from the n-substrate far away from the junction after the slice was diffused with Zn at 800°C for 6 hours. The data have been corrected for the wavelength response of the detection system.

we note that the intensity of band 1 (I_1) decreases about 370 times while the intensity of band 2 (I_2) increases only about 6 times. This observation implies the creation of an appreciable concentration of defects which can capture most of the injected holes. Although I_1 and I_2 are practically constant for all slices before the diffusion, figure 3 shows the large variations which are found after diffusion. We see that I_1 and I_2 both vary in the same manner for each slice, indicating that recombination through the defects formed during the diffusion dominates. Scanning of the excitation light spot across the sample shows that I_1 and I_2 vary less than 10% for each slice. The values in figure 3 represent the average values. I_1 and I_2 measured at 77°K also exhibit the same variation as that shown in figure 3. Thus, the recombination process which dominates at 20°K also dominates at 77°K.

4. A theoretical model

In this section, our purpose is to relate I_{th} to I_1 which is the photoluminescence intensity of band 1 in the bulk n-type substrate after the diffusion. The model will be constructed on the basis of experimental observations stated in the previous section.

4.1. Bulk n-substrate photoluminescence

In the bulk n-type substrate, the Fermi level is located somewhere in the conduction band. Hence, all the levels in the energy gap are filled in the dark. Thus, a donor defect will act

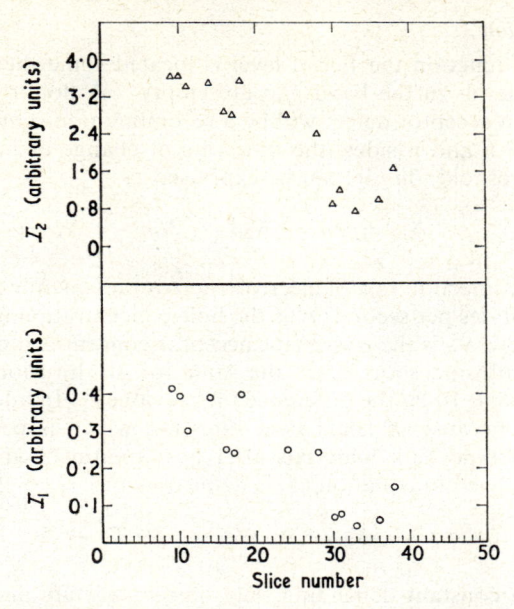

Figure 3. Photoluminescence intensities at 20°K of band 1 (I_1) and band 2 (I_2) for each n-substrate after the diffusion.

as a recombination centre while an acceptor will be a trapping centre. However, if the ionization energy of the trapping centre is greater than $4\,kT$, the trapping centre can be considered as a recombination centre because the holes captured at the centres necessarily recombine with electrons from the conduction band and are not released thermally into the valence band. From the peak position of band 2, we can regard the centres responsible for band 2 as recombination centres both at 20°K and 77°K. If we assume that the defect formed during the diffusion introduces a level (either donor or acceptor) with ionization energy of more than $4\,kT$ (this assumption will be discussed in section 5) and that the optical excitation is small, the time rate of change of the injected hole concentration in the valence band, dp/dt, can be expressed as

$$\frac{dp}{dt} = F - \beta_1 n p - \beta_2 N_2 p - \beta_t N_t p - \beta_x N_x p, \tag{2}$$

where F is the generation rate of the electron-hole pairs and is a constant for constant excitation; β's are recombination or capture probabilities per second, depending on whether the centre is a recombination centre or a trapping centre; n is the electron concentration in the conduction band; N_2 is the concentration of the centres responsible for band 2; N_t is the concentration of the defects formed during the diffusion and, as will be seen later, these defects probably involve the donor (Te) impurity; N_x is the sum of concentrations of other deep centres not associated with the donor impurity. Before the diffusion, $N_t = 0$ and the fact that both I_1 and I_2 remain essentially unchanged for all slices indicates that each slice contains nearly equal n, N_2, and N_x. After the diffusion, the overall radiative recombination is greatly reduced. We must have

$$\beta_t N_t \gg \beta_1 n + \beta_2 N_2 + \beta_x N_x. \tag{3}$$

Thus, under steady state conditions ($dp/dt = 0$), $F \simeq \beta_t N_t p$ and I_1 can be expressed as

$$I_1 \propto \beta_1 n p \simeq \frac{\beta_1 n F}{\beta_t N_t}. \tag{4}$$

4.2. *p-side of the junction*

On the p-side of the junction the Fermi level is located somewhere in the valence band tail. Hence, all deep levels in the band gap are empty. A donor defect will thus act as trapping centre while an acceptor defect will be a recombination centre. If we assume that N_t is the same on both n and p sides, the time rate of change of injected electrons in the conduction band at threshold, dn'/dt, can be expressed as

$$\frac{dn'}{dt} = N - c_1 p' n' - c_2 N_2 n' - c_t N_t n' - c_x N_x n' \tag{5}$$

where $N \simeq I_{th}/qAd$ is the injection rate of electrons across the junction; c's are the recombination or capture probabilities per second; p' is the hole concentration in the valence band and is equal to $N_A - N_D$, where N_A is the active zinc acceptor concentration in the active region of the junction. If the diffusion sources are the same for all diffusion runs, p' for all slices should differ no more than 10%, the difference in N_D values. In addition, we may assume that $c_1 p' \gg c_2 N_2 + c_x N_x$ because the near edge emission which is proportional to $c_1 p' n'$ is the dominant one in p-type bulk materials after heat treatment at 800°C (Hwang 1967). The electron lifetime τ given in equation (1) is defined as

$$\frac{1}{\tau} \equiv c_1 p' + c_2 N_2 + c_x N_x + c_t N_t \simeq H(T) + c_t N_t, \tag{6}$$

where $H(T) \simeq c_1 p'$ is a constant depending only on temperature and is approximately the same for all diodes made from different slices. Since F is constant for constant optical excitation, equations (1), (4), and (6) give

$$I_{th} \simeq A(T) + B(T) \left(\frac{1}{I_1}\right) \tag{7}$$

where $A(T)$ and $B(T)$ are temperature dependent constants.

4.3. *Comparison with experiment*

In figure 4, values of $1/I_1$ at 20°K and 77°K are plotted as a function of slice number in the following manner. The values of I_{th} at 77°K taken from table 1 are plotted as vertical bars.

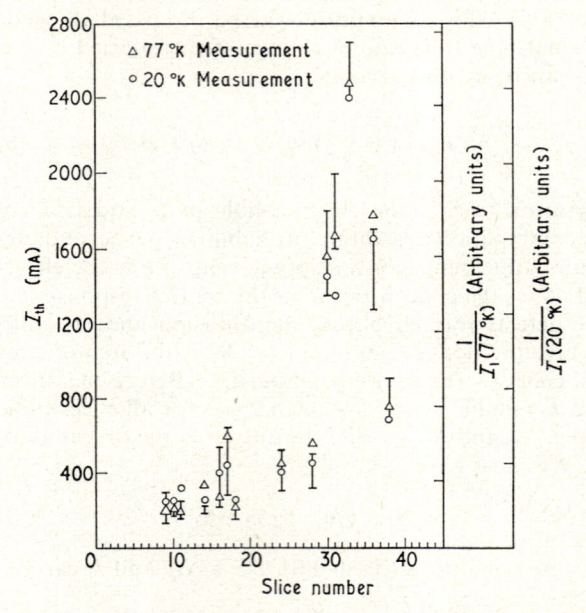

Figure 4. Comparison of the laser threshold currents with $1/I_1$ from data measured both at 20°K and 77°K. The vertical scales for $1/I_1$ are adjusted for the best fit to the threshold data.

Each bar represents the threshold range of the lasers made from one particular slice. The vertical scales for $1/I_1$ at 20°K and 77°K are adjusted so that the best fit of $1/I_1$ with the vertical bars is obtained. It is seen that the fit is in good agreement with the functional dependence predicted by equation (7). We, therefore, conclude that the defects formed during the diffusion indeed play a dominant role in determining I_{th}.

5. Discussion

Since the defects formed during the diffusion were detected in the region of the n-substrate far from the Zn diffusion front, the defects are actually produced by the annealing of the crystal and are not associated with the indiffused Zn. The defect can be either an acceptor or a donor species. In either case, some of the injected electrons will be captured and consequently the laser threshold will increase. If the recombinations through these defects are predominantly radiative, the defect eneryy level must lie more than 0·4 ev into the energy gap from a band edge since no new photoluminescence response besides I_1 and I_2 was detected in the photon energy region between 1·5 ev and 1·1 ev. If recombinations through the defects are predominantly non-radiative, the energy released must be dissipated through a multiphoton process. This process is most likely to occur when the interaction between the defect centre and its neighbouring atoms is strong. Since the stronger the interaction the larger the ionization energy of the centre (Williams 1967), this defect is also likely to be a deep centre. Hence, the assumption that the ionization energy of the defect is $\geqslant 4\,kT$ which was made in the previous section appears quite reasonable.

Let us now discuss our assumption of section 3 in which we assumed that the hole quasi-Fermi energy was constant. The actual Zn acceptor concentration N_A in the active region is not exactly known. However, a value of $N_A - N_D \simeq 3 \times 10^{18}$ cm^{-3} should be appropriate for an injection laser (Stern 1966). Using N_D values from figure 1, N_A should vary less than 10% for all our slices. The corresponding variation in the hole quasi-Fermi energy is probably less than 1 mev as can be seen from the calculation of Stern (1966). The presence of defects formed during diffusion is not expected to alter the hole quasi-Fermi level since the defect concentration is much smaller than N_A. For example, if we assume a one-to-one correspondence between the loss of active Te donors and the formation of defects, the maximum defect concentration will be 3×10^{17} cm^{-3} from figure 1.

We also find that the largest decreases in N_D occur for the higher N_D values and longer diffusion times. These results are consistent with the annealing experiments of Fuller and Wolfstirn (1963) which show that this behaviour should occur for $N_D \geqslant 2 \times 10^{18}$ cm^{-3}. Their annealing kinetics suggest that complexes containing the Te atom are probably formed during diffusion. It is likely that these defects capture electrons in the p-region and increase I_{th}. Complexes such as GaTe$_3$ are known to exist (Hulliger and Mooser 1963). The experiments of Fuller and Wolfstirn (1963) also indicate that about 10 hours are required to achieve the equilibrium defect concentration at 794°C. Thus, in our experiments, I_{th} should also depend on the diffusion time. Table 1 seems to show such a trend in addition to the dependence on N_D.

In our experiments, we found no correlation between I_{th} and the prediffusion parameters of a slice. The present results coupled with those of Hwang (1967) show that defects can be both created and annealed at high temperatures. These comments should apply equally well to diffused electroluminescence lamps although Brice (1967) has reported a correlation between the lamp efficiency and the As vapour pressure under which the original crystal was grown. Decreases in efficiency were attributed to increases in the number of As vacancies. Since these diodes were made by Zn diffusion at high temperatures and since the defects associated with As vacancies are known to depend strongly on the annealing temperature (Potts and Pearson 1966), this correlation seems to have little significance as was pointed out previously (Hwang 1968). Wright et al. (1967) also reported a correlation between lamp efficiency and the presence of a certain trapping level in the material. This correlation is even less certain because, in this case, not only were the diodes fabricated by Zn diffusion but the defects were detected in samples compensated with Cu.

6. Summary and conclusion

In this paper we report the first experimental correlation between I_{th} and the near edge photoluminescence intensity (I_1) measured in the bulk n-substrate *after* Zn diffusion. Of the three parameters ρ_0, E_0, and τ which determine I_{th}, we show that only decreases in the total electron lifetime (τ) can account for the increases in I_{th} found in some slices. Large reductions in I_1, after diffusion indicate that considerable concentrations of defects are formed during diffusion. By assuming that these defects capture injected electrons on the p-side of the junction, an expression relating I_{th} and I_1 is derived (equation (7)). This theoretical dependence is in good agreement with the experimental data of figure 4. The defect centres produced during diffusion probably involve complexes containing Te atoms similar to the proposal of Fuller and Wolfstirn (1963).

If thermally generated defects control I_{th}, the correlation described in this paper should enable much time to be saved in selecting good crystals for laser fabrication. However, under certain circumstances, ρ_0 and E_0 might provide the dominant contribution in equation (1). Therefore, in order to achieve the lowest possible I_{th} values, compromises must be made in selecting the doping levels and diffusion times.

Acknowledgments

The authors wish to thank R. L. Brown for performing the infrared reflectance measurements and assisting in the photoluminescence and electroluminescence experiments, and R. E. Mayer for laser fabrication. They are also indebted to L. A. D'Asaro, G. E. Smith and E. I. Gordon for valuable comments.

References

BRICE, J. C., 1967, Solid State Electronics, **10**, 335–7.
BURSTEIN, E., 1954, *Phys. Rev.*, **93**, 632–3.
DOUSMANIS, G. C., and STAEBLER, D. L., 1966, *J. Appl. Phys.*, **37**, 2278–80. Note that τ in this reference is the radiative lifetime because a unity quantum efficiency is assumed.
DYMENT, J. C., 1967, *Appl. Phys. Lett.*, **10**, 84–6.
DYMENT, J. C., and D'ASARO, L. A., 1967, *Appl. Phys. Lett.*, **11**, 292–4.
FULLER, C. S., and WOLFSTIRN, K. B., 1963, *J. Appl. Phys.*, **34**, 2287–9.
HILL, D. E., 1964, *Phys. Rev.*, **133A**, 866–72.
HULLIGER, F., and MOOSER, E., 1963, *J. Phys. Chem. Solids*, **24**, 283–95.
HWANG, C. J., 1967, *J. Appl. Phys.*, **38**, 4811–7.
HWANG, C. J., 1968, *J. Appl. Phys.*, **39**, 1654–9.
MOSS, T. S., 1954, *Proc. Phys. Soc.*, **B67**, 775–82.
NELSON, D. F., GERSHENZON, M., ASHKIN, A., D'ASARO, L. A., and SARACE, J. C., 1963, *Appl. Phys. Lett.*, **2**, 182–4.
POTTS, H. R., and PEARSON, G. L., 1966, *J. Appl. Phys.*, **37**, 2098–103.
SPITZER, W. G., and WHELAN, J. M., 1959, *Phys. Rev.*, **114**, 59–63.
STERN, F., 1966, *Phys. Rev.*, **148**, 186–94.
WILLIAMS, E. W., 1967, *Brit. J. Appl. Phys.*, **18**, 253–62.
WRIGHT, H. C., HUNT, R. E., and ALLEN, G. A., 1967, *Solid State Electronics*, **10**, 633–9.

Theory of Q-switching and time delays in GaAs junction lasers

J. E. RIPPER

Bell Telephone Laboratories, Incorporated, Murray Hill, New Jersey, U.S.A.

Abstract. In previous publications, the existence of a double acceptor type trap was proposed to account first for the characteristics of the long delays t_d between the beginning of the current pulse and the onset of stimulated emission and second for the existence in some diodes of a Q-switched light pulse after the end of the injection pulse.

In the present paper, an outline of a mathematical development of this model is presented. Computer-generated curves show that the model can account for all the known properties of the delays and Q-switching. An anomalous delay behaviour predicted by the theory was experimentally observed with very good agreement between the theoretical and experimental curves.

1. Introduction

Some GaAs junction lasers when operated near room temperature present a long delay ($\sim 10^{-7}$ sec) between the beginning of the current pulse and the onset of stimulated emission (Winogradoff and Kessler 1964; Konnerth 1965; Fenner 1967). In a study of the temperature dependence of these delays, it was proposed that this effect can be explained by the presence near the active region of a double-acceptor type trap (Dyment and Ripper 1968). It was also proposed that the newly discovered internal Q-switching behaviour of some GaAs laser diodes can be accounted for by the same model (Ripper and Dyment 1968).

The purpose of this paper is threefold: first to outline a theory based on the double-acceptor trap model; second to present results of computer calculations using the theory; and third to compare these results with experimental data, some not previously published.

2. Proposed model

As described previously (Dyment and Ripper 1968), we postulate the existence of a double-acceptor trap in the active region of the laser. This trap can be in three different states depending on the number of electrons it contains, figure 1. In its first state Tr_1, the trap can capture an electron at an energy E_{12} and thus go into the second state Tr_2. This transition is reversible and will be controlled by the relative position of E_{12} and the valence band quasi-Fermi level E_F, since E_{12} is assumed to be close enough to the valence band and nearby impurity states to be in thermal equilibrium with them. Having captured one electron and being in the state Tr_2, the trap can capture a second electron at an energy E_{23}

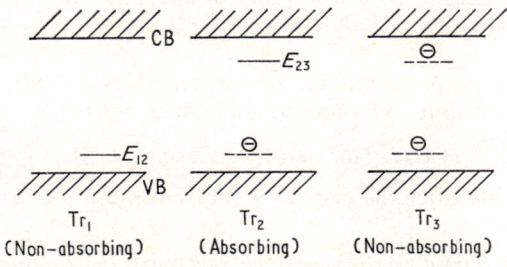

Figure 1. States of the double acceptor trap. $Tr_1 + e^- \rightleftharpoons Tr_2$; $Tr_2 + e^- \rightleftharpoons Tr_3$. State Tr_2 is labelled absorbing, because a photon of energy close to the gap can be captured as an electron goes from the valence band to the level E_{23}.

and go to the state Tr3 where it cannot capture another electron in our range of interest. Since E_{23} is assumed near the conduction band, the transition into Tr3 can occur in two ways: by trapping an injected electron directly from the conduction band or by capturing the electron from the valence band, with the absorption of photon accounting for the energy difference. Because of this last mechanism the trap, when in state Tr2, is optically absorbing for photons of energy near the gap. The transition Tr2-Tr3 corresponds to the trap proposed by Fenner (1967).

3. Outline of the theory

The objective of the theory is to calculate the laser gain G and losses L as a function of time so that the delay will be given, at any heat sink temperature and injection current level, by:

$$G(I, t_d) = L(I, T_{HS}, t_d). \qquad (1)$$

The symbols of this and subsequent equations are defined in table 1.

Table 1. Definition of symbols used in section 3

B	Capture probability per unit time of injected electrons by Tr2
E_F	Valence band quasi-Fermi level
E_ρ	Centre of distribution $\rho(E-\rho)$
$G(I, t_d)$	Laser gain
I	Injection current level
k	Boltzmann constant
$L(I, T_{HS}, t_d)$	Laser losses
L_0	Constant
$n_i(t)$	Effective number of traps on state Tri
n_ρ	Total effective number of traps $= \int_{-\infty}^{\infty} \rho(E-E_\rho)\, dE$
n_{21}	Number of effective traps that decay from Tr2 to Tr1 at the end of the current pulse due to a change in $E_F - E_\rho$
N	Number of injected electrons
t	Time from the beginning of the pulse
t_d	Delay
t_p	Pulse length
T	Laser junction temperature
T_{HS}	Heat sink temperature
T_0	Constant
T_t	Transition temperature
β	Constant
ϵ	Constant
$\rho(E-E_\rho)$	Effective density of level E_{12} weighted by the normalized photon density
τ	Time constant of build-up of population inversion
τ_e	Emptying time of the second electron of the traps
τ_{sp}	Spontaneous recombination time of injected electrons
τ_{th}	Thermalization time of level E_{12}

The gain is assumed proportional to the current (Pilkuhn and Rupprecht 1963) and builds up with a time constant τ (Konnerth and Lanza 1964).

$$G(I, T_{HS}, t) = \beta I \{1 - \exp(-t/\tau)\}. \qquad (2)$$

If lasing has not occurred, after the end of the current pulse the carriers decay with a time constant τ_{sp}. Then:

$$G(I, T_{HS}, t) = \beta I \{1 - \exp(-t_p/\tau)\} \exp\{-(t-t_p)/\tau_{sp}\}. \qquad (3)$$

The losses should be divided in two parts; first the normal laser losses in absence of the traps, which in our diodes are exponential with temperature; second a term proportional to the number of traps in the absorbing state Tr_2:

$$L(T_{HS}, I, t) = L_0 \exp(T/T_0) + \epsilon n_2(t). \tag{4}$$

Our problem is then reduced to calculating $n_2(t)$. During the current pulse it will be controlled by two transitions; first between the states Tr_1 and Tr_2 controlled by the relative position of E_{12} and E_F:

$$n_2(0) = n_\rho - n_1(0) = \int_{-\infty}^{+\infty} \rho(E - E_\rho) \frac{1}{1 + \exp\{(E - E_F)/kT\}} dE \tag{5}$$

where

$$E_F - E_\rho = F(T, I); \tag{6}$$

and second by the capture of a second electron from the conduction band making the transition to Tr_3, which, as shown by Fenner can be described by:

$$\frac{dn_3(t)}{dt} = n_2(t) BN - \frac{n_3(t)}{\tau_e}. \tag{7}$$

This equation can be solved assuming B to be constant for all traps, and $n_3(0) = 0$.

After the end of the current pulse some of the traps in Tr_2 decay into Tr_1 with a fast thermalization time τ_{th} due to a change in $E_F - E_\rho$. So for $t > t_p$

$$n_2(t) = n_2(t_p) - n_{21}[1 - \exp\{-(t - t_p)/\tau_{th}\}]. \tag{8}$$

These equations allow us to solve equation (1) and divide the temperature-current space in three regions:

(a) Normal lasing region where equation 1 has a solution with $t_d < t_p$. This region is limited by the threshold curve.

(b) Q-switching region, where equation 1 has a solution *only* for $t_d > t_p$. This region does not exist for most diodes.

(c) Spontaneous emission region where equation 1 has no solution for any t.

4. Theoretical results

In order to compare our model with the experimental results, a computer program was set up based on the theory outlined and taking heating during the pulse into account.

Figure 2 shows computer-generated curves of the threshold current I_{th} and the delay

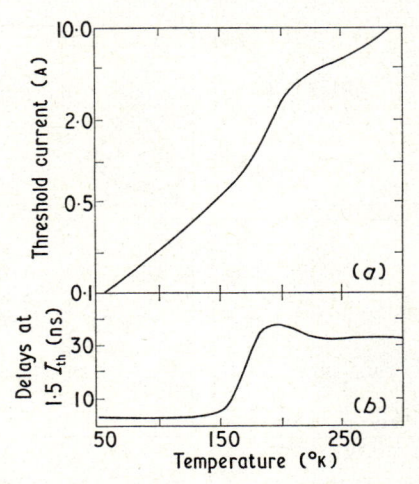

Figure 2(a). Threshold curve calculated for one diode with transition temperature 180°K. (b) Delay calculated for currents 50% above threshold for the same diode.

at $I = 1 \cdot 5\, I_{th}$ for a particular diode. The following characteristics, previously found experimentally, can be readily observed:

(a) Two temperature regions of short and long delays respectively, separated by a transition region centred at a transition temperature T_t. A small change in the parameters used in the calculation produces a large change in T_t, consistent with the wide variation experimentally found in the value of T_t. The transition temperature can occur well above 300°K, explaining the absence of long delays on some diodes at all temperatures tested.

(b) There is a 'kink' in the threshold curve at approximately T_t.

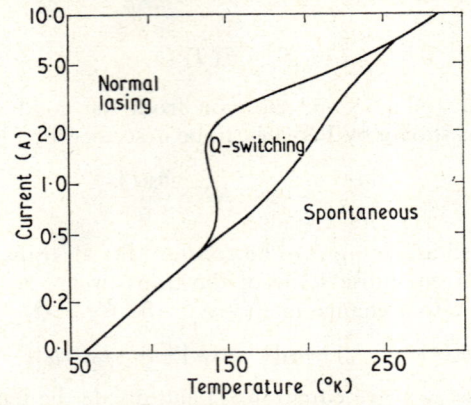

Figure 3. Calculated normal lasing, Q-switching, and spontaneous emission region for a diode with a transition temperature about 140°K.

Figure 3 shows the calculation of the Q-switching region for a diode with a very low transition temperature. This was calculated assuming $\tau_{th} = 0$ in equation (8); this approximation and the assumption of an infinitely sharp current pulse fall-off tend to increase the Q-switching region in the low current limit, since it neglects any gain reduction and superestimate the loss reduction at the time the Q-switching pulse is generated. This explains the small discrepancy with the experimental curve (Ripper and Dyment 1968), but otherwise the same general behaviour is observed, including a temperature region where a double threshold is found.

Finally on the examination of the delay behaviour at a fixed temperature, far below or above T_t, the delays were calculated to decrease with increasing currents as previously

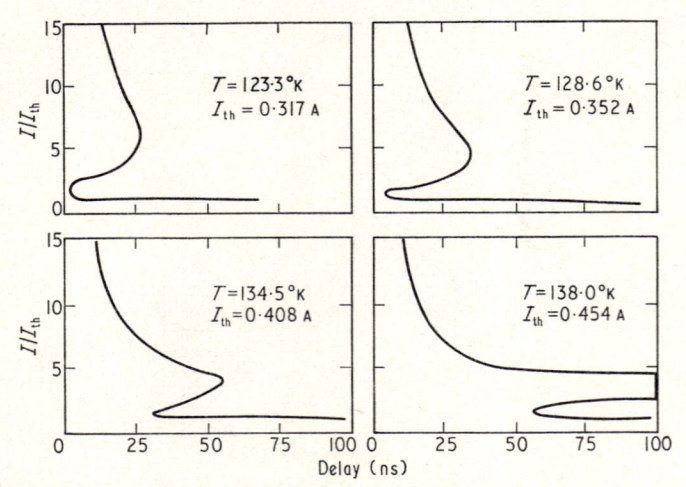

Figure 4. Abnormal delay behaviour calculated for the diode of figure 3. Last curve corresponds to the beginning of the Q-switching region.

experimentally observed. However, on diodes with low T_t, the delay can have an abnormal behaviour just below T_t as shown in figure 4. In this region, as the current is increased, the delays initially decrease rapidly to a minimum value, only to increase again as traps go into the absorbing state. After reaching a maximum value the delays again decrease normally with current; when this maximum value tends to be bigger than the pulse length t_p, Q-switching will occur (last curve). This abnormal delay behaviour was later confirmed experimentally as shown in figure 5.

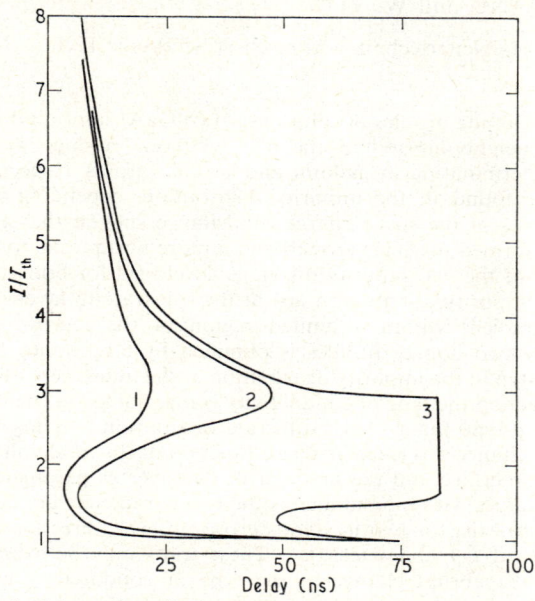

Figure 5. Experimental delays for diode L-137-10 showing the abnormal delay behaviour. 84 nsec pulses.

	Temp (°K)	I_{th} (A)
1	124	0·54
2	125	0·59
3	129	0·82

5. Conclusions

The computer-generated curves show that the double-acceptor trap model can not only account for the properties of delays and Q-switching in GaAs diodes. The anomalous delay behaviour also predicted by it was later verified experimentally.

The problem of the identity of the trap remains unsolved. There is a strong possibility that it is a defect introduced during the zinc diffusion.

Acknowledgment

The author would like to thank J. C. Dyment, T. H. Zachos, and C. J. Hwang for stimulating discussions during the development of this theory.

References

DYMENT, J. C., and RIPPER, J. E., 1968, *J. Quant. Electronics*, **QE-4**, 155–60.
FENNER, G. E., 1967, *Solid State Electronics*, **10**, 753–64.
KONNERTH, K., 1965, *IEEE Trans. Electron Devices* (1965 Solid State Device Research Conference Abstracts), **ED-12**, 506.
KONNERTH, K., and LANZA, C., 1964, *Appl. Phys. Lett.*, **4**, 120–1.
PILKUHN, M., and RUPPRECHT, H., 1963, *Proc. IEEE*, **51**, 1243–4.
RIPPER, J. E., and DYMENT, J. C., 1968, *Appl. Phys. Lett.*, **12**, 365–7.
WINOGRADOF, N. N., and KESSLER, H., 1964, *Solid State Commun.*, **2**, 119–22.

Doping profiles of solution grown GaAs injection lasers

H. BENEKING and W. VITS

Institut für Halbleitertechnik Rheinisch-Westfälische Technische Hochschule, Aachen, Germany

Abstract. Doping profiles of solution-grown GaAs laser diodes are investigated in the immediate neighbourhood of the p–n junction. Using ⟨111⟩ and ⟨$\overline{111}$⟩ oriented substrates, terminating in gallium and arsenic atoms respectively, considerable differences are found in the impurity distribution, depending on substrate orientation. Measurements of the space charge capacitance suggest that a nearly linearly graded junction is formed for [$\overline{111}$] growth and a more abrupt one for [111] growth.

A model of the net impurity profile is developed for both orientations, assuming a step in the impurity distribution just at the substrate-to-layer interface and a constant impurity gradient within a limited region of the regrown layer. The difference between the two doping profiles is obtained by taking into account an orientation-dependent step in the impurity distribution at the interface. If the crystal is orientated in [$\overline{111}$] direction the type of conductivity is maintained at the step, whilst it is changed from n- to p-type for a ⟨111⟩ substrate orientation. Using these models the space charge capacitance was calculated as a function of the total voltage, taking the impurity step at the interface and the gradient in the layer as parameters. By comparing the calculated values with measurements the two parameters are determined. In addition traps seem to exist for p–n junctions grown in [111] direction as indicated by a hump in the forward I–V-characteristic. These traps may be explained by lattice defects caused by the abrupt change in the type of conductivity at the substrate-to-layer interface.

1. Introduction

As recently reported, it has been found that the properties of solution-grown GaAs injection lasers are strongly affected by substrate orientation (Beneking and Vits 1968). In particular, the losses per unit length turned out to be the highest for diodes grown in [$\overline{111}$] direction and to differ almost by a factor of two, when compared to the [111] growth direction. Calculations concerning wave-guide effects in p–n junctions carried out by several authors (see for example Yariv and Leite (1963), McWhorter (1963), Stern (1964)) indicate, that the impurity distribution in the neighbourhood of the p–n junction has a considerable influence on laser losses.

Because the formation of the doping profile is affected by the performance of the epitaxial process the method used will be briefly described (Beneking and Vits 1966). The substrates are 10^{18} cm^{-3} selenium-doped single crystals. The melt contains 4 g Ga, 0·14 g Zn and 0·8 g n-type GaAs. At 850 °C the melt is brought into contact with the substrate. Within the next 5 minutes the temperature is raised by 5 °C in order to clean the surface by dissolution. Then the system is cooled down to 450 °C at a rate of 10 °C min^{-1}.

2. Doping profile at the p–n junction

2.1. *Space charge capacitance*

We measured the space charge capacitance as a function of the applied voltage for both types of diodes with an accuracy of $\pm 2\%$. The results obtained are shown for one example in figures 1 and 2. Here the square and the cube of the reciprocal capacitance per unit area are plotted versus the applied voltage. As can be seen from these diagrams the p–n junction grown in [$\overline{111}$] direction (figure 1) has a nearly linearly graded impurity distribution indicated by the straight line of the $(C_s/A)^{-3}$ plot. For the diode grown in [111] direction

PLATE IX—*J. J. Tietjen et al.: paper 9*

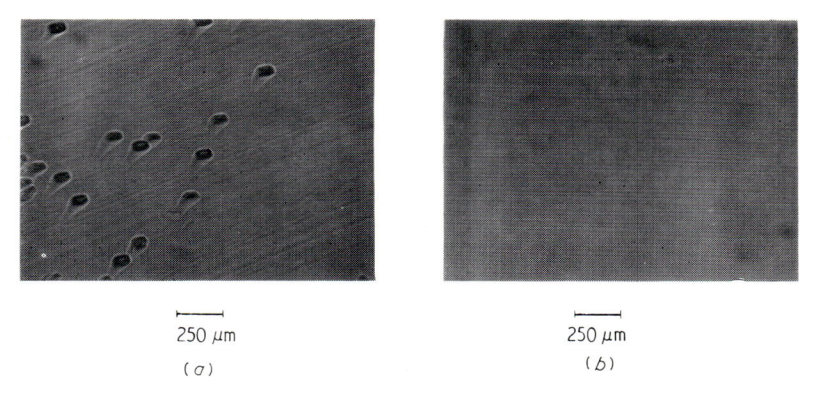

250 μm
(a)

250 μm
(b)

Figure 4. The effect of impurity striations in the substrate on the perfection of a GaAs epitaxial layer.

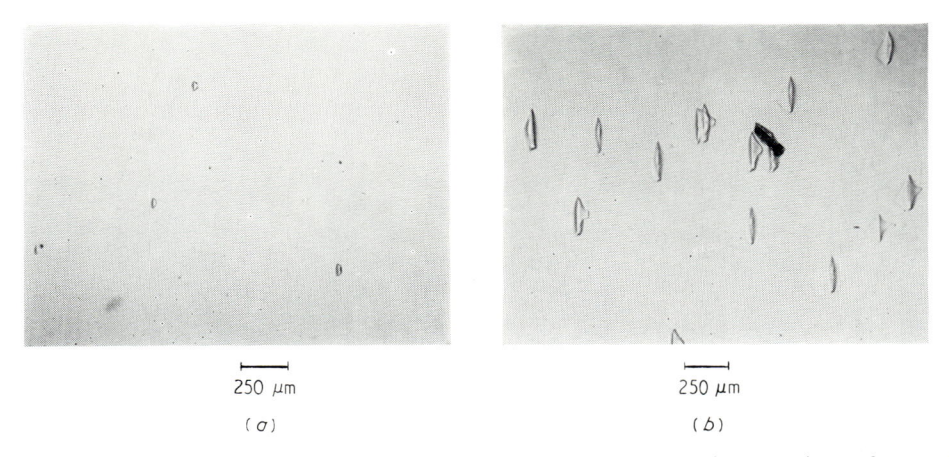

250 μm
(a)

250 μm
(b)

Figure 5. The effect of exposing the substrate to the laboratory atmosphere on the surface perfection of the deposited GaAs layers.

PLATE X—J. J. Tietjen et al.: paper 9

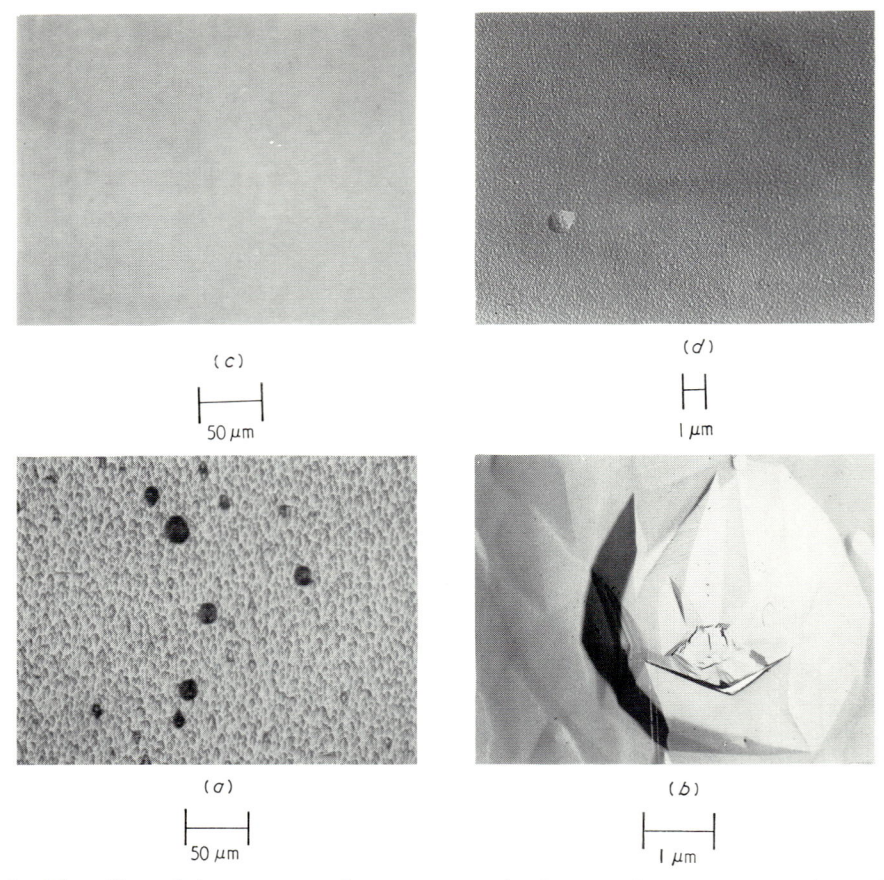

(c) 50 μm

(d) 1 μm

(a) 50 μm

(b) 1 μm

Figure 6. The effect of the presence of water vapour in the growth apparatus on the perfection of a GaAs epitaxial layer.

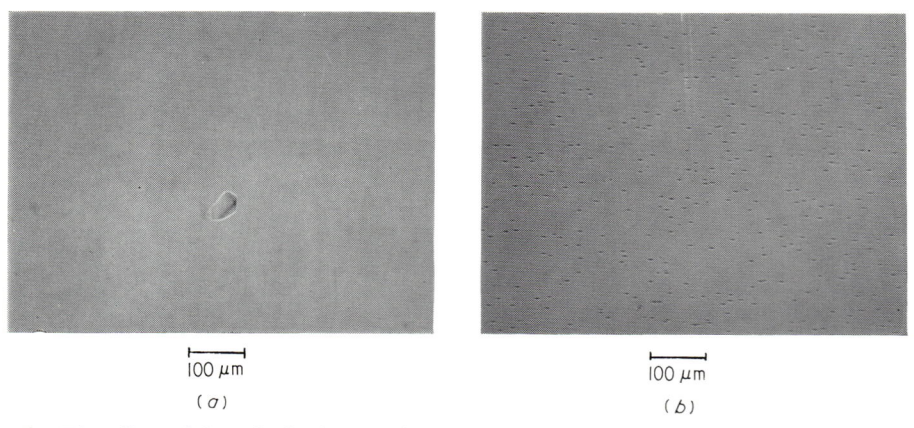

(a) 100 μm

(b) 100 μm

Figure 7. The effect of deposits in the reaction tube on the perfection of the deposited GaAs layer.

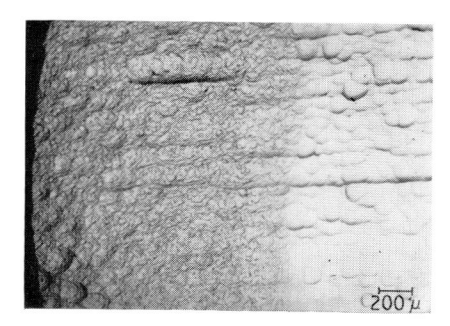

Figure 1. Distribution of precipitations (as revealed by etching) at the periphery of a Te-doped GaAs slice cut perpendicular to the (111)-growth axis. Note the sharp boundary between regions of high- and low-precipitation density.

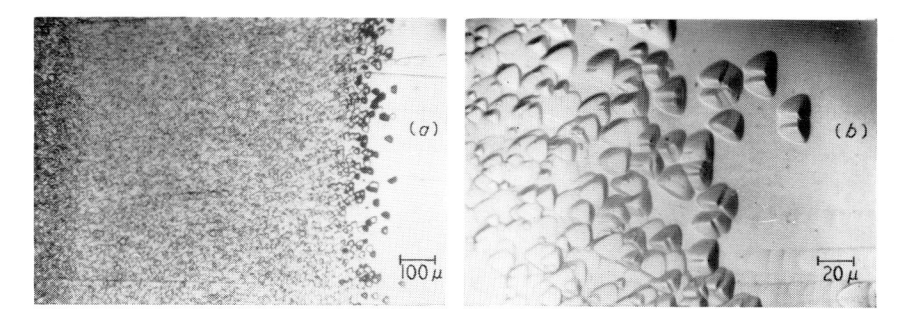

Figure 2. (a) The same part of the (111) slice as shown in figure 1, but after annealing at 1100°C for four hours. Stacking faults surrounded by closed, partial dislocation loops have been formed in the region of the steepest precipitation gradient. (b) Magnified view of the transition region. Pairs of etch pits are visible, connected by notches along one of the three $\langle 110 \rangle$ directions lying in the $\{111\}$ plane.

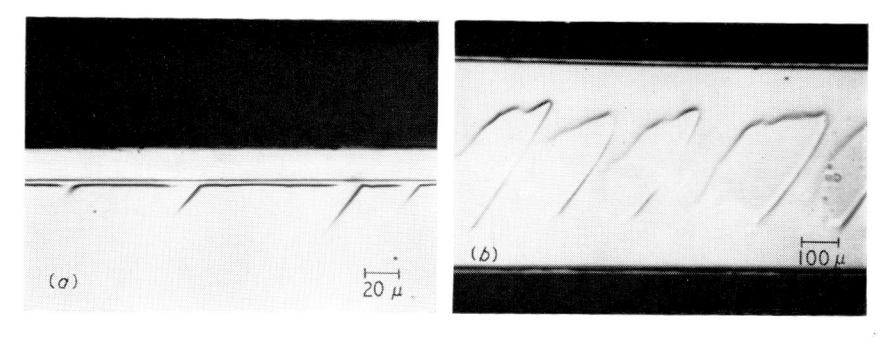

Figure 3. Thermal conversion effect. Vertical section of an n-type slice. (a) Early stage: the upper continuous line is due to regular Be-diffusion. The lower interrupted line is due to the diffusion of a parasitic acceptor which is extremely rapid on one side of some preferential $\{111\}$ planes. (b) Later stage: only some isolated regions in the centre of the sample are still n-type.

PLATE XII—*H. R. Winteler and A. Steinemann*: paper 12

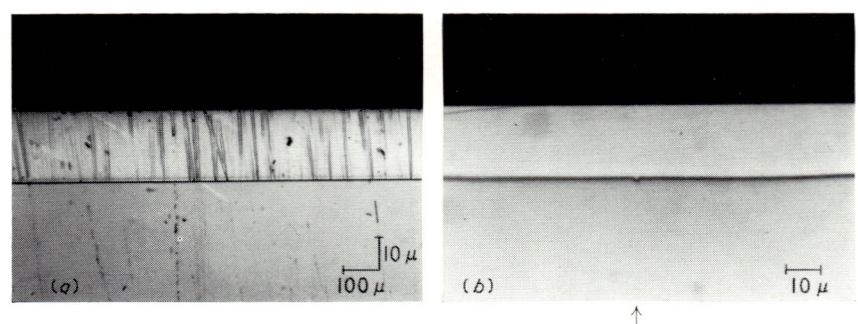

Figure 4. Perfect diffusion of Zn at 750°C with high surface concentration. (*a*) In dislocation-free n-GaAs (oblique section). (*b*) In GaAs with 10⁴ dislocations/cm² (vertical section). The spike in the p–n junction is due to a single dislocation locally increasing the diffusion rate.

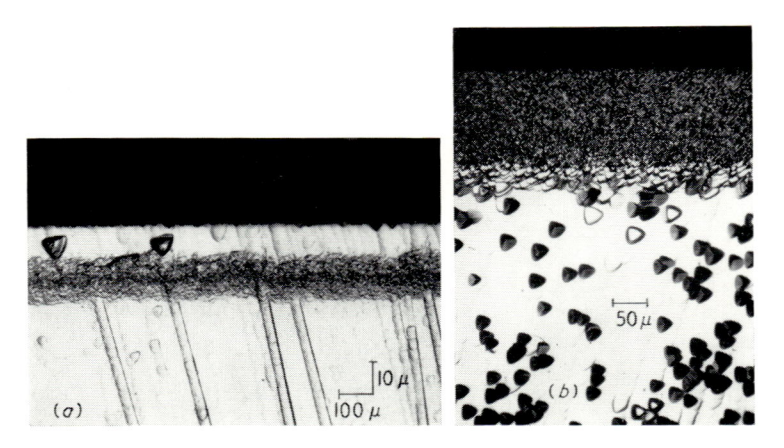

Figure 5. Imperfect diffusion of Zn at 850°C in dislocation-free GaAs with high surface concentration and excess As. (*a*) First appearance of precipitations around the p–n junction at a depth of ca 15 μ (angle lapping). (*b*) For large penetration depths (about 100 μm), the diffused zone is completely filled with precipitations. In the bulk material dislocations are formed (vertical section).

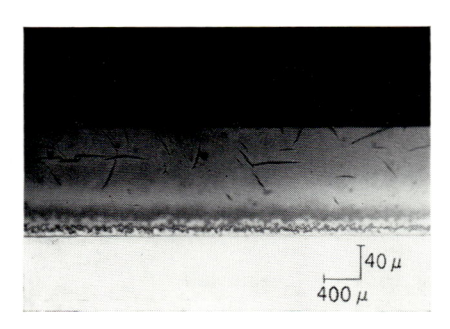

Figure 6. Vertical section through a deeply Zn-diffused, homogeneously Te-doped, dislocation-free GaAs sample without excess As. The junction is nearly flat, although the diffused region is full of precipitations.

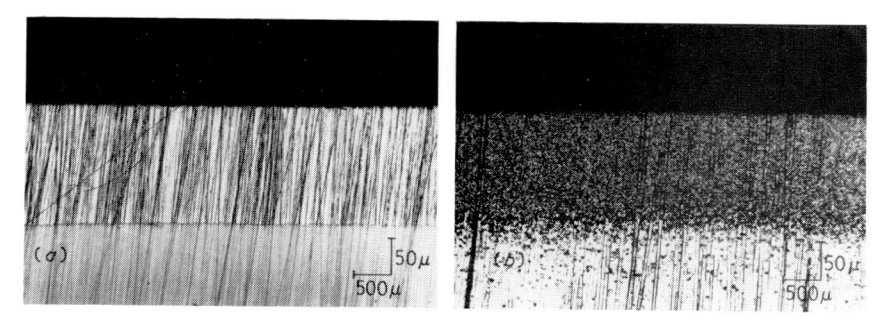

Figure 7. Oblique section through a deep Mn-diffused sample of dislocation-free, homogeneously Te-doped, GaAs. (*a*) Junction-etching. (*b*) Dislocation-etching. The diffused zone is full of precipitations.

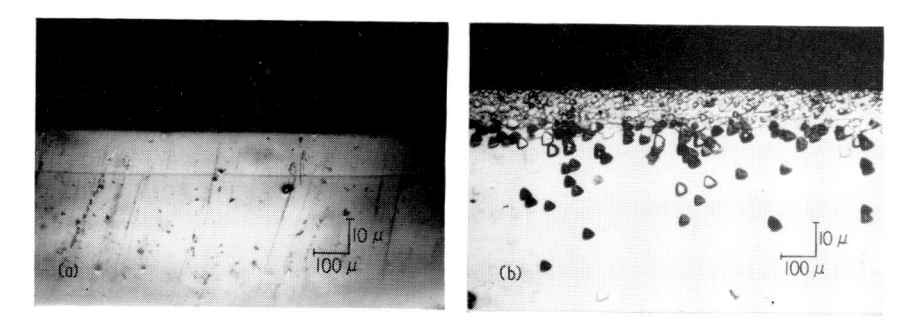

Figure 8. Oblique section through an Si-diffused, originally dislocation-free p-type sample (Zn doping). (*a*) Junction etching; the diffusion front is nearly flat. (*b*) Dislocation etching: Precipitations in the diffused region and dislocations in the bulk material are being created.

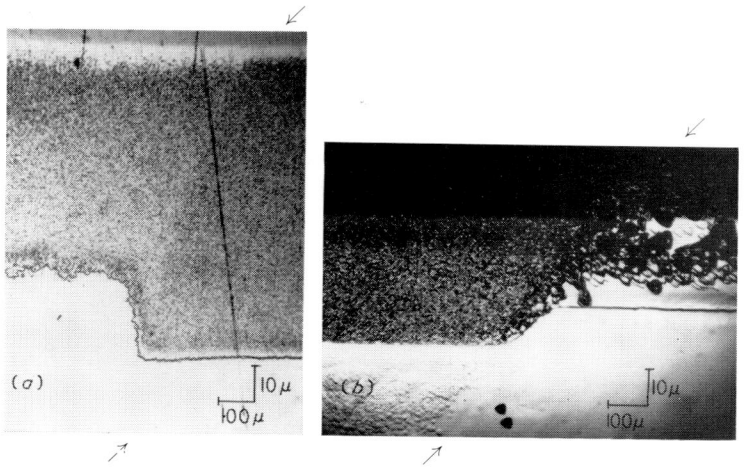

Figure 9. Oblique section through Zn-diffused layers (with high surface concentration of Zn) in dislocation-free GaAs. (*a*) Without As addition. The crystal portion to the right of the arrows is full of as-grown precipitations. The junction is nearly flat and the penetration depth is greater than in the left-hand portion, where only a few precipitations are visible in the substrate. The diffusion front, however is very wavy. (*b*) With excess As. In the left-hand part of the sample, where a large density of as-grown precipitations are present, the diffusion zone is full of new precipitations and the p–n junction lies much deeper than in the right-hand part where no precipitations are detectable and where only a few dislocations in the centre of the diffusion zone have been formed.

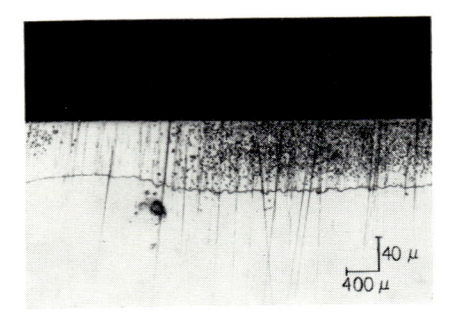

Figure 10. Variations around the mean penetration depth of a Mn-diffused layer due to the inhomogeneous formation (depending on the local crystal properties) of precipitations and dislocations. Oblique section through the sample.

PLATE XV—E. Mohn: paper 17

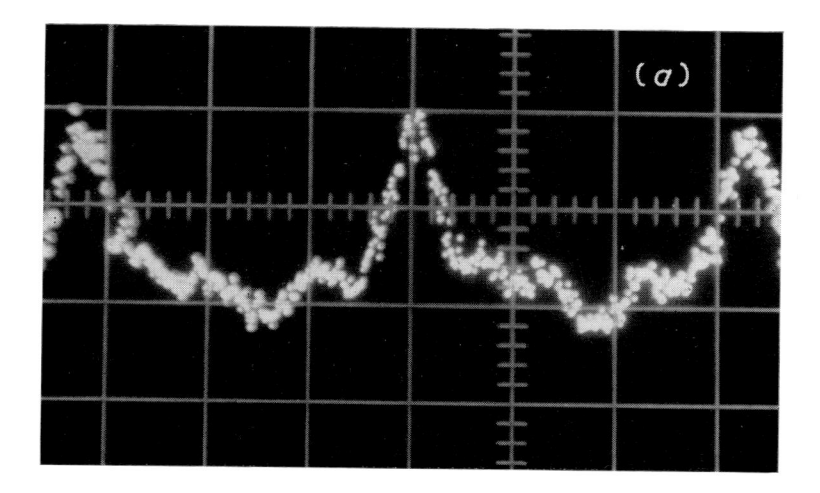

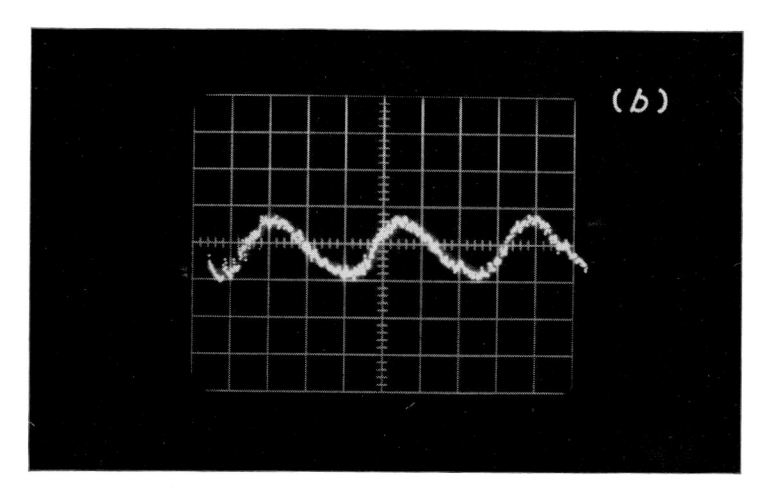

Figure 6(a). Output from gain-modulated laser with $\Omega = \pi c/L$. 5 mv/div. vertically, 1 ns/div. horizontally. Pulse-width about 600 ps, repetition period 3·3 ns. (b) Resonator length detuned by 3 cm. Same scales as in (a).

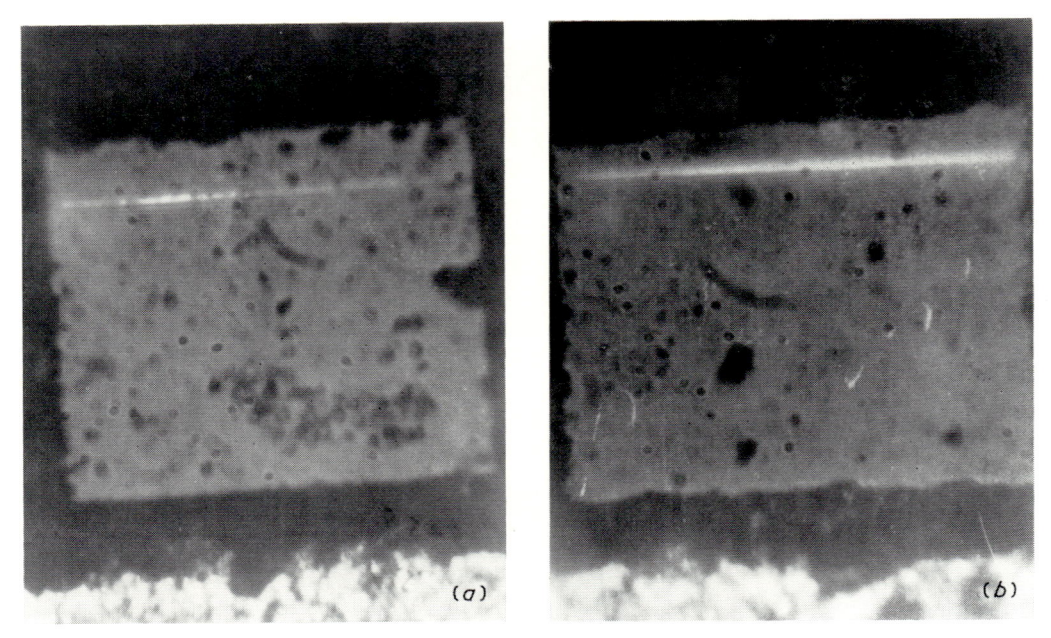

Figure 1. Near field pattern of (a) a typical 'spotty' lasing diode (b) a typical 'uniform' lasing diode.

(figure 2), however, neither $(C_s/A)^{-2}$ nor $(C_s/A)^{-3}$ give an exact proportionality to the applied voltage, although the $(C_s/A)^{-2}$ plot is closer to a straight line. This suggests that the impurity distribution in this type of junction is discontinuous rather than linearly graded.

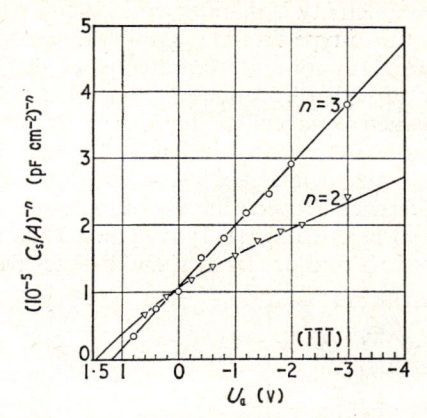

Figure 1. Square and cube of the reciprocal space-charge capacitance per unit area as a function of the applied voltage (growth direction: $[\overline{1}\overline{1}\overline{1}]$).

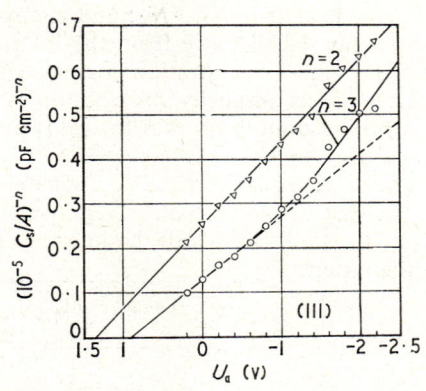

Figure 2. Square and cube of the reciprocal space-charge capacitance per unit area as a function of the applied voltage (growth direction: $[111]$).

2.2. Model of the net impurity distribution

The following considerations concerning the formation of the p–n junction during epitaxial growth are related to the deposition of zinc-doped, p-type layers on to selenium-doped substrates. The doping profile of solution-grown p–n junctions is generally determined by impurity segregation from the liquid phase and a simultaneous solid-to-solid diffusion after deposition. If, in the present case, the final structure of the doping profile had been mainly affected by solid diffusion, the concentration of zinc in the $[\overline{1}\overline{1}\overline{1}]$ grown layer should be higher than that of selenium in the substrate, in order to obtain a linearly graded junction. On the other hand, for the more abrupt profile the doping level in the $[111]$ grown layer should be lower, compared to the level of the substrate. Thus a higher zinc concentration ought to be found in the case of $[\overline{1}\overline{1}\overline{1}]$ growth. Measurements of the width of the space charge region indicate, however, quite the reverse. For diodes grown in $[\overline{1}\overline{1}\overline{1}]$ direction the zero-bias space charge width was nearly twice as large as for diodes grown in the opposite direction. From this it follows that the final doping profile is mainly affected by impurity segregation during epitaxial growth. The formation of linearly graded or more abrupt p–n junctions depending on substrate orientation can be explained by the following assumptions concerning the incorporation of impurities into the epitaxial layer.

1. At the very beginning of the epitaxial process a considerable amount of selenium dissolved in the melt is still available in the neighbourhood of the substrate-to-melt interface, due to dissolution of the substrate surface prior to epitaxial growth. The selenium concentration in the regrown layer, however, decreases rapidly with increasing layer thickness.

2. The zinc segregation, probably affected by the selenium concentration, is relatively low at the substrate-to-layer interface, but increases rapidly with distance from the substrate within a small region.

3. The distribution coefficients of zinc and selenium are orientation-dependent in the following way: growing in $[\overline{1}\overline{1}\overline{1}]$ direction the distribution coefficient of selenium is greater than for the $[111]$ direction. For zinc, however, the opposite must be valid.

Distribution coefficients similarly dependent on orientation have been found by Williams (1964) for zinc and tellurium using vapour-phase epitaxy and by Willardson and Allred (1966) for tellurium during crystallization from the stoichiometric melt.

Following these assumptions an idealized net doping profile can be derived for either growth direction. At the substrate-to-layer interface donors as well as acceptors are

E

incorporated. Because the distribution coefficient of selenium is smaller than unity the doping level for donors must be lower in the regrown layer compared to the substrate, causing a step in the net impurity concentration. Owing to orientation-dependent distribution coefficients of zinc and selenium the type of conductivity is maintained at the step for [$\overline{1}\overline{1}\overline{1}$] growth, that is $N_D > N_A$, and changes from n- to p-type for [111] growth ($N_D < N_A$). Within a limited distance from the interface the selenium concentration decreases rapidly whilst the zinc concentration increases for either growth direction. In our model this variation of net impurity distribution is approximated by a constant impurity gradient. The resulting doping profiles for [111] and [$\overline{1}\overline{1}\overline{1}$] growth are shown in figure 3. They are characterized by a concentration step from N_0 to N_0^* at the interface and a linear decrease of the net doping level $N_D - N_A$. The essential difference between the two doping profiles is in the sign of N_0^*, which is negative for [111] and positive for [$\overline{1}\overline{1}\overline{1}$] growth. Thus the p–n junction coincides with the interface in the first case and occurs at some distance from it in the latter.

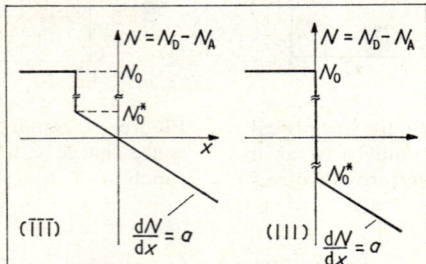

Figure 3. Model of the net impurity distribution at the p–n junction for [111] and [$\overline{1}\overline{1}\overline{1}$] growth.

2.3. *Determination of the doping profile*

By solving the Poisson equation in the space-charge approximation the width of the space charge region was calculated as a function of the total voltage for either doping profile of figure 3, taking $K = N_0^*/N_0$ and $a = dN/dx$ as parameters. The space charge capacitance per unit area is then given by

$$\frac{C_s}{A} = \frac{\epsilon}{w(U_j)}$$

where A is the junction area, ϵ the dielectric constant, w the space charge width, and U_j the total junction voltage.

By plotting the calculated values of C_s/A versus the total voltage and the measured versus the applied bias in two separate diagrams the theory can be compared with experiments by

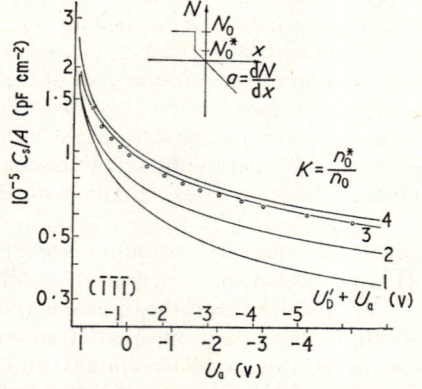

Figure 4. Calculated and experimental space charge capacitance per unit area as a function of voltage (growth direction [$\overline{1}\overline{1}\overline{1}$]). Curve 1, $a = -10^{22}$ cm^{-4}, $K = 10^{-2}$; curve 2, $a = -5 \cdot 10^{22}$ cm^{-4}, $K = 10^0$; curve 3, $a = -5 \cdot 10^{22}$ cm^{-4}, $K = 10^{-1}$; curve 4, $a = -10^{23}$ cm^{-4}, $K = 10^0$.

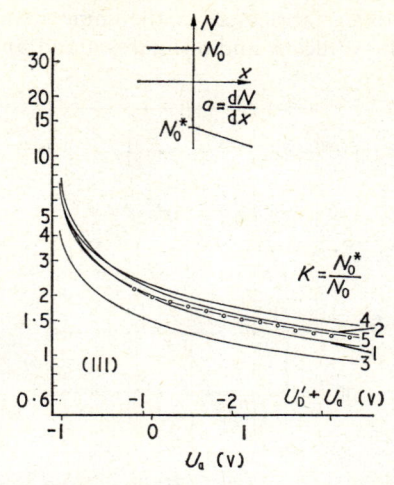

Figure 5. Calculated and experimental space charge capacitance per unit area as a function of voltage (growth direction [111]). Curve 1, $a = -10^{23}$ cm^{-4}, $K = -10^0$; curve 2, $a = -10^{24}$ cm^{-4}, $K = -10^0$; curve 3, $a = -10^{23}$ cm^{-4}, $K = -10^{-1}$; curve 4, $a = -10^{24}$ cm^{-4}, $K = -10^{-1}$; curve 5, $a = -10^{24}$ cm^{-4}, $K = -10^{-2}$.

the following procedure. The diagrams are superposed and displaced against one another along the voltage axis until the measured curve coincides with one of the calculated curves. Thus the parameters $K = N_0^*/N_0$ and $a = dN/dx$ can be determined within an order of magnitude. In figures 4 and 5 the theoretical and experimental curves are shown after having been adjusted. From this it turns out that the real doping profiles can fairly well be approximated by the simple model. The results obtained are summarized in table 1.

Table 1. Impurity concentrations and gradients at the p–n junction for [111] and [$\bar{1}\bar{1}\bar{1}$] growth

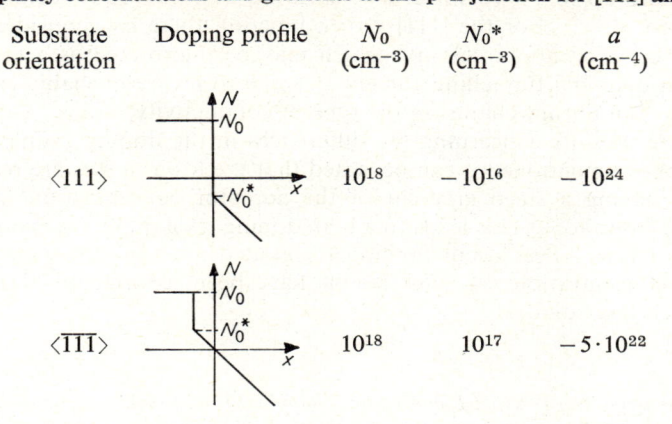

Substrate orientation	Doping profile	N_0 (cm^{-3})	N_0^* (cm^{-3})	a (cm^{-4})
$\langle 111 \rangle$		10^{18}	-10^{16}	-10^{24}
$\langle \bar{1}\bar{1}\bar{1} \rangle$		10^{18}	10^{17}	$-5 \cdot 10^{22}$

For a $\langle 111 \rangle$ oriented substrate a junction is formed having an abrupt change in the type of conductivity at the interface from $N_0 = 10^{18}$ cm^{-3} to $N_A \simeq -N_0^* = 10^{16}$ cm^{-3} and a very steep gradient of the acceptor concentration in the regrown layer amounting 10^{24} cm^{-4}.

Growing in [$\bar{1}\bar{1}\bar{1}$] direction there is a discontinuity in the donor concentration from $N_0 = 10^{18}$ cm^{-3} to $N_0^* = 10^{17}$ cm^{-3} at the interface and a linear decrease of the net impurity concentration with a gradient $a = dN/dx = -5 \cdot 10^{22}$ cm^{-4} in the epitaxial film.

It must be noted here, that these data are limited to a small voltage region, in which the measurements were carried out, corresponding to a maximum width of the space charge region of 0·12 μm for [111] growth and 0·21 μm for [$\bar{1}\bar{1}\bar{1}$] growth. Similar results were found for all investigated diodes fabricated in the same manner. The accuracy of the matching process to obtain the values K and a (figures 4 and 5) is within one order of magnitude.

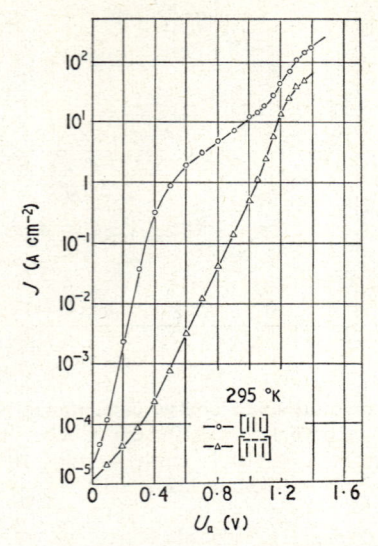

Figure 6. Current densities as a function of forward bias for [111] and [$\bar{1}\bar{1}\bar{1}$] grown p–n junctions.

As already mentioned, the junction is formed within the epitaxial layer growing in [$\bar{1}\bar{1}\bar{1}$] direction and coincides with the substrate-to-film interface using a $\langle 111 \rangle$ orientation. In the latter case lattice defects are expected to be found at the interface, which may become effective in the electrical behaviour of these junctions. Figure 6 shows the I–V-characteristics of both types of junction. We found that the slope of these curves is almost independent of temperature up to 1 v forward bias, showing that mainly tunnelling processes are involved in current flow. For the [111] grown junction there is a considerable change in the slope of the curve at about 400 mv, which may be interpreted as a transition from a trap-assisted to a diagonal tunnelling current. These traps are probably caused by lattice defects induced by an abrupt change in the type of conductivity.

Summarizing the results concerning the differences in the doping profile with regard to the laser losses above mentioned it can be stated that the lower losses are related to a more abrupt junction having a steep gradient of the acceptor concentration within a limited distance from the interface. This leads to a high doping level in the regrown layer. On the other hand higher laser losses occur for linearly graded junctions showing a smaller slope of the impurity concentration. Similar results have been reported by Karnaukhov *et al.* (1967) for diffused laser diodes.

References

Beneking, H., and Vits, W., 1966, *IEE conference public.*, **27**, 429 (6th international conference on microwave and optical generation and amplification, Cambridge, U.K., 1966).
Beneking, H., and Vits, W., 1968, *IEEE J. Quant. Electronics*, QE-4, 201.
Karnaukhov, V. G., Kryukova, I. V., and Petrov, A. I., 1967, *Soviet Physics Semiconductors*, **1**, 1133.
McWhorter, A. L., 1963, *Solid-State Electronics*, **6**, 417.
Willardson, R. K., and Allred, W. P., 1966, *Proc. Intern. Symp. on GaAs*, Conf. Series No. 3, 35.
Williams, F. V., 1964, *J. electrochem. Soc.*, **111**, 886.
Stern, F., 1964, *Radiative Recombination in Semiconductors*, p. 165, Dunod, Paris.
Yariv, A., and Leite, R. C., 1963, *Appl. phys. lett.*, **2**, 55.

Properties of a GaAs laser coupled to an external cavity

E. MOHN

Institute of Applied Physics, University of Berne, Berne, Switzerland

Abstract. A GaAs laser diode has been coupled to an external cavity by lapping one end of the diode at Brewster's angle with respect to the junction plane. The total losses introduced by the external resonator and the coupling mechanism amount to about 83%. Most of this is lost at the Brewster surface due to the small apparent thickness of the active region seen by the incident radiation. Threshold current densities are between 1800 and 4800 A cm^{-2} at 84°K for a resonator length of 50 cm and a reflectance of the external mirror of 90%. Mode selective effects at the Brewster window as a result of diffraction phenomena are only apparent below a current of 1·5 to 2 times threshold. By exciting the laser with pulses of duration less than the cavity round-trip time, the photon lifetime has been measured as $\tau_p = 1\cdot6$ ns and an estimate has been made for the Einstein coefficient B. Direct gain modulation by application of a r.f. component on top of the rectangular pump current pulse with a frequency Ω equal to the axial mode spacing has resulted in partial mode-locking, the output consisting of a train of short pulses with 600 ps half-width and a repetition frequency equal to Ω.

1. Introduction

For many studies and applications it is desirable to be able to place additional elements within the laser cavity. With the exception of a saturable absorber formed by an unpumped region of the junction it is essential to use an external cavity. The extra degrees of freedom afforded by the latter allow the inclusion of such items as an electro-optic modulator (Nelson 1967) and nonlinear devices, as well as the ability to modify the spectral and spatial properties of the emission by control of the geometry of the external cavity.

One of the main difficulties in coupling a diode laser to an external cavity is the suppression of oscillatory modes within the diode itself. The very high gain within the active region (~ 100 cm^{-1}) requires that the reflectance of the GaAs surface be reduced to a value of considerably less than one per cent. Generally this is achieved by evaporation of a film of dielectric m... ive index $n' = \sqrt{n_{\text{GaAs}}}$, a quarter wavelength in thickness. Both ...condition and have been used successfully for this purpose by a ...nd 1963; Crowe and Craig 1964; Crowe and Ahearn 1968). ...in this work, is to lap one of the diode surfaces at Brewster's ...e junction plane (Kosonocky 1968; Mohn et al. 1967). Light ...iode at ϕ_B, and reflection is reduced to zero for a particular ...e method has the advantage that any light reflected back into ...m the active region into the lossy unpumped parts of the diode. ...n reflectance requirement is not so stringent as with anti-... normal to the junction plane. On the other hand, the losses ...ted from the external mirror enters the GaAs at a glancing ...hich the active region presents an even smaller aperture than

...s work was to study some of the properties of a GaAs diode ...particular, the coupling losses, the spectral behaviour, and ...ut for current pulses shorter than the cavity round-trip time ...de current.

2. Properties of the external cavity

2.1. *Experimental arrangement*

The basic arrangement has already been described by Mohn *et al.* (1967) and so only a brief summary will be given here. As is shown in figure 1 the diode (D) was fixed to a heat sink (S) within a small cryostat fitted with anti-reflection coated windows (W). A cooled copper rod (C) maintained the diode at the temperature of liquid air, about 84°K. The light emitted from the Brewster window was collected by a 26 mm focal length, f/0·94 aperture, Videcon camera objective (CL) whose support was rigidly attached to the cryostat.

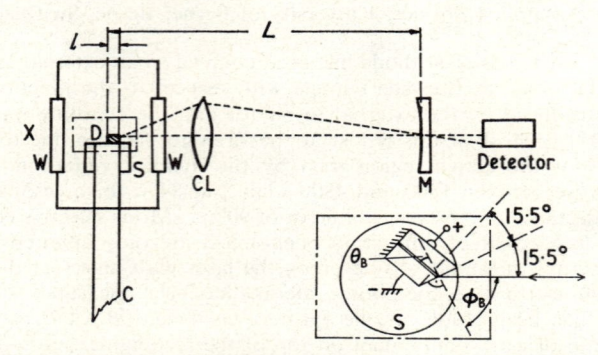

Figure 1. Arrangement of Brewster diode D with lens CL and external mirror M. The diode is fixed to the heat sink S maintained at 84°K by the cooled copper rod C. The windows W are anti-reflection coated. L is the total resonator length and l the diode length. A monitor photocell may be placed at X. Inset: Plan view of dashed region round diode showing orientation of Brewster window with respect to junction and axis of external cavity.

Initially a plane, dielectric coated mirror (M) was used but this caused difficulties with alignment and stability. Very much improved stability was obtained with a curved mirror of 100 cm radius coated with a gold film of 90% reflectance. All measurements reported here were made with this mirror and, unless otherwise stated, with a cavity length of 50 cm. The laser output and cavity alignment was usually monitored by means of a photo-electric cell placed at the position X to receive the light transmitted through the rear surface of the diode (perpendicular to the junction plane and forming the second cavity mirror). Spectral measurements were made on the light transmitted through the external mirror by means of a Jarell–Ash 0·5 m grating monochromator with a maximum resolution of 0·2 Å. In the experiments with r.f. modulation and short drive pulses the output was focused with a 12 mm, f/1·3 objective onto the sensitive area of a Philco Type L 4501 photodiode connected either to a Tektronix Type 661 sampling oscilloscope or to a spectrum analyser. The diode and coupling circuit had a frequency response extending into the Ghz region making the overall resolution time of the detection system between 0·1 and 0·2 nsec.

2.2. *Threshold and loss measurements*

For a series of Brewster diodes prepared from the same slice of material in one fabrication process, dimensions and threshold current densities are given in table 1.

Threshold measurements on a conventional Fabry-Perot diode (No. 6) in the external resonator allowed determination of the losses arising from scattering and absorption at the Dewar window, the objective, and the external mirror, as well as from incomplete coupling between the diode and the external cavity. This was done by calculating an effective reflectance R_{eff} for the external mirror, which includes the losses mentioned, and by comparing R_{eff} to the measured value R of this mirror. We start from the well-known threshold relation

$$\beta j_{\text{th}} = \alpha + l^{-1} \ln (R_1 R_2)^{-1/2}, \tag{1}$$

Table 1. Diode specifications for external cavity

Diode number	Length l (μm)	Width (μm)	Threshold j_{th} (A cm^{-2})	Reflectance R_1 of cleaved mirror (%)
				Brewster diodes
1	475	130	3360	32
2	500	150	2000	90
3	520	150	4780	32
4	515	200	1830	32
5	515	180	1820	32
6	350	320	805	Fabry-Perot diode
				32

in which R_2 is defined as the total fraction of light reflected back into the diode from the external cavity and α is the bulk loss within the diode. α/β is the threshold current density j_{th} for infinite diode length l and R_1 the reflectance of the cleaved mirror. R_2 may be calculated from the measured ratio

$$k = \frac{\beta j_{th} \text{ (with ext. mirror)}}{\beta j_{th} \text{ (without ext. mirror)}} = \frac{\alpha + l^{-1} \ln (R_1 R_2)^{-1/2}}{\alpha + l^{-1} \ln R_1^{-1}}, \quad (2)$$

if β is assumed independent of current. The threshold of diode No. 6 was 980 A cm^{-2} without the mirror, and fell to 805 A cm^{-2} within the aligned resonator, from which we obtain $R_2 = 0.58$. Allowing for multiple reflections between the inner diode mirror and the external mirror yields the following relation

$$R_{eff} = \frac{(R_2 - R_1)}{1 - 2R_1 + R_1 R_2} = 0.47 \quad (3)$$

In fact, the reflectance R of the external mirror is approximately equal to 90%, and the difference $R - R_{eff} = 0.43$ may be accounted for by the losses inherent in the external cavity.

Fabry-Perot diodes made from the same material and with dimensions similar to those of the Brewster diodes listed in table 1 are expected to have thresholds near 1000 A cm^{-2}. The above-calculated loss of 43% should still result in a decrease of threshold for the Brewster diodes. The measured thresholds (table 1), however, are strongly increased, so there must be another source of loss, which evidently is the Brewster window itself.

Losses due to misorientation of the Brewster window are less than 5%, provided that the error in the angle is less than $\pm 0.4°$. Deviations from the internal Brewster angle $\theta_B (= 15.5°)$ to higher values are very critical since the latter is within 1° below the angle for total internal reflection.

A much more serious loss mechanism arises from the shallow angle at which the reflected light strikes the GaAs surface, reducing the effective thickness of the active region, as seen from the external mirror, by a factor $\cos \phi_B$ to less than 0.5μ. In addition the use of an objective with circular symmetry results in a poor match with the highly unsymmetric aperture of the active region, causing laser action only in a narrow strip and consequent diffraction loss in the junction plane. This effect could be observed at currents slightly above threshold. The position of the strip could be moved in the junction plane by adjustment of the lens or mirror. This additional difficulty could be avoided by using a cylindrical lens and mirror but such items, of the necessary high quality, are difficult to obtain.

A direct estimate of the cavity losses with a Brewster diode is also made. The intensity of the laser beam after a complete round trip in the cavity may be written as

$$I = I_0 T^2 R_1 R_{eff} \exp \{(g - \alpha) 2l\} \quad (4)$$

where g is the gain per unit length and T an effective transmittance of the Brewster window accounting for all the aforementioned losses not yet incorporated in R_{eff}. At threshold, $I = I_0$ and therefore

$$g = \beta j_{th}(B) = l^{-1} \ln (T^2 R_1 R_{eff})^{-1/2} + \alpha.$$

Again taking the ratio of the threshold $j_{th}(B)$ of the Brewster diode with that of a Fabry-Perot diode made from the same material and without external cavity, $j_{th}(FP)$, the average transmittance may be estimated from

$$T = (R_1 R_{eff})^{-1/2} \exp\left[-l\left\{\frac{j_{th}(B)}{j_{th}(FP)}(l^{-1}\ln(R_1^{-1})+\alpha)-\alpha\right\}\right]. \quad (5)$$

For $j_{th}(B)/j_{th}(FP) = 2$, $R_{eff} = 0.47$, $l = 5 \times 10^{-2}$ cm and $\alpha = 15$ cm^{-1}, T is about 13%. This is the average of the two-way transmission but of course the greater loss is for light entering the diode. Beside the corresponding single-pass loss of 87% all other effects may be neglected.

One further loss mechanism should be mentioned. Hatz (1967) has shown that diodes carefully prepared from homogeneous dislocation-free starting material emit with a constant intensity over a large fraction of the junction width and with TM polarization (Hatz and Mohn 1967). For the present work such diodes were used. The orientation of the Brewster window was correspondingly chosen to transmit TM modes without reflection. But if the light generated in the active region were to contain a component of TE polarization caused, for example, by inhomogeneities, there would be an additional loss since the reflectivity of the Brewster window is 70% for this latter polarization.

2.3. Spectral properties

All Brewster diodes listed in table 1 showed qualitatively similar results.

The features of the radiated spectra of the Brewster diodes in an external cavity are illustrated in figure 2a–c taken from diode No. 5 at various current densities.

(a) Below a well defined threshold current, which was considerably lower (a factor of about two) than the threshold for super-radiance without external mirror, there was one broad line due to spontaneous emission.

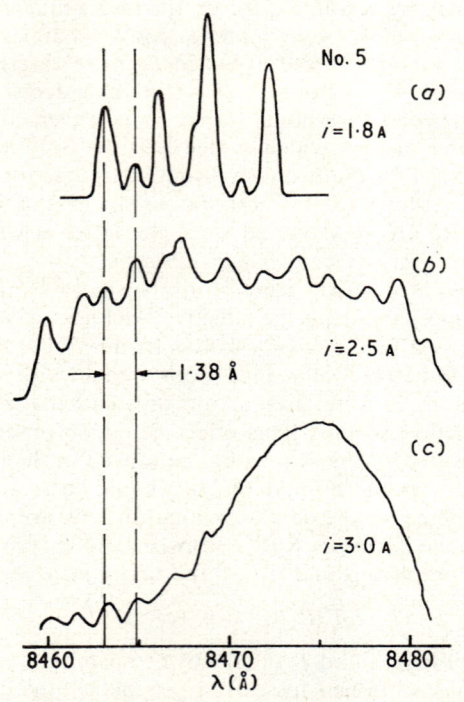

Figure 2. Spectra of diode No. 5 within the 50-cm external resonator. Current increasing from threshold 1·8 A (a) to 2·5 A (b) and 3·0 A (c). Vertical scales are arbitrary.

(b) At or slightly above threshold the spectrum exhibited a number of modes whose separation of 1·38 Å was, within the error limits, in agreement with the value of 1·31 Å calculated from the diode length. It is important to notice that these modes disappeared when the resonator was misaligned or blocked, showing that they really were oscillating within the external resonator, but picked out of the many possible eigenoscillations of the latter by the mode-selecting action of the Brewster window (figure 2a). All these modes were TM polarized according to the orientation of the Brewster surface.

(c) By further increasing the current the spectrum became broader and mode selection less pronounced as may be seen in figure 2b.

(d) Finally, at 1·5 to 2 times threshold current, the spectrum consisted of only one, 10 to 20 Å wide line (figure 2c), indicating that more than 1000 axial modes were probably oscillating within the 50-cm external resonator. The separation of 7×10^{-3} Å of these modes was of course far beyond the resolving power of our monochromator. The radiation was entirely TM polarized.

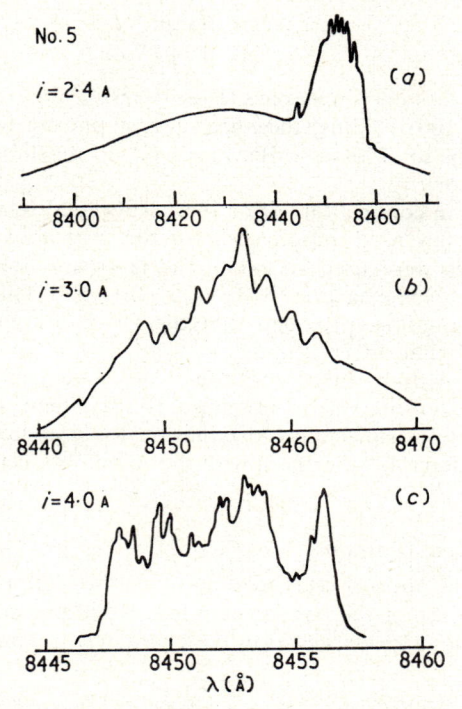

Figure 3. Spectra of diode No. 5 without external resonator. Injection current 2·4 A (a), 3·0 A (b), and 4·0 A (c). Note the change in wavelength scale from (a) to (c).

For comparison the spectra of the same diode but without the external cavity are shown in figure 3 for three values of current. Line narrowing is clearly present but the mode structure is highly irregular, having no correspondence between the diode length and the mode separation. Measurement of the angular distribution of the far field intensity in a plane perpendicular to the junction plane and parallel to the diode length axis showed a maximum in the intensity at an angle of 59° to the axis, in agreement with the refraction angle through the Brewster window. No subsidiary maximum was seen at 0°, as would be the case if a small part of the Brewster surface were damaged stepwise along a cleavage plane, thus forming a perfect resonator within the diode.

The mode-selective action of the Brewster window is thought to be due to higher diffraction losses suffered by those modes which do not have a node of the electric field along the edge formed by the p–n junction plane and the Brewster surface. Calculations of the emitted far-field show that the radiation pattern is spatially more extended for the case of an antinode on the aperture, thus explaining the higher diffraction losses in this case. But, due to the fact that the Brewster surface is not parallel to the phase fronts of the two light beams, travelling in opposite directions, it is not possible for a node to exist over the entire Brewster window. In other words, the condition of zero electric field over the whole window cannot be fulfilled by the usual TE and TM solutions of Maxwell's equations. Therefore at higher drive current and hence at higher gain, where the electric field extends farther into the bulk material, this mode-selecting mechanism may be expected to contribute a smaller fractional change to the total gain. This probably accounts for the disappearance of internal modes at moderately high currents.

3. Transient behaviour

3.1. *Measurement of photon lifetime*

The photon lifetime within a laser resonator is a function of the total losses and is given by

$$\tau_p = \frac{L+l(n-1)}{ln} \cdot \frac{ln}{c\epsilon} \simeq \frac{L}{c\epsilon} \tag{6}$$

where ϵ is the fractional loss per single pass (Röss 1966), L the total resonator length and l the diode length ($L/ln > 100$). Thus, measurement of photon lifetime is an independent method of determining the total losses. Because $\epsilon \leqslant 1$, τ_p must be $\ll L/c$, i.e., $\tau_p \geqslant 1\cdot33$ ns for a 40 cm long resonator.

A charged coaxial line in connection with a mercury-wetted relay produced short, rectangular current pulses of a few ns duration and, for the part above threshold, a decay time $\leqslant 0\cdot 1$ nsec. These pulses were used to excite the laser and the decay time ($1/e$) of the corresponding light pulses was measured to be $(1\cdot 6 \pm 0\cdot 1)$ ns. This time may be considered identical to the photon lifetime, provided that the latter is considerably longer than the overall recombination lifetime τ.

A separate experiment with modulation of the diode at frequencies up to 900 MHz showed that at the same current densities τ was less than 10^{-10} s, hence the effect of τ upon the measurement may be safely neglected and equation (6) used to estimate ϵ. This yields a value of $\epsilon = 0\cdot 84$, in reasonable agreement with the value obtained from threshold measurements in section 2.2.

3.2. *Characteristic damping constant*

It is possible to solve the appropriate rate equations (see section 4) in a linear approximation to describe the response of the laser system to a small perturbation (Röss 1966). For a laser resonator with strong mode selection the characteristic damping time ($1/e$) is shown to be

$$\theta = L/(BW\tau_p ln) \tag{7}$$

where W is the pump rate and B the Einstein coefficient for stimulated emission.

In order to measure the time constant θ, the laser was excited with a long rectangular current pulse of several μs duration, upon which a very short ($< 2L/c$) pulse from the mercury relay was superposed. The short pulse is reflected several times by the external mirror with decreasing amplitude. From the intensity ratio between two successive echoes it is possible to evaluate θ. This experiment was repeated for several resonator lengths and the corresponding values for θ are plotted versus L in figure 4. According to equation (7) the points may be approximated by a horizontal straight line, since τ_p itself is also proportional to L.

If η is the internal quantum efficiency, d the thickness of the active region, F the area of the p–n junction and i the pump current, the pump rate W may be written as

$$W = \eta i/edF \quad (e = 1\cdot 6 \times 10^{-11} \text{ A s}),$$

hence

$$B = \frac{LedF}{\theta \eta i \tau_p ln}. \tag{8}$$

Numerically this yields $B=9\cdot 10^{-7}$ cm³ s⁻¹, which is of the same order of magnitude as the value calculated from the spontaneous recombination lifetime τ_s. The following numerical values have been used: $L=40$ cm, $d=3\mu$, $F=7\cdot 5\times 10^{-4}$ cm² (diode No. 2, table 1), $\theta=3\cdot 9$ nsec (figure 5), $\eta=0\cdot 7$ (Unger 1967) $i=2$ A, $\tau_p=1\cdot 6$ ns (section 3.1), $l=0\cdot 05$ cm. Time-resolved spectral measurements showed that the emitted spectrum was the same in each of the reflected short pulses and was typical for a Brewster diode pumped slightly above threshold. Of course this stems from the fact that the long current pulse was somewhat higher than threshold current, whereas the high current short pulse did not influence the spectrum because its length was shorter than the cavity round-trip time.

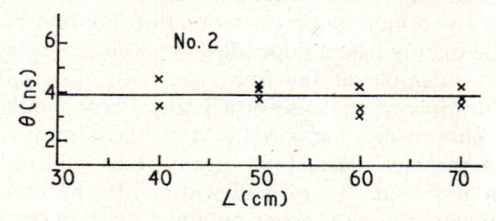

Figure 4. Damping time θ as a function of resonator length L. The straight line is a mean least square fit to the experimental points.

4. Direct gain modulation

Conventional rate equations may be applied to describe the temporal behaviour of the injection laser. In the following form, which is similar to that used by Lasher (1964), they are adequate to give the time dependence of photon number S and electron density N of a diode laser within an external cavity:

$$\frac{dN}{dt} = \frac{j\eta}{ed} - \frac{N}{\tau_s} - Sg(\omega, N)\zeta \qquad (9a)$$

$$\frac{dS}{dt} = VSg(\omega, N)\zeta - \frac{S}{\tau_p} + Vf\frac{N}{\tau_s}. \qquad (9b)$$

$g(\omega, N)$ is the stimulated emission rate per photon of energy $\hbar\omega$ into the mode ω, and per unit volume, V the volume of the active region, f the fraction of excited to total cavity modes, and ζ accounts for the ratio of optical path lengths in the diode and external resonator,

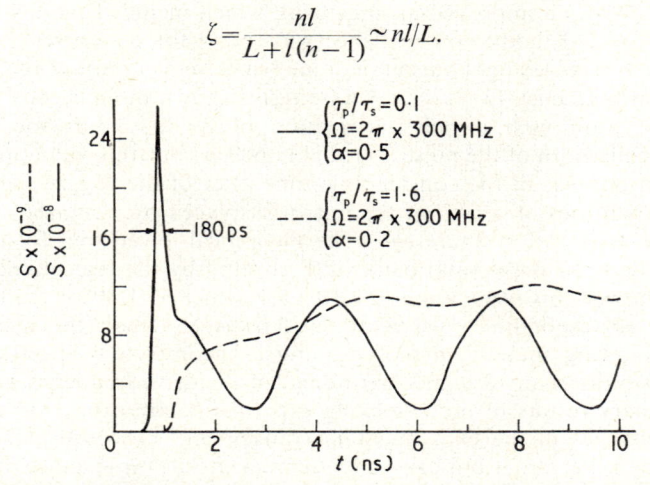

$$\zeta = \frac{nl}{L+l(n-1)} \simeq nl/L.$$

Figure 5. Computer solutions of rate equations (9) for photon number S vs. time near threshold. Solid line is characteristic for $\tau_p \ll \tau_s$, dashed line for $\tau_p \simeq \tau_s$. Modulation frequency is 300 MHz in both cases.

Equations (9) have been written for one single mode. It is assumed that the frequency ω is at the centre of the gain profile. Because of the large number of modes fitting into the external resonator, this is certainly allowed. The pump current density is taken to be

$$j(t) = j_0(1 + \alpha \sin \Omega t) \tag{10}$$

with α and Ω being the modulation depth and frequency, respectively. Unger (1967) has given an analytic approximation for the gain factor $g(\omega, N)$ valid within an error limit of a few per cent up to room temperature and including the effect of band tailing. Using this approximation and the appropriate parameter values for GaAs, equations (9) have been solved by computer. The time dependence of the solutions is extremely sensitive to the ratio of photon lifetime to spontaneous recombination lifetime τ_p/τ_s. Two characteristic types of solution may be distinguished depending on whether τ_p/τ_s is much less than or of the order of unity. An example of the first type with $\tau_p/\tau_s = 10^{-1}$ is given in figure 5 (solid line), which might apply to the case of a Fabry–Perot diode within an external resonator, where the mean photon lifetime is reduced by those photons not leaving the diode. Before the photon number has reached its steady-state sinusoidal variation, there is a pronounced spike with less than 0·5 ns half-width. If the diode is operated very near threshold it is likely that steady state is never obtained but that the light output consists of a train of spikes having a repetition frequency equal to the modulation frequency of the pump current. Indeed such spiking operation, in a Fabry–Perot diode modulated near threshold at 300 MHz, has been observed by the author and it was confirmed that this does not depend on the presence of an external mirror. Similar effects were observed by Roldan (1968) and Magalyas et al. (1967) in diodes with non-uniform distribution of inversion. Details will be published in a forthcoming paper.

The other type of solution, which is more relevant to the situation of a Brewster diode with an external mirror, is also illustrated in figure 5 (dashed line) corresponding to a ratio $\tau_p/\tau_s = 1·6$. The photon number rises smoothly until steady-state conditions are established. Note, that the small modulation depth is partially due to the smaller value chosen for α in this case. The experimental dependence of relative modulation depth of the light output on resonator length has been previously given by Broom and Mohn (1968). There it was shown that the results could be qualitatively described by the assumption of partial phase coherence between individual cavity modes. Experimentally the modulation depth was strongly enhanced when the cavity length was related to the modulation frequency by $L = m\pi c/\Omega$, where m is a positive integer. This effect cannot be predicted by the above analysis, based on a single mode or equivalently, assuming many modes with random relative phases. With an improved arrangement, which included the fast detection system described in section 2.1, additional evidence was found for the presence of partial mode-locking. The time-resolved light output of diode No. 2 for a current of 16% above threshold and a modulation frequency $\Omega = \pi c/L$ ($= 300$ MHz) is shown in figure 6a. It consisted of a continuous pulse train with a pulse separation of $2\pi/\Omega = 3·3$ ns for resonator length of 50 cm. The half-width of the pulses is roughly 600 ps but this value may be considered as an upper limit because of the considerable time jitter of the trigger signal. In contrast, with the same conditions of drive but the mirror displaced by 3 cm, the output was nearly a pure sine wave as shown in figure 6b. The theoretical lower limit for the pulse width is equal to $1/\Delta\nu$ where $\Delta\nu$ is the total bandwidth covered by the laser oscillations. Spectral measurements showed this to be about 2×10^{11} s^{-1} which is 1/30 of the total spontaneous linewidth. The corresponding pulse width of 5 ps is less than the measured one by a factor of 120, showing that the mode coupling is incomplete. Spectral analysis of the light pulse yields more than 10 higher harmonics of the drive frequency Ω.

These preliminary results of mode-locking experiments are expected to be improved by some obvious changes in the experimental arrangement. The use of a c.w. diode laser might help to reduce perturbations caused by thermal effects, for instance by variations of the refractive index. Anyway, it is possible to produce sub-nanosecond pulses by direct gain modulation of diode lasers, though it is doubtful whether it will prove possible to have the total spontaneous linewidth participating in the locking mechanism.

Acknowledgment

The author thanks Prof. K. P. Meyer for the opportunity to perform this work and for his continuous encouragement. Special thanks are due to R. F. Broom for his help with some experiments and with the manuscript and, together with Ch. Deutsch, J. Hatz, and Ch. Risch, for many discussions and critical comments.

References

BROOM, R. F., and MOHN, E., 1968, *J. Appl. Phys.*, **39**, 4251–2.
COUPLAND, M. J., HAMBLETON, K. G., and HILSUM, C., 1963, *Phys. Lett.*, **7**, 231–2.
CROWE, J. W., and CRAIG, R. M. JR., 1964, *Appl. Phys. Lett.*, **5**, 72–3.
CROWE, J. W., and AHEARN, W. E., 1968, *IEEE J. Quant. Electronics*, **QE-4**, 169–72.
HATZ, J., 1967, *IEEE J. Quant. Electronics*, **QE-3**, 643–4.
HATZ, J., and MOHN, E., 1967, *IEEE J. Quant. Electronics*, **QE-3**, 656–62.
KOSONOCKY, W. F., and CORNELY, R. H., 1968, *IEEE J. Quant. Electronics*, **QE-4**, 125–31.
LASHER, G. J., 1964, *Solid-State Electronics*, **7**, 707–16.
MAGALYAS, V. I., PLESHKOV, A. A., RIVLIN, L. A., SEMENOV, A. T., and TSVETKOV, V. V., 1967, *JETP Lett.*, **6**, 68–70.
MOHN, E., BROOM, R. F., DEUTSCH, CH., and HATZ, J., 1967, *Phys. Lett.*, **24A**, 561–2.
NELSON, D. F., 1967, *IEEE J. Quant. Electronics*, **QE-3**, 667–74.
RÖSS, D., 1966, *Laser Lichtverstärker und -Oszillatoren* chap. 12.7., 318–20. (Akademische Verlagsgesellschaft, Frankfurt am Main, 1966).
ROLDAN, R., 1967, *Appl. Phys. Lett.*, **11**, 346–8.
UNGER, K., 1967, *Z. Physik*, **207**, 322–41.

Filamentary lasing and delay time in GaAs laser diodes

H. B. KIM[†]

General Electric Company, Syracuse, New York, U.S.A.

Abstract. This report shows that scattered values in threshold current density of zinc diffused GaAs laser diodes were found in those devices where the filamentary lasing phenomenon resulted in 'spotty' near-field patterns, and that the lower threshold laser diodes with a fixed cavity length and process displayed more 'spotty' near-field patterns. With the use of the threshold current density values of the 'uniform' lasing diodes selected by the observation of near-field patterns, reliable values of the absorption coefficient and the gain constant of the active region of the laser diode were determined.

The position of the dislocation band induced by zinc diffusion into GaAs, was shown to be a sensitive measure of the heat-treatment temperature. Significant reductions in delay time of the stimulated emission were found in the heat-treated laser diodes. The dislocation band may be the source of the optical absorbing centres of Fenner's model as indicated by correlation of reduction of delay time with the band position after heat treatment.

1. Introduction

The filamentary lasing phenomenon in GaAs laser diodes has been reported by several workers: Fenner and Kingsley (1963), Kingsley and Fenner (1963), and Abrahams and Buiocchi (1966). The phenomenon is considered as detrimental to the potential applications of GaAs laser diodes in areas where spatial coherency is required such as beam steering (Fenner 1966), the master oscillator for phased array amplifications, (Vuilleumier *et al.* 1967), and in applications where high power is required. The near-field patterns of the filamentary lasing phenomenon appear 'spotty' for filament sizes of a few microns and 'uniform' for lasing across the entire junction (50 to 100 μm). The 'spotty' lasing patterns were observed at temperatures ranging from 77° to 300°K most commonly in diode junctions formed by diffusion processes.

This report presents results of experimental observations on the filamentary lasing phenomenon usable in determining reliable values of the absorption coefficient, α, and the gain constant, β, of the active region of the laser diode. Correlation between the process-dependent reduction in delay time of the stimulated emission and the process-dependent position of the dislocation band, (Kim 1965), induced by Zinc diffusion into GaAs, was observed. The position of the dislocation band with respect to the p–n junction was found to depend on the diffusion conditions (Kim 1965) and the heat-treatment processes (Carlson 1967). The band is considered as a probable source of the optical absorbing centres of Fenner's model, (Fenner 1967), as indicated by correlation of reduction of delay time with the band position after the heat treatment.

2. Experimental

2.1. *Junction formation processes*

The p–n junctions were formed by a 3-hour diffusion of Zn into n-type GaAs (Sn : $1 \cdot 2 \times 10^{18}$ donors cm^{-3} or Se : $1 \cdot 9 \times 10^{18}$ donors cm^{-3}), followed by a 3–4-hour heat treatment at temperatures ranging from 900° to 1000°C. The substrate wafers were oriented in either {111} or {100} crystallographic directions. Each group of laser diodes was fabricated from one wafer and mounted with the lasing beam directed perpendicular to the top surface of conventional TO-46 packages. Each group yielded both 'spotty' and 'uniform' diodes.

[†] Now with Monsanto Company New Enterprise Division St. Louis, Missouri.

2.2. *Near field pattern and delay time measurements*

The optics were erected on a vibration-free optical bench. The cryogenic Dewar, which contained the pulser and the cold finger, was an integral part of the microscope stage. For observation of the filamentary lasing phenomenon, the emission from the laser diode was magnified through the optics, and the near-field pattern was displayed on both an infrared image converter (Varo 5500C) and Polaroid film (IR413). Figure 1(a) shows a typical 'spotty' lasing pattern and figure 1(b) shows a 'uniform' lasing pattern. For measurement of the stimulated emission delay time, the same emission was also detected by a silicon photodiode (EGG SGD-444A), and the input current pulse and the output of the detector were displayed on a Tektronix 661 sampling oscilloscope. The rise time of the input pulse was about 30 ns with a pulse width of 300 ns.

3. Results and discussion

3.1. *Determination of α and β*

At a given temperature, for a laser cavity of length L, the threshold condition can be expressed as

$$R \exp\{(G(J_t) - \alpha)L\} = 1 \tag{1}$$

with R the reflectivity coefficient of the cavity mirrors, $G(J_t)$ the gain within the cavity, and α the absorption coefficient of the cavity medium. If the gain is assumed as linearly dependent on the threshold current density, $G(J_t) = \beta J_t$, then the threshold current density can be expressed from equation (1) as

$$J_t = \frac{1}{\beta}\left(\alpha + \frac{1}{L}\ln\frac{1}{R}\right) \tag{2}$$

where β is the gain constant per unit length.

For a given GaAs substrate, the threshold current density depends on the diode operation temperature, the p–n junction formation processes and the cavity length. At a fixed operation temperature the relation between the threshold current density and the cavity length expressed in equation (2), therefore, should yield the cavity parameters, α and β, if the process and the substrate remain constant. Figure 2 shows a typical plot of the threshold current density vs. the cavity length. The scatter of the points was typical even among adjacent diodes fabricated from the same cleaved bar.

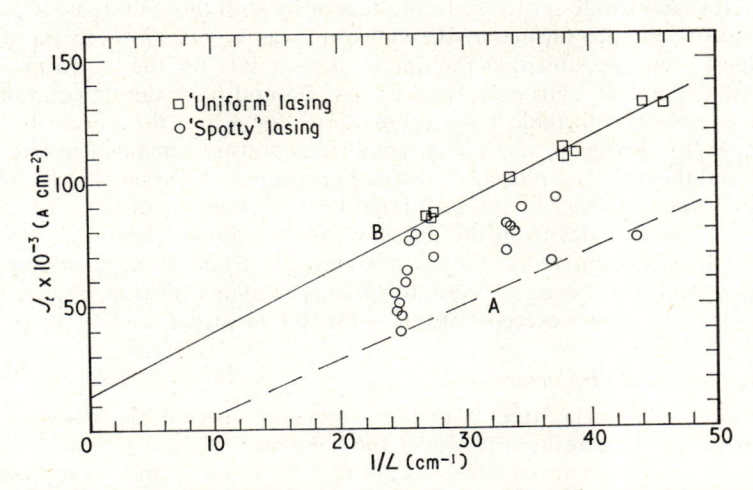

Figure 2. Room temperature threshold current density vs. $1/L$ with respect to near field pattern; a dashed line A for arbitrarily selected 'spotty' lasing diodes; a solid line B for the 'uniform' lasing diodes.

Equation (2) is derived only for a one-dimensional cavity and applies to the 'uniform' lasing cavity. For the 'spotty' lasing cavity the lasing area and the threshold current must be determined for each filament, which was not attempted here because of its inherent difficulty. A demonstration of the valid range of application of equation (2) follows: a dashed line A represents data points with 'spotty' lasing; a solid line B represents data points with 'uniform' lasing. The calculation of the absorption coefficient α and the gain factor β from figure 2 can be misleading if filamentary lasing is not taken into account. To illustrate the point, α and β from the two lines A and B in figure 2 are calculated. The dashed line A was drawn arbitrarily through those points with low threshold current density values observed for each fixed cavity length. Ranges of many α values can be obtained by arbitrary selection of other data points which yield misleading values of α. A positive value of 6·5 cm^{-1} for α in the active region was obtained from the solid line B, whose data points represented 'uniform' lasing diodes. Only those data points for 'uniform' lasing diodes are unique and usable in determination of reliable α and β. Figure 3 shows the threshold

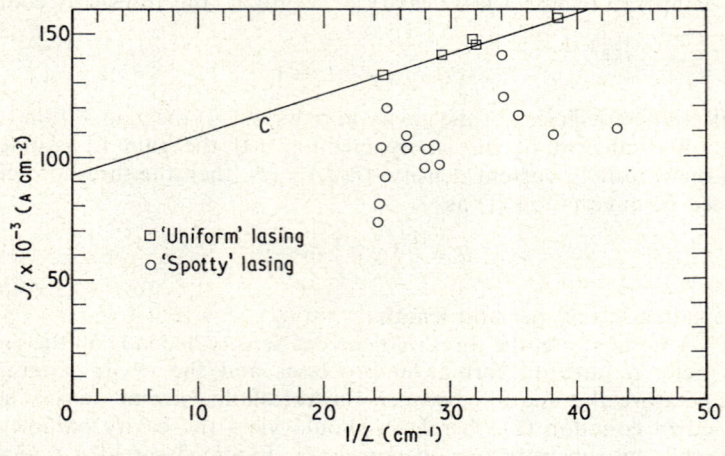

Figure 3. Room temperature threshold current density vs. $1/L$ with respect to near field pattern; line C for the 'uniform' lasing diodes; lapped surface condition.

vs. the length for laser diodes processed simultaneously with those of figure 2 under identical conditions except that the surface of the substrate was lapped prior to the diffusion run. Again the line C demonstrates that the unique data points for the 'uniform' lasing diodes yield the reliable α and β. The calculated values of α and β are tabulated in table 1. Also included for comparison in table 1 are those values for α and β reported by Pilkuhn and Rupprecht (1967). The gain factor β at room temperature remained nearly constant for the diffusion and the diffusion plus heat treatment processes; however, a reduction in absorption coefficient α was realized by the additional heat treatment process. The lack of α and β values for the laser diodes from the diffusion process alone (see line 3, table 1) was the result of difficulties encountered in finding low enough thresholds at room temperature and also having a 'uniform' lasing pattern for the particular substrate selected here. The estimated α for these diodes exceeds 100 cm^{-1} for this substrate.

3.2. *Process controlled parameters*

(*a*) *Delay time.* The delay time τ_t of the stimulated emission decreases with increasing input current above the threshold (Carlson 1967; Fenner 1967). A typical reduction ratio of the delay time τ_{2t}/τ_t observed here was about 1/2, where τ_t and τ_{2t} represent the delay time measured at the threshold and at twice the threshold current respectively. No definite correlations could be observed between the delay time τ_t and the variations in the observed near-field pattern of the filamentary lasing, if the laser diodes were processed under the same

Table 1

Process	Curve	$T(°K)$	τ_t(ns)	$\Delta d(\mu)$	α(cm^{-1})	β(cm A^{-1})	Near field pattern
D+A (970°c) MP	B	298	0–20	7·5	6·5	$4·6 \times 10^{-4}$	Uniform
D+A (970°c) LS	C	298	0–20	9·0	70·0	$7·5 \times 10^{-4}$	Uniform
D		298		1·0	~100·0		
D (Pilkuhn, '67)		296			20·0	$5·7 \times 10^{-4}$	Unknown
E+A (Pilkuhn, '67)		296			92·0	$3·84 \times 10^{-3}$	Unknown

τ_t Delay time measured at the threshold
Δd Distance between the dislocation band and the p–n junction
α Absorption coefficient of the active region
β Gain constant of the active region
D   Diffusion
A   Heat treatment
E   Epitaxy
MP   Polished surface
LS   Lapped surface

conditions. With the rise time of the input pulse of 30 ns, the delay time ranged from 0 to 20 ns for all the lasers presented in figures 2 and 3 regardless of the values of the threshold current density J_t, the observed variations in the near field pattern, and the differences in the surface treatment conditions prior to diffusion.

The delay time τ_t remained constant over the temperature range 250° to 300°K measured at the heat sink. The reduction rate in τ_t as the operation temperatures decreased below 250°K differed slightly from diode to diode, but there were not enough data available to see any trend in correlating the reduction rate in τ_t to the near-field pattern. At temperatures below 200°K, τ_t was zero. The reduction in delay time with increasing heat treatment temperature observed in this work agreed well with that reported by Carlson (1967).

(b) *Dislocation band.* It was reported by Kim (1965) that the dislocation density was highest in the 'knee' region of the Zn concentration profile in GaAs, because the steep gradient in Zn concentration at the 'knee' region induced a large strain which in turn was relieved by the formation of dislocations. This region of high dislocation density was delineated as an etch pit band by etching the cross section of the Zn diffused wafer in Schell (1957) etch. The spread of the band and its distance from the p–n junction increase with increasing heat-treatment temperature. Figure 4 shows the change of the band position in relation to the p–n junction with respect to the heat-treatment temperature.

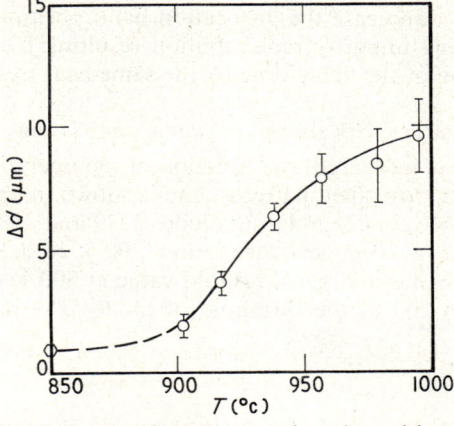

Figure 4. Changes of the band distance from the p–n junction with respect to the heat-treatment temperature.

Fenner's model (1967) explains the delay time in GaAs laser diodes by postulating a moderate number of optical absorbing centres ($> 10^{16}$ cm^{-3}) which cause optical losses and suppress the stimulated emission until these centres are filled by electrons injected into the p-side of the diffused junction; this in effect makes these centres optically transparent or 'bleached'. We attempted to identify the possible optical absorbers, as postulated by Fenner (1967) to be located in the vicinity of the active region. The effects of the optical absorbers are probably sensitive to changes within and near the active region of the laser diode, which was estimated to extend approximately 3 μm on the p-side of the p–n junction.

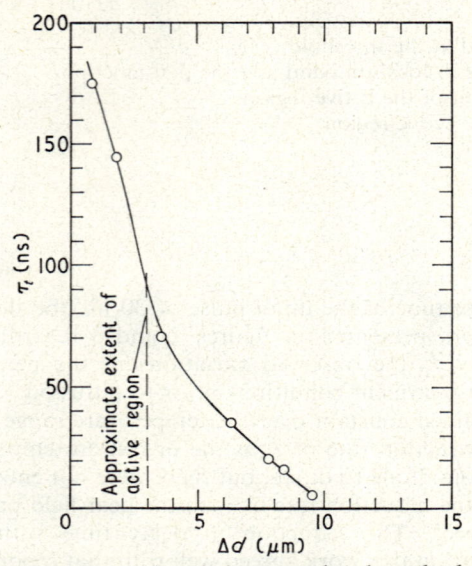

Figure 5. Delay time of the stimulated emission correlated to the band distance from the p–n junction.

Figure 5 is the plot of the delay time vs. the band distance from the p–n junction. Rapid reduction of the delay time was noted up to near the band position corresponding to the estimated extent of the active region. This observation indicates that the dislocation band may be the major source of the optical absorbing centres of Fenner's model. The characteristics of the dislocation band as an optical 'trap' of this kind are not entirely clear. It could be that the optical centres are distributed throughout the bulk and on heat treatment precipitate out to the surface of the wafer or the dislocation band and hence vanish as effective delay agents. In any case the dislocation band position appears to be a sensitive measure of the dopant and impurity redistribution resulting from the heat treatment and correlates to the reduction of the delay time by the same heat treatment.

3.3. *Threshold vs. temperature with respect to filamentary lasing*

The lasing threshold was measured as a function of the operation temperature for diodes with 'uniform' and 'spotty' near-field patterns, and is shown in figure 6. The threshold has a T^3 dependence for the two 'uniform' lasing diodes 5279 and 5289, while the 'spotty' lasing diode 5286 changed from a T^3 dependence below 200°K to a T^2 dependence near room temperature. Diode 5289 has a larger threshold value at 300°K than diode 5286; however, at temperatures less than 200°K, the threshold of diode 5289 is actually less than that of diode 5286.

4. Conclusion

The scatter in the threshold current density data for identically processed laser diodes was shown to depend on the filamentary lasing characteristics of the diodes. Diodes which

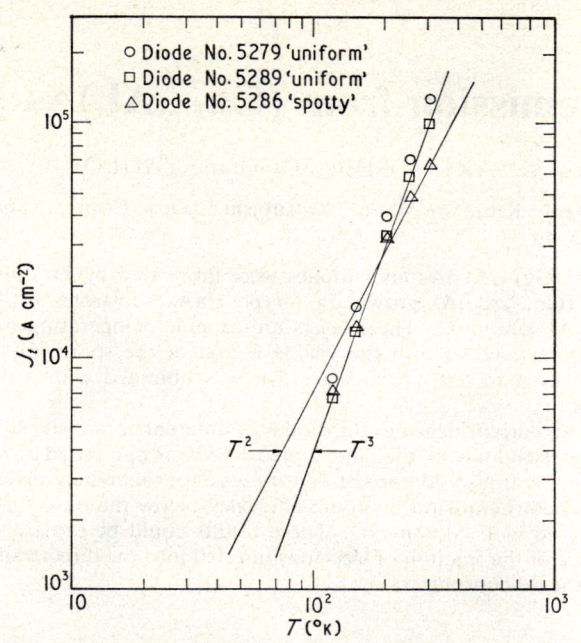

Figure 6. Threshold current density vs. operation temperature for laser diodes 5279 and 5289 with 'uniform' lasing and 5286 with 'spotty' lasing.

display 'spotty' near-field patterns at the threshold were found to have lower threshold current density values than those diodes with the 'uniform' lasing pattern. The values of the absorption coefficient α and the gain constant β of the active region of the laser diode depended on the selection of proper data points and were shown to be reliable only for the 'uniform' lasing diodes. The position of the dislocation band induced by Zn diffusion into GaAs, was shown to be a sensitive measure of the heat-treatment process. Significant reductions in delay time of the stimulated emission were realized on the heat-treated laser diodes. The dislocation band may be the source of the optical absorbing centres of Fenner's model as indicated by correlation of reduction of delay time with the band position after heat treatment.

Acknowledgment

I wish to thank B. A. Pokol and W. G. Horton for their skillful assistance in measuring and processing the laser diodes; F. E. Gentry and R. I. Scace for their support and encouragements; and R. O. Carlson for valuable discussions.

References

ABRAHAMS, M. S., and BUIOCCHI, C. J., 1966, *J. Appl. Phys.*, **37**, 1973.
CARLSON, R. O., 1967, *J. Appl. Phys.*, **38**, 661.
FENNER, G. E., and KINGSLEY, J. D., 1963, *J. Appl. Phys.*, **34**, 3204.
FENNER, G. E., 1966, *J. Appl. Phys.*, **37**, 4991.
FENNER, G. E., 1967, *Solid State Electronics*, **10**, 753.
KIM, H. B., 1965, Carnegie I. of Tech., Ph.D. Thesis.
KINGSLEY, J. D., and FENNER, G. E., 1963, *Bull. Am. Phys. Phys. Soc.*, **8**.
PILKUHN, M. H., and RUPPRECHT, H., 1967, *J. Appl. Phys.*, **38**, 5.
SCHELL, H. A., 1957, *Z. Metall.*, **48**, 158.
VUILLEUMIER, R., et al., 1967, *Proc. IEEE*, **55**, 1420.

Stimulated emission from $(Ga_{1-x}Al_x)As$ junctions

WATARU SUSAKI, TOSHIO SOGO and TAIJI OKU

R & D Dept., Kitaitami Works, Mitsubishi Electric Corp., 1. Shugaike, Ojika, Itami, Hyogo, Japan

Abstract. $(Ga_{1-x}Al_x)As$ laser diodes were fabricated by Zn diffusion into Te doped layers of $(Ga_{1-x}Al_x)As$ grown on n-type GaAs substrates by liquid phase epitaxy from Ga–Al solutions. These lasers are capable of operating over a broad range of wavelengths extending into the visible region of the spectrum. The shortest wavelength achieved to date, 6380 Å at 77°K was obtained with a diode composition of 0·40 AlAs.

Threshold current densities for diodes of different compositions but fabricated in the same manner and having the same dimensions were compared at 77 and 300°K. It was found that the threshold current density was approximately inversely proportional to the spontaneous emission quantum efficiency below the onset of stimulated emission, and increased with Al content. These results could be explained by considering the dependence of the fraction of electrons injected into the direct valley of the conduction band on the Al concentration.

1. Introduction

The search for an injection laser which will emit coherent radiation in the visible spectrum has led to p–n junctions made from $Ga(As_{1-x}P_x)$ alloys as the most promising material since the first announcement by Holonyak and Bevacqua in 1962. On the other hand, Black and Ku (1966) have shown that $(Ga_{1-x}Al_x)As$ alloys have a direct band gap up to about 0·50 mole fraction of AlAs. They also reported that $(Ga_{1-x}Al_x)As$ p–n junctions fabricated by vapour phase epitaxy could emit radiation in the visible spectrum (Ku and Black 1966). However, coherent radiation was not obtained in this material until the later part of last year (Rupprecht et al. 1967a). This was attributed (Ku and Black 1966) partially to the difficulty of preparing single-crystal $(Ga_{1-x}Al_x)As$ alloys having a high degree of chemical homogeneity and purity.

Recently, a liquid-phase growth method of preparing epitaxial deposits of $(Ga_{1-x}Al_x)As$ alloys has been reported and high purity and homogeneity of these materials has been demonstrated by Rupprecht et al. (1967). They also reported coherent radiation from p–n junctions made of these materials by liquid phase epitaxy at 77°K (Rupprecht et al. 1967b). The shortest lasing wavelength observed by them was 7500 Å, and was obtained in a diode with a threshold current density of 2×10^4 A cm^{-2}.

We have also reported coherent emission from $(Ga_{1-x}Al_x)As$ p–n junctions fabricated by diffusing Zn into liquid-phase, epitaxially grown, layers at room temperature as well as at 77°K (Susaki et al. 1968).

In this paper, we wish to describe more detailed results obtained in this kind of diffused p–n junction. In particular, we wish to report the effects of AlAs composition on the threshold current densities of $(Ga_{1-x}Al_x)As$ p–n junction lasers.

2. Experimental

The p–n junctions studied in this experiment were formed by diffusing Zn into n-type $(Ga_{1-x}Al_x)As$ epitaxial layers which were grown on n-type $\langle 100 \rangle$ GaAs substrates by liquid-phase epitaxy (Nelson 1963). Zn diffusion was performed in a sealed quartz tube from a $ZnAs_2$ source (1 mg cm^{-3}) 850°C for 2·5 h. The n-type $(Ga_{1-x}Al_x)As$ epitaxial layers studied were doped with Te either to 1×10^{18} electrons cm^{-3} or 4×10^{18} electrons cm^{-3}.

All of the diodes studied had Fabry–Perot structures with the cleaved ends about 100 μm wide and the etched edges about 400 μm long. The planarity of the junctions was examined by cleaving and staining the junction region by using an etch (10H$_2$O : 1H$_2$O$_2$: 1HF) and was found to be excellent for both series of diodes used.

The AlAs composition of diodes was analysed with an electron probe microanalyser. Many of the epitaxially grown layers were of somewhat graded composition along the growth axis but homogeneous along planes parallel to the substrate–epitaxial layer interface. The composition of the diodes was therefore carefully determined by probing the p–n junction with an electron beam focused to a diameter of about 1–2 μm.

Each diode was tested at 77°K with 1·0 μs current pulses at a repetition rate of 10^3 pulses per second. At room temperature a mercury relay switch was used to test the diodes with a 5-Ω pulse forming network to form nearly square-topped pulses of 100 ns durations at a repetition rate of 60 pulses per second. At 77°K, each diode was also tested with 100 ns current pulses at a repetition rate of 60 pulses per second. The laser threshold was determined by measuring the emission with a fast silicon photodiode as a function of diode current or by observation of the spectral pattern of the emitted radiation in a Jarrel–Ash 0·5 metre grating spectrometer. The threshold current was not changed by applying 100 ns pulses or 1 μs pulses at 77°K. The external efficiency was measured with the aid of a modified integrating sphere similar to that used by Cheroff et al. (1963).

3. Results

When biased in the forward direction, these diodes emitted light from the p-region in a layer about 2 μm in thickness adjacent to the p–n junction as observed directly under a microscope. Figure 1 gives spontaneous emission spectra for a series of diodes having donor levels of 1×10^{18} cm^{-3} with different compositions. These spectra were taken at a current density of about 100 A cm^{-2} at 77°K. The emission lines of these diodes were generally about 200 Å wide at current densities from 10 to 10^3 A cm^{-2}. Raising the temperature to 300°K resulted in a peak shift toward longer wavelengths, and spreading of the line width up to about 350 Å.

Figure 2 gives the variation of the peak photon energy of spontaneous emission with

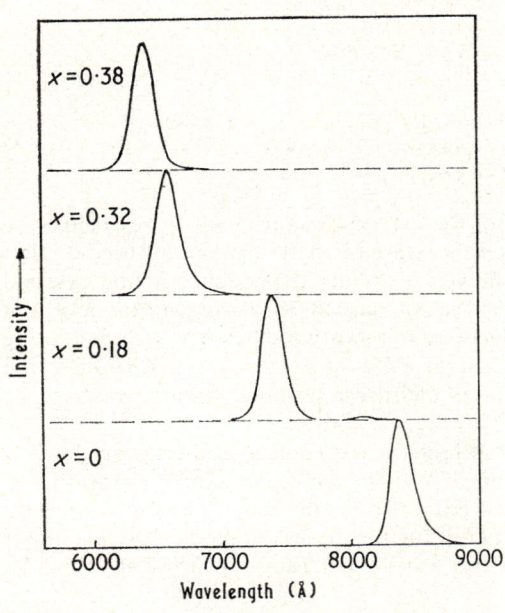

Figure 1. Spontaneous emission spectra at 77°K for a series of diodes ($N_d = 1 \times 10^{18}$ electrons cm^{-3}) of different AlAs composition.

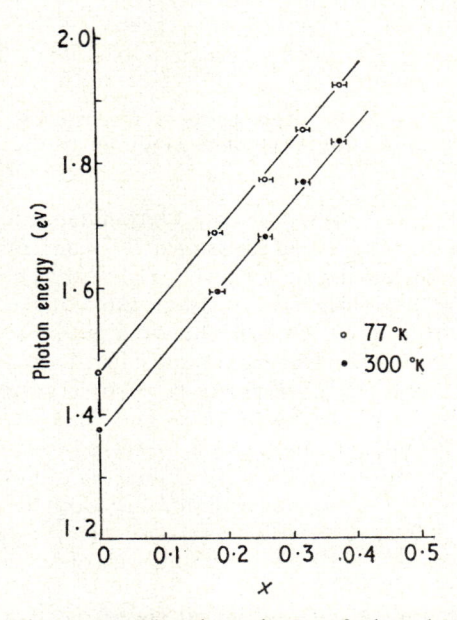

Figure 2. The dependence of the photon energy of the (Ga$_{1-x}$Al$_x$)As diodes ($N_d = 1 \times 10^{18}$ electrons cm^{-3}) on the AlAs composition at 77 and 300°K.

alloy composition at 77°K and room temperature. The peak energy was measured at a current density of about 100 A cm^{-2} at both temperatures. It was found that the peak photon energies of spontaneous emission were lower than the corresponding 77°K values by 0·45 ev as the AlAs component x was increased from zero to 0·40 at both temperatures. This result is in good agreement with the experimental data first reported by Ku and Black (1966).

The external quantum efficiencies of these diodes were measured at 77°K and room temperature. The values of external quantum efficiency below the onset of stimulated emission depends on the amount of reabsorption of the emitted light, i.e., on the geometry of the diodes and on the impurity distribution near the p–n junction.

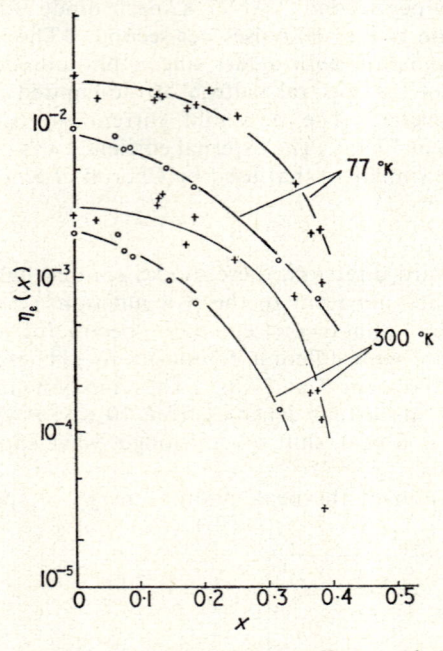

Figure 3. The dependence of the external quantum efficiency below the onset of stimulated emission on the AlAs composition for two series of diodes at 77 and 300°K. + : 1×10^{18} e cm^{-3}; ○: 4×10^{18} e cm^{-3}

Figure 3 shows the composition dependence of the external quantum efficiency below the onset of stimulated emission at 77°K and room temperature for both series of diffused diodes. Diodes having donor levels of 1×10^{18} cm^{-3} showed a smaller dependence of the external quantum efficiency on composition (up to 0·38 AlAs) than those having donor levels of 4×10^{18} cm^{-3}, but both showed a progressive decrease in quantum efficiency with increasing Al content. The maximum efficiencies obtained at 77°K and at room temperature are 1·7 and 0·3%, respectively. These results resemble those obtained with Ga(As$_{1-x}$P$_x$) epitaxial diodes by Maruska and Pankove (1967).

Stimulated emission was observed by applying high current pulses in the forward direction. In the series of diodes having donor levels of 4×10^{18} cm^{-3}, stimulated emission was observed in the wavelength range from 8500 to 6400 Å ($0 \leqslant x \leqslant 0.39$) at 77°K and from 9050 to 7800 Å ($0 \leqslant x \leqslant 0.15$) at room temperature, while in the series of diodes having donor levels of 1×10^{18} cm^{-3} lasing was observed in the wavelength range from 8450 to 6380 Å ($0 \leqslant x \leqslant 0.40$) at 77°K.

Figure 4 gives a typical spectrum of a (Ga$_{0.61}$Al$_{0.39}$)As diode having donor levels of 4×10^{18} cm^{-3} at 77°K. The spectrum was taken with 1·0 μs current pulses at a 10^3 c/s repetition rate. The threshold current is 1·5 A, which corresponds to a current density of 4.5×10^3 A cm^{-2}. The spectrum exhibits the familiar Fabry–Perot cavity mode structure with a mode spacing of 1·0 Å, which corresponds to a value for $(n_0 - \lambda_0\, dn/d\lambda)$ of 5·4, where n_0 is the index of refraction, λ_0 is the photon wavelength, and $dn/d\lambda$ is the dispersion.

Figure 5 shows the composition dependence of threshold current density for two series of diodes with different donor levels. The experimental points are somewhat scattered but a

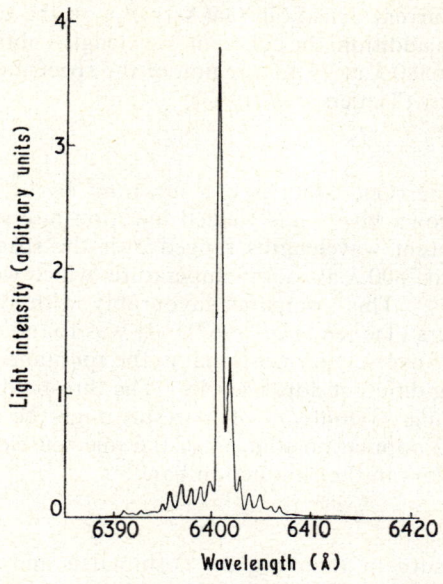

Figure 4. Stimulated emission spectrum from a $(Ga_{0.61}Al_{0.39})As$ diode at 77°K.

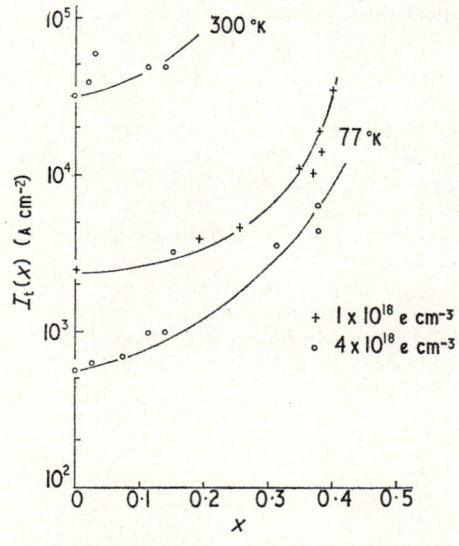

Figure 5. Threshold current density as a function of the AlAs composition for two series of diodes at 77 and 300°K.

clear distinction between the two series of diodes can be seen. In the series of diodes having donor levels of 1×10^{18} cm^{-3} the threshold current density changed little with composition up to $x = 0.38$, where it increased more rapidly. On the other hand, in the series of diodes having donor levels of 4×10^{18} cm^{-3} the threshold current density gradually increased as the AlAs composition increased. The solid lines in figure 5 indicate the inverse of the external quantum efficiency below the onset of stimulated emission. The scales are so adjusted that the threshold current density dependence coincides with the AlAs composition points. There is fairly good agreement for both series of diodes having different donor levels at 77°K.

The dependence of threshold current density on the AlAs composition also resembles the dependence of threshold current density in $Ga(As_{1-x}P_x)$ on the GaP composition reported by Tietjen et al. (1967). In addition, the coherent wavelengths obtained in our $(Ga_{1-x}Al_x)As$ diodes range from 8500 to 6380 Å at 77°K, a region of the spectrum close to that obtained in $Ga(As_{1-x}P_x)$ injection lasers (Tietjen et al. 1967).

4. Conclusions

The method of preparing $(Ga_{1-x}Al_x)As$ p-n junctions by diffusing Zn into the n-type liquid-phase epitaxially grown layers has yielded injection lasers having very good device characteristics. The coherent wavelengths ranged over the spectral region from 8500 to 6380 Å at 77°K, and 9050 to 7800 Å at room temperature, which were obtained with remarkably low current densities. This compares favourably with the best data obtained in $Ga(As_{1-x}P_x)$ injection lasers (Tietjen et al. 1967). It was found that the threshold current density was approximately inversely proportional to the spontaneous quantum efficiency for two series of diodes having different donor levels. The threshold current density decreased progressively with increasing Al content. This result might be explained qualitatively by considering the intervalley population sharing of the injected electrons between the direct valley and the indirect valleys in the conduction band.

Acknowledgments

The authors take pleasure in thanking S. Yoshimatsu and K. Fujibayashi for their interest in this work. They are indebted to H. Miki for preparation of GaAs crystals, M. Ishii for growing $(Ga_{1-x}Al_x)As$ layers by liquid-phase epitaxy, and A. Yoshitome for performing the *AlAs* compersition measurements.

References

BLACK, J. F., and KU, S. M., 1966, *J. Electrochem. Soc.*, **113**, 249.
CHEROFF, G. et al., 1963, *Rev. Sci. Instrum.*, **34**, 1138.
HOLONYAK, N., and BEVACQUA, S. F., 1962, *Appl. Phys. Lett.*, **1**, 82.
KU, S. M., and BLACK, J. F., 1966, *J. Appl. Phys.*, **37**, 3733.
MARUSKA, H. P., and PANKOVE, J. I., 1967, *Solid State Electronics*, **10**, 917.
NELSON, H., 1963, *RCA Rev.*, **24**, 603.
RUPPRECHT, H. et al., 1967a, *Appl. Phys. Lett.*, **11**, 81.
RUPPRECHT, H. et al., 1967b, *IEEE Semiconductor Laser Conference*.
SUSAKI, W. et al., 1968, *IEEE J. Quantum Electronics*, **QE-4**, 422.
TIETJEN, J. J. et al., 1967, *Trans. Met. Soc. A.I.M.E.*, **239**, 385.

CHAPTER 4

Spontaneous emission

Investigation of liquid-epitaxial GaAs spontaneous light-emitting diodes

K. L. ASHLEY and H. A. STRACK

Southern Methodist University and Texas Instruments Incorporated, Dallas, Texas, U.S.A.

1. Introduction

It will be demonstrated in this paper that the light emitter characteristics of solution-grown GaAs diodes, doped amphoterically with silicon (solution diodes) can be characterized by an energy parameter $E_0 = kT_0$, which is related to the band structure of solution-grown GaAs. We will show that the experimental data are consistent with a model based on this characteristic energy (or characteristic temperature).

The existence of a band tail on the conduction band under certain doping conditions and the associated spectral peak shifting has been well established (Pankove 1960; Gershenzon *et al.* 1962). Hayashi (1968) has obtained a solution to the varying manner in which the states are occupied under various conditions of temperature and injection level. In the following, his conclusions are used in connection with solution diodes to explain certain optical and electrical characteristics of the devices. Of primary significance is the characteristic temperature T_0 which separates two regions of behaviour; above this temperature the radiative current is $J \sim \exp(qV/kT)$ whereas below this temperature the exponent becomes temperature independent.

The solution diodes have been observed to exhibit an external quantum efficiency of about 14% at 77°K. This is reduced by a factor of 3.5 to about 4% at room temperature. Room temperature efficiencies as high as 20% have been observed in dome configurations.

An attempt has been made to determine the mechanism responsible for the reduction in external quantum efficiency with increasing temperature. It is found that the reduction is due at least partially to the onset of absorption with increasing temperature.

2. Device model

The major features of a model for solution diodes concern the conduction-band tail, the recombination centre levels and the manner in which they affect the light-emission energy spectrum. Mechanisms for spectral peak shifting with current level and the relatively low energy of light emission, typically 0.1 ev less than bandgap energy, must be accounted for.

The tendency for the peak shifting to decrease with increasing temperature, as well as other characteristics of the device, are consistent with considerations outlined by Hayashi (1968) concerning the effect of energy band shapes on carrier distributions. Hayashi defines a transition temperature $T_0 = E_0/k$ above which the band tail tends to be completely filled at all levels of injection with the opposite prevailing below T_0.

The parameter E_0 characterizes the shape of the band tail, i.e., the electron states in the tail are distributed according to

$$\rho = \rho_0 \exp(E/E_0)$$

where E is the electron energy. Below T_0, the spectral peak is roughly at the quasi-Fermi level, hence the peak shift is proportional to the applied voltage. As the current level is increased, a saturation effect occurs corresponding to the onset of the filling of the tail. At temperatures above T_0, the band is filled at all levels and no shifting appears.

E_0 is expected to increase with increasing doping level (Kane 1963). The solution diodes studied are known to be graded out to a point well away from the junction and probably throughout the bulk region of the device. Capacitance measurements on solution diodes have shown that the junction has a linearly graded impurity concentration with a slope of 10^{21} cm^{-4}. The net impurity concentration in the vicinity of the junction is in the low 10^{16} cm^{-3} range. The top of the p-layer has a net acceptor concentration of about 10^{18} cm^{-3}, as determined from capacitance measurements of a shallow diffused n-type layer at the top surface. Because of a graded impurity distribution, we can expect the following: (a) the radiative recombination is taking place over a region of graded doping, and (b) the value for E_0 varies with the difference from the junction. We will, however, discuss our results in terms of an average value for E_0 and an average value for the peak emission.

At temperatures higher than T_0 where the electron states are occupied above the conduction band energy E_0, one would expect the radiative current voltage relation to obey the functional relation $J \sim \exp(qV/kT)$. Below T_0 the considerations which lead to this relation are not valid. The current will follow the relation $J \sim \exp(qV/E_0)$. This conclusion is based on the following reasoning: let us consider minority-carrier electrons, $n(x)$, injected into a p-region. The radiative current is

$$J_R = \int_0^w n(x)/\tau_n \, dx \sim n(0)$$

where w is the width of the p region and τ_n is the electron lifetime. Assume that electrons fill the band up to the quasi Fermi level E_F, and that $E_F \sim qV$ where V is the applied voltage. Then

$$n(0) \simeq \int_0^{qV} \rho_0 \exp(E/E_0) \, dE \sim \exp(qV/E_0)$$

and

$$J_R \sim \exp(qV/E_0)$$

One other conclusion can be drawn. The internal absorption of light will be different for temperatures below and above the characteristic temperature. At temperatures $T < T_0$, tail states below the quasi-Fermi level are filled and absorption requires the participation of states above the Fermi level. When recombination takes place from states below the Fermi level, light is not absorbed. At high current levels, the quasi-Fermi level moves close to the conduction band and absorption will take place. At room temperature, transitions occur from near the conduction band and absorption can be expected.

The validity of the model should manifest itself through experimentally obtained voltage–current, integrated light emission–voltage characteristics, and emission spectra taken over a range of temperatures, including the characteristic temperature T_0.

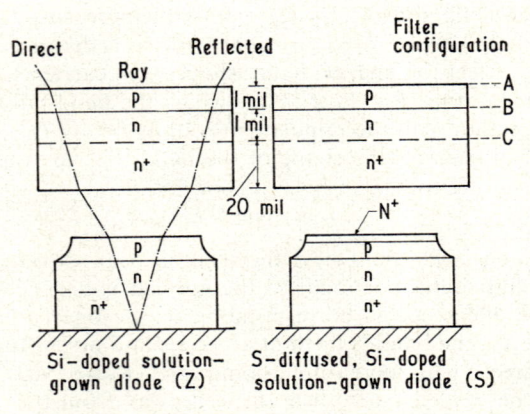

Figure 1. Configurations of diodes and filters.

3. Device fabrication

Solution diodes were fabricated by a process similar to the one described by Rupprecht *et al.* (1966). A silicon-doped slice with a donor concentration of $\sim 10^{17}$ cm^{-3} was inserted into a Ga melt saturated at 950°C with GaAs and doped with silicon. The melt was cooled down at a rate of 1°C min^{-1} for 2 hours. The p-type layer obtained was about 1 mil. A 20-min zinc diffusion was then performed at a temperature of 600°C. Individual 6×7 mil^2 diodes were formed by mesa etching (figure 1, Z-Diode).

To study the properties of the top region of the p-layer, a sulphur diffusion was performed. A slice was enclosed, together with 1 mg Ga$_2$S$_3$, in an evacuated quartz ampoule (volume 15 cm^3) and diffused for 1 hour at 925°C. Diodes were fabricated again by mesa etching (figure 1, S-diode). This structure is equivalent to a transistor and use was made of the transistor-like properties of the n-p-n device.

In some experiments a solution-grown slice was used as a filter to study the absorption characteristics of the solution-grown material. The filter was used in 3 configurations, as indicated in figure 1: (*a*) filter consisting of the p–n–n$^+$ structure, (*b*) filter with p-layer etched off, and (*c*) filter with the p- and n-layer etched off.

4. Experimental results

In figure 2 a plot is shown of the logarithm of the integrated light intensity versus voltage of a Z-diode, obtained at temperatures between 79°K and 300°K. Figure 3 is a logarithmic plot of light intensity versus current at 77°K and 300°K. The departure from the line with a slope of 1 at decreasing current is due to parallel, non-radiating currents. At 77°K, a wide range of current exists over which the non-radiative current is small compared to the radiative current. At 300°K, the characteristic does not exhibit this condition except in the upper range of currents.

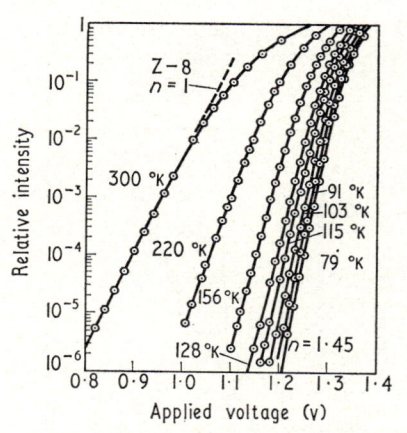

Figure 2. Logarithmic plot of relative intensity versus applied voltage for various temperatures.

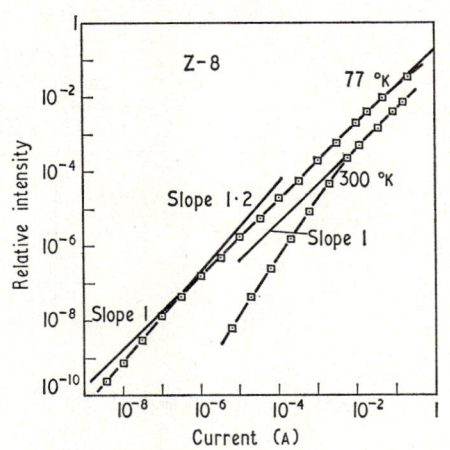

Figure 3. Logarithmic plot of relative intensity versus current at 77°K and 300°K.

The low temperature curve at low currents approaches a condition where the light intensity varies with slope 1·2. This slope is n_n/n_r where the radiative component of current is varying as $\exp(qV/n_r kT)$ and the non-radiative parallel current is related to voltage as $\exp(qV/n_n kT)$. From figure 2 one finds $n_r \simeq 1\cdot45$ such that $n_n \simeq 1\cdot8$. (In this low range of current, the total current also goes as $I_T \sim \exp(qV/n_n kT)$ because the radiative contribution to current is negligible.)

The temperature dependence of the slopes of the plots in figure 2 is more apparent in figure 4 where the quantity of nkT is plotted versus temperature. In the upper temperature range, the curve follows a slope 1 curve corresponding to $n=1$. As the temperature decreases, nkT approaches a constant, temperature-independent value. The gradual transition indicates that the mechanism is changing with decreasing temperature as opposed to having a competing radiative current component increasing in respect to the kT current.

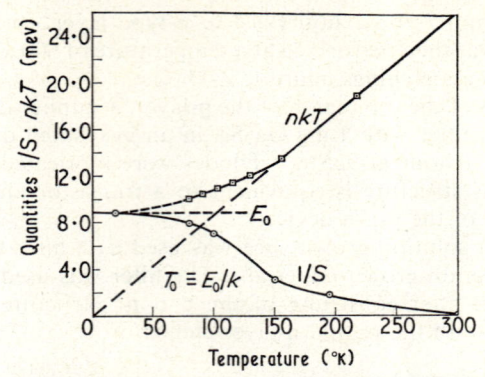

Figure 4. Quantities $1/S$ and nkT versus temperature.

The spectral response of the diode at 77°K is shown in figure 5 for various currents. The curves exhibit a peak shift which saturates at higher currents. At room temperature the peak emission occurs at 1·33 ev independent of current level. Figure 6 is a plot of the logarithm of the current versus the spectral peak for various temperatures. An exponential relation of current and peak energy and a saturation effect at higher currents are noticed in the 77° curve. The quantity

$$1/S \equiv \Delta E / \Delta \ln I$$

derived from curves as in figure 6, is plotted in figure 4. The parameter $1/S$ changes from zero at room temperature to a constant value $E_0 = 9$ mev below liquid-nitrogen temperature.

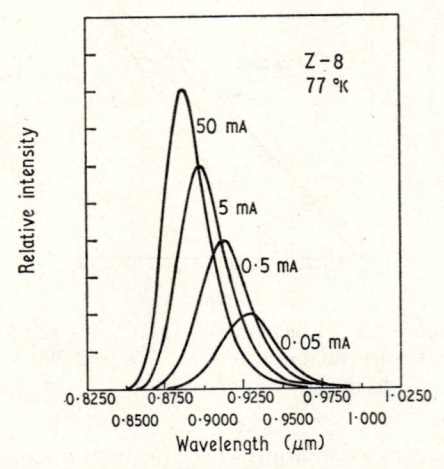

Figure 5. Emission spectra for a Z-diode at 77°K for various currents.

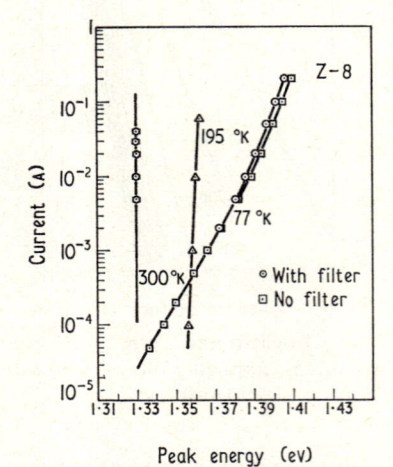

Figure 6. Logarithmic plot of current versus peak energy for various temperatures. The 77°K data are taken with filter configuration A and without filter.

At higher currents, a saturation effect is observed. The saturation can be due to a combination of effects: (a) transitions occur from states high in the tail near the conduction band, (b) heating, (c) absorption. Heating could be excluded by pulsing the diode and varying the pulse length. To study the effect of absorption, the light through filter configuration A, being at the same temperature as the device, was observed. The influence of absorption on the peak shift is shown in figure 6. This also explains the slight decrease from the line with a slope of 1 of the light intensity versus current plot (figure 3) at high currents.

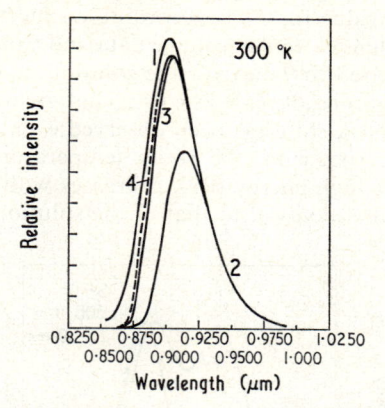

Figure 7. Emission spectra of Z-diode at room temperature using various filter configurations: curve 1, no filter; curve 2, filter configuration A; curve 3, filter configuration B; and curve 4, filter configuration C.

The region mostly responsible for the absorption is the p-type region. This is demonstrated in figure 7. Curve 1 shows the spectral distribution of the light with no filter; curve 2, the same diode through a complete wafer (configuration A); curve 3, through a wafer with the p-layer etched off (configuration B); and curve 4, through the substrate only. It is noted that absorption is decreased considerably when the p-layer is removed. All measurements were made with the diode and filter at room temperature.

The effect of the absorption is reduced at low temperature as can be seen in figure 8, where the external quantum efficiency is shown for constant current as a function of temperature for a diode without and with filter (configuration A). The current is chosen so that at all temperatures the light intensity was proportional to the current. The curves are arbitrarily normalized at 80°K to demonstrate the stronger influence of absorption at higher temperatures than at lower temperatures. Also, emission spectra in the low current range at 77°K have been obtained with and without filtering and identical spectral shape was observed. At high currents, a slight amount of absorption occurs, as evidenced in figure 6.

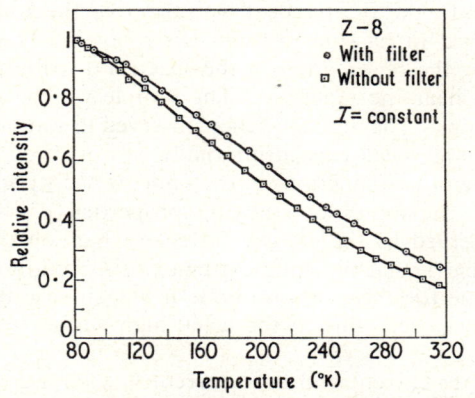

Figure 8. Relative intensity at constant current as a function of temperature. Data taken without filter and with filter configuration A.

So far, we have discussed the electro-optical properties in the vicinity of the solution-grown junctions only. To study the properties of the material closer to the surface, a shallow sulphur diffusion was performed in the top surface (S-diode in figure 1). In this diode, electron injection will take place into a p-type material more heavily doped than in the solution diode junction region. The compensating donor concentration is enhanced due to sulphur atoms. A different value for E_0 is accordingly expected. The value of E_0 calculated from integrated light intensity versus voltage data and from peak shift data obtained at 77°K is $E_0 \simeq 16$ mev as compared to 9 mev for the solution diode. The larger value for E_0 gives a characteristic temperature of $T_0 \simeq 185°$K. This greater value of T_0 is consistent with the fact that a spectral peak shift has been observed even at room temperature.

Unlike the solution diode spectrum, the room temperature spectrum for the S-diode exhibits a double peak. The high energy peak decreases with decreasing temperature and the 77°K spectrum is essentially identical to that of the solution diode.

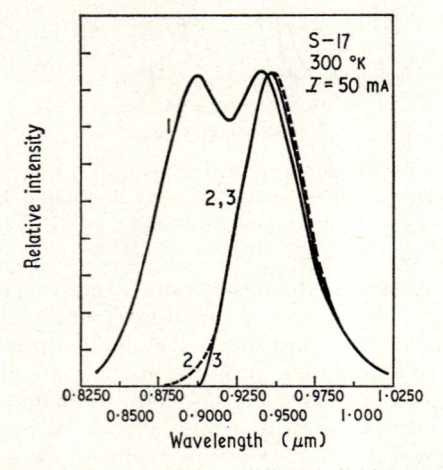

Figure 9. Emission spectra of: curve 1, sulphur diffused diode; curve 2, sulphur diffused diode through filter configuration A; curve 3, solution diode.

The high energy peak is likely to be affected by internal absorption. In figure 9 the spectral distribution of light emitted from the S-diode is shown (curve 1). The high energy portion of this spectrum is strongly absorbed by the filter of configuration A (curve 2). Also shown is the solution diode spectrum of the same device (curve 3). The similarity of this spectrum to that of the filtered spectrum could suggest that the internal spectrum of the solution diode is similar to the double-peak spectrum of the S diode and that the high energy portion is absorbed in passing through the p-layer. The absence of an internal double-peak is verified by the observation of the spectral distribution of light produced by electron-beam excitation near the junction. The sample was bevelled for the purpose of probing the junction region. The high efficiency observed in solution diodes also would not be possible if the high energy peak existed internally.

Because of the presence of two junctions in the same device structure, transistor measurements provided another possibility of studying properties of the solution diode. The solution-grown n-layer served as the emitter. Reverse bias on the S-diode established a condition of zero excess carriers at the outer extreme of the solution-grown p-layer. This is probably a good analogue for the solution diode in which case the excess carrier density near the surface approaches zero due to the usual high surface recombination velocity of GaAs.

Measurements of emitter current and base current yielded a common base current gain of 0·6. This corresponds to a diffusion length of 20 μm for a base width of 1 mil. This gain did not vary much between 77°K and 300°K. The internal quantum efficiency therefore must be only slightly dependent upon temperature if we exclude the unlikely possibility that the separate temperature dependence of the base transport factor and the emitter efficiency exactly compensate one another.

5. Conclusion

The behaviour of various diodes demonstrated above is consistent with the model described in section 2. The current-voltage characteristic changes from a $J \sim \exp(qV/kT)$ behaviour to a $J \sim \exp(qV/E_0)$ characteristic. The value of E_0 obtained from figure 4 is $E_0 \simeq 9$ mev. The corresponding value for the transition temperature is $T_0 \simeq 104°$K, which falls within the transition region of figure 4.

Both the low-temperature saturation value and the high-temperature value of the peak energy correspond to transitions to an acceptor level about 0·1 ev less than the bandgap energy. Below saturation (in a peak shift region), recombination occurs from the band tail to the 0·1 ev acceptor level. Radiation has been observed due to electron transitions from the tail states at least 0·09 ev below the conduction band at 77°K. At room temperature and in the saturation region, transitions occur from near the conduction band. The transitions at the surface of the diode originate from tail states about 0·02 ev below the conduction band, and terminate in the 0·1 ev level and in a shallower 0·03 ev level. The shallow level agrees with the value for silicon acceptors (Kressel et al. 1968).

A first-order estimate of the magnitude of the absorption can be obtained by comparing the filtered spectrum of a diode with an unfiltered spectrum. The estimate is based on the assumption that

$$f_1(E)/f_2(E) \simeq \exp[\alpha(E) w]$$

where $f_1(E)$ and $f_2(E)$ are the spectral intensities, without and with filter (for example, curves 1 and 2 of figure 7), E is the emission energy, $\alpha(E)$ is the absorption coefficient, and w is the thickness of the p-layer. Anticipating an exponential dependence of $\alpha(E)$ upon energy, of the form $\alpha(E) \sim \exp(E/E_\alpha)$, $\ln(f_1/f_2)$ is plotted on a logarithmic scale against the energy of emission. The result is shown in figure 10. The straight line behaviour demonstrates the exponential dependence of $\alpha(E)$ upon energy. E_α is calculated to be $E_\alpha \simeq 20$ mev. The absolute value scale computed for $\alpha(E)$ in the figure is based on an absorbing layer thickness of 1 mil.

The external quantum efficiency changes by a factor of 3·5 from 77°K to 300°K and by a factor of 4·6 when a filter is used. Though the effect of the filter is not very pronounced, the major cause for the reduction in efficiency is thought to be due to absorption. The absorption is quite selective, as is evident from the data in figure 10. Changes in internal

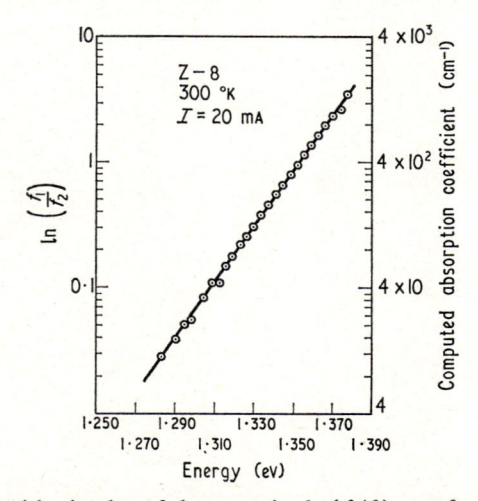

Figure 10. Logarithmic plot of the quantity $\ln(f_1/f_2)$ as a function of energy.

quantum efficiency with temperature are likely to be small in view of the transistor data. Dome-shaped units are expected to change less with temperature since the light exists through the n-layer with a smaller absorption effect. The smaller absorption is partially offset by the increased thickness of the dome configuration. Typical decreases in quantum efficiency for dome shaped units are of the order of 2·5. Mesa units exhibit external quantum efficiencies of 4% at room temperature. The high efficiencies are due partly to the geometry of the device and to the participation of back-reflected rays (see figure 1).

To summarize the observations, we note that the device may be characterized according to whether the temperature is above or below a characteristic value T_0. The transition temperature is related to the band structure of the material. Below this temperature, peak shifting occurs and the exponential term in the diode equation becomes temperature independent. The absorption becomes negligible. Above the characteristic temperature, absorption is significant. The diode current is the conventional diffusion current and no peak shifting is observed.

References

GERSHENZON, M., NELSON, D. F., ASHKIN, A., D'ASARO, L. A., SARACE, J. C., 1962, *Bull. Am. Phys. Soc.*, **8**, 202.
HAYASHI, I., 1968, *J. Quant. Electron.*, **QE4**, 113–8.
KANE, E. O., 1963, *Phys. Rev.*, **131**, 79–88.
KRESSEL, H., DUNSE, J. V., NELSON, H., and HAWRYLO, F. Z., 1968, *J. Appl. Phys.*, **39**, 2006–11.
PANKOVE, J. I., 1960, *Phys. Rev. Lett.*, **4**, 20–1.
RUPPRECHT, H., WOODALL, J. M., KONNERTH, K., and PETTIT, D. G., 1966, *Appl. Phys. Lett.*, **9**, 221–3.

A visible light source utilizing a GaAs electroluminescent diode and a stepwise excitable phosphor

S. V. GALGINAITIS and G. E. FENNER

General Electric Research and Development Center, Schenectady, N.Y., U.S.A.

Abstract. A stepwise excitable phosphor and an infrared-emitting electroluminescent diode have been combined to form a system for converting infrared radiation into visible light. The diode is made from Si-doped GaAs. The phosphor consists of LaF_3 activated by Er^{3+} and sensitized by Yb^{3+}. The primary visible output of the device is centred at a wavelength of 0·54 μm. Devices have been made with a brightness of 200 ft-L for a diode current of 50 mA. Visible output varies approximately quadratically with diode current. The decay time constant for the visible radiation was ~1·1 ms. Diodes made from InP, while providing a better wavelength match to the phosphor, did not prove as effective a source as Si-doped GaAs, because of their lower external efficiency.

1. Introduction

Continuous generation of photons with bandgap energy utilizing successive absorption of infrared quanta was first demonstrated in sulphide phosphors by Halsted, Apple, and Prener (1959). Following a suggestion by Bloembergen (1959), who proposed the use of rare-earth impurities in a host lattice to provide appropriate levels for conversion of infrared to higher energy radiation, Porter (1961) demonstrated stepwise excitation in La : $PrCl_3$. Brown and Shand (1963, 1964) demonstrated similar behaviour in Pr- and Er-doped fluorides of La, Sr, Ca, and Ba, while Esterowitz and Noonan (1965) observed comparable effects in Er-doped CdF_2. In all these cases, the fluorescent output was at visible wavelengths. Since the phosphors in question exhibit low absorption in the infrared region of the spectrum, their conversion efficiency was low. It was then found, by Johnson Geusic, and Van Uitert (1966), by Auzel (1966), and by Ovsyakin and Feofilov (1966), that the addition of Yb to the system resulted in a marked improvement in conversion efficiency. The enhanced response resulted from the facts that Yb^{3+} exhibits reasonable absorption in the infrared, and that efficient energy transfer takes place between the excited Yb^{3+} and the rare-earth ions in the host lattice.

In a study of a large number of rare-earth-activated materials, Hewes and Sarver (1968) discovered that Yb^{3+} is particularly effective in La, Gd, and Yb fluorides activated by Er^{3+}. These workers have developed a LaF_3 phosphor in which Er^{3+} and Yb^{3+} concentrations have been optimized. This phosphor emits primarily in the green when excited by infrared radiation centred at 0·975 μm. The energy level scheme for the system according to the data of Krupke and Gruber (1963), is shown in figure 1. The level diagram is simplified in that only relevant energy levels of the Er^{3+} ion are shown. It will be noted that both upward transitions involved in the stepwise excitation of the Er-doped phosphor involve energy intervals which are equal or nearly equal. Therefore, regardless of the specific details of the mechanisms involved in energizing the phosphor, only a single wavelength is required for excitation. This suggests the possibility of using an infrared-emitting diode as a convenient source to drive the phosphor.

2. The diode-phosphor combination
2.1. Si-doped GaAs as diode material

Curve 1 of figure 2 represents the visible luminescent output of a typical phosphor used in this work as a function of photon energy of the exciting radiation. (Taken from Hewes and

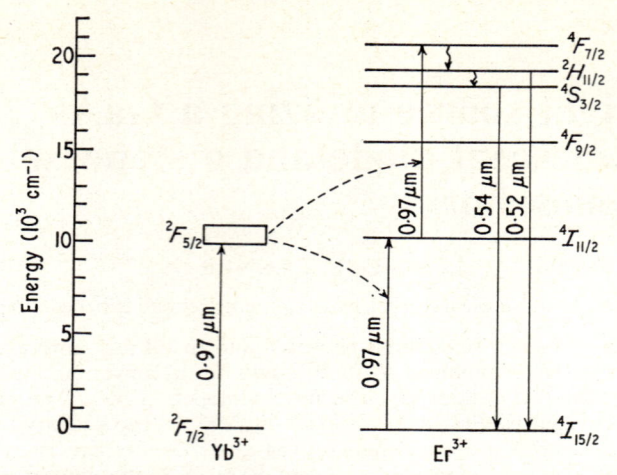

Figure 1. Energy level scheme for Er^{3+} in LaF$_3$, sensitized with Yb^{3+}. Only the relevant energy levels of the Er^{3+} ion are shown.

Sarver 1968). The curve exhibits a sharp peak at 1·273 ev, with a broad shoulder at 1·295 ev. It is evident that standard diffused GaAs diodes, whose emission peaks at about 1·38 ev at room temperature, will be ineffective as a source for exciting the phosphor. However, GaAs diodes made by means of solution regrowth, using Si as an amphoteric dopant (Rupprecht et al. 1966) emit in a region which overlaps the excitation spectrum of the phosphor. In curve 2 of figure 2 is shown a typical output spectrum of such diodes. In curve 3 is shown the relative response of the phosphor to the diode characterized by curve 2. If we assume maximum possible response of the phosphor to the diode radiation, then about 26% of the output of the diode will be effective in producing visible light. If curve 2 is shifted along the energy axis so that the output spectrum of the diode more nearly matches the response of the phosphor, a maximum possible utilization of about 38% is obtained. Curve 4 (inset in figure 2) illustrates how the fraction of diode output utilized varies with peak position of the

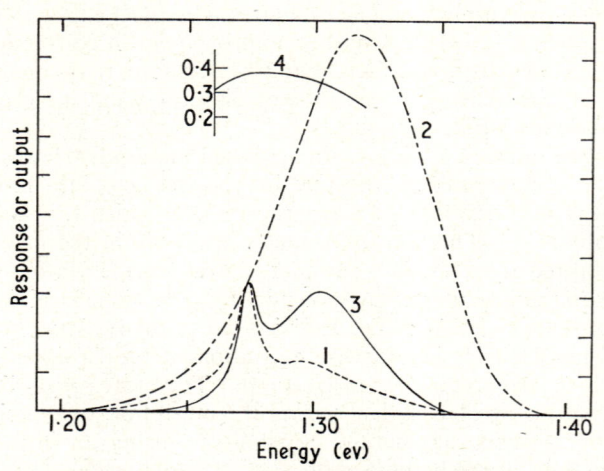

Figure 2. Curve 1—Excitation spectrum for the green luminescence of a typical phosphor used in this study. Curve is normalized to constant energy input. Curve 2—Output spectrum of a typical Si-doped GaAs diode used in this work. Diode current was 60 mA. Curve 3—Relative response of phosphor to the diode characterized by curve 2. Curve 4 (use vertical scale in inset)—Fraction of diode output utilized in phosphor vs. position of peak output of diode. It is assumed that all the radiation at 1·273 ev is utilized in the phosphor.

diode spectrum. From this it is clear that the emission characteristics of Si-doped GaAs represent a reasonable wavelength match. This fact, together with the high external efficiency of these diodes, should make them good sources for the excitation of LaF$_3$ two-step phosphors.

In the design of a diode-phosphor combination, it is important to insure that as much radiant energy as possible be incident on unit area of the phosphor. Since the diode and the phosphor are in proximity, it is necessary that the diode have a high radiance. Carr (1966) has evaluated a number of diode geometries in terms of their photometric properties, and his results indicate that a structure utilizing 'wide aperture end emission' will exhibit the highest average radiance. Consequently, diodes for use as exciting sources were of flat geometry, with cleaved sides, and as small in cross-section as possible. With very careful control of conditions during solution regrowth, diodes could be made with external quantum efficiencies of 4·5–5% without any anti-reflection coating.

2.2. System properties—experimental

The short-circuit current of a cluster of calibrated Si solar cells was measured to obtain the external efficiency of the diodes. Brightness measurements of the visible output of the devices were made by means of a Gamma Model 721 photometer, using a photometric microscope with a fibre optics probe. Relative output measurements and some of the time response measurements were made with a 1P21 photomultiplier tube, with a Corning 4–97 filter placed between diode and detector. A Jarrel-Ash 1/2 meter spectrometer was used for recording spectra. For visible spectra, the signal from a photomultiplier tube with an S-20 surface was amplified with a PAR HR-8 lock-in amplifier. For near-infrared spectra, the signal from a photomultiplier with an S-1 surface was, after preamplification, passed on to a sampling oscilloscope for display and recording.

The phosphor was applied to the surface of the diode with an appropriate binder. For the binder, polystyrene cement proved to be convenient.

2.3. System properties—visible output

The visible output spectrum for the device when operated at room temperature is displayed in figure 3. The radiation in the red part of the spectrum represents about 5% of the total visible energy output of the device.

The infrared output of the diode is a linear function of current, if heating is absent. Hence, the visible light output of the device as a function of diode current is a direct measure of visible light output vs. excitation. For d.c. operation, brightness is a convenient photo-

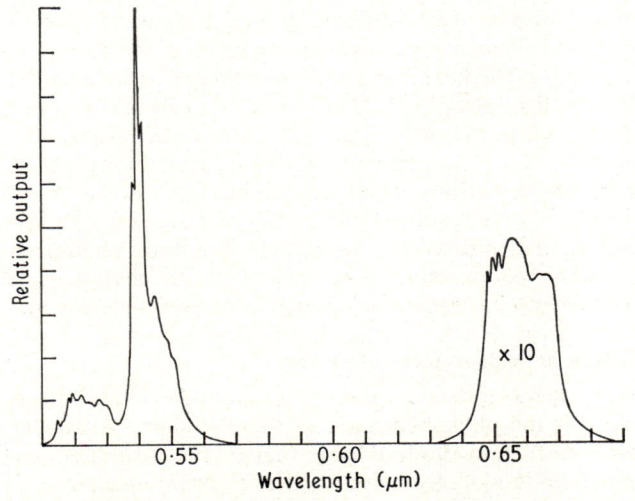

Figure 3. Visible light output of device, operated at room temperature.

metric quantity by means of which to express the visible output. In figure 4 is plotted the brightness of one of the devices as a function of diode current. At currents low enough to preclude heating, the brightness varies approximately as the square of diode current. For the device represented in figure 4, effects of heating appear at 40–50 mA. For pulsed operation, relative output was measured by means of a photomultiplier tube. For 8 ms pulses, approximately quadratic dependence of output on diode current extended to at least 300 mA.

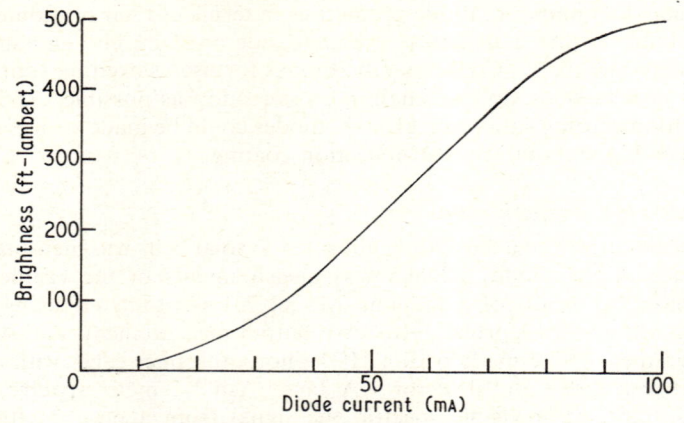

Figure 4. Brightness of device vs. diode current.

2.4. *System properties—time reponse*

Response to a 50 μs pulse is shown in figure 5. The decay time constant for the visible proves to be $\sim 1\cdot 1$ ms. It can also be noted that an output maximum occurs about 250 μs after the pulse is turned off. Response to longer pulses is shown in figure 6. The delay in maximum output persists at longer pulses, although the magnitude of the delay decreases with increasing pulse length. At a given current, a steady-state value of light output is reached in 6–7 ms.

2.5. *System properties infrared output*

The radiative output of the system includes a substantial amount of infrared. A portion of this infrared undoubtedly represents diode radiation which has not been absorbed, but it also includes radiation which has been re-emitted by the phosphor. In figure 7 are shown spectra of this output, sampled at two different points in time. Curve 1 represents the near infrared output at a time when current was flowing through the diode and the visible output had reached steady state. The height of curve 1 in figure 7 is roughly 1/5 that of curve 2 in figure 2. The effect of absorption at 1·273 ev is quite evident. Curve 2 in figure 7 represents the near infrared output shortly after the current was turned off and primary diode emission had gone to zero. Therefore, it represents transitions which occur exclusively within the phosphor. Observations on Yb-containing LaF$_3$ powders with and without the addition of Er indicate that the results shown in figure 7 are due primarily to the Yb. The decay time constant for the infrared was $\sim 2\cdot 2$ ms when both Yb and Er were present, and $\sim 2\cdot 0$ ms when Yb alone was present. The reason for the change in decay time with the addition of Er, particularly the direction of change, is not yet understood.

3. Possible modification or improvement of system

InP, with a room temperature energy gap of about 1·27 ev would, in principle, provide a nearly perfect match for the phosphors used in these studies. However, when the energy of the radiation emitted from a diode is near that of the forbidden gap, then absorption within the bulk of the diode will seriously limit the external efficiency. For diodes of this type, one could probably expect a limiting efficiency no better than the best seen in standard

Zn-doped GaAs diodes—approximately 1%. For the present state of the art in InP, a more nearly practical value for external efficiency would be about 0·1%. Since the response of the phosphor decreases approximately quadratically as the excitation decreases, one might expect that the radiant output from InP diodes would not be high enough to excite the phosphor to a useful brightness. These conclusions were borne out in a series of trials with InP diodes.

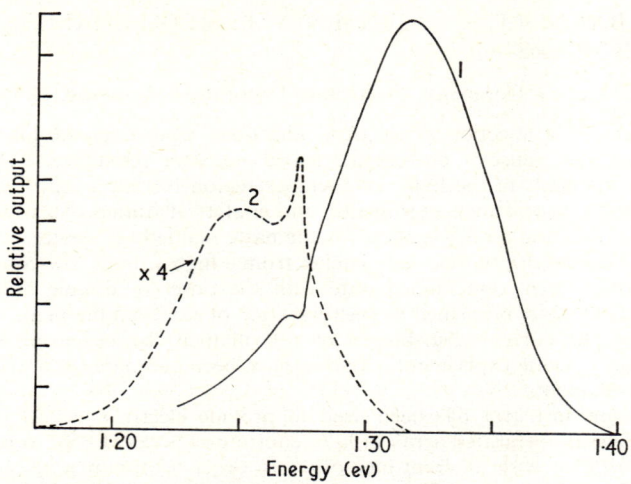

Figure 7. Near infrared output spectrum of device. Curve 1—output while current flows through the diode, and visible output has reached steady state. Curve 2—output shortly after current has been turned off.

Other possible semiconductors which, when used in electroluminescent diodes, would provide a good match for the phosphor, are $GaIn_xAs_{1-x}$ and $GaSb_xAs_{1-x}$, where x is of the order of several per cent. If one were to adjust the energy gap in these mixed crystals to about 1·27 ev, then limitations on external efficiency due to internal absorption would again come into play. However, for use in the system under discussion, only small additions of In or Sb to GaAs would be required. Hence, one can possibly expect that Si might behave amphoterically in these mixed crystals, and yield high-efficiency diodes as in GaAs. In this event, one can expect to increase the visible output of the diode-phosphor combination by about a factor of 2 over the values obtained in this work.

Acknowledgments

The phosphors used in this study were supplied by R. A. Hewes. The Si-doped junction layers were grown by F. K. Heumann. We are indebted to R. J. Connery for diode fabrication.

References

AUZEL, F., 1966, *Compt. Rend.*, **262B**, 1016–9.
BLOEMBERGEN, N., 1959, *Phys. Rev. Lett.*, **2**, 84–5.
BROWN, M. R., and SHAND, W. A., 1963, *Phys. Rev. Lett.*, **11**, 366–8.
BROWN, M. R., and SHAND, W. A., 1964, *Phys. Rev. Lett.*, **12**, 367–9.
CARR, W. N., 1966, *Infrared Physics*, **6**, 1–19.
ESTEROWITZ, L., and NOONAN, J., 1965, *Appl. Phys. Lett.*, **7**, 281–3.
HALSTED, R. E., APPLE, E. F., and PRENER, J. S., 1959, *Phys. Rev. Lett.*, **2**, 420–1.
HEWES, R. A., and SARVER, J. F., 1968, *Bull. Am. Phys. Soc.* Series II, **13**, 687–8.
JOHNSON, L. F., GEUSIC, J. E., and VAN UITERT, L. G., 1966, *Appl. Phys. Lett.*, **8**, 200–2.
KRUPKE, W. F., and GRUBER, J. B., 1963, *J. Chem. Phys.*, **39**, 1024–30.
OVSYAKIN, V. V., and FEOFILOV, P. P., 1966, *J.E.T.P. Lett.*, **4**, 317–8. (*Zh.E.T.F. Pisma* **4**, 471–4).
PORTER, J. F. JR., 1961, *Phys. Rev. Lett.*, **7**, 414–5.
RUPPRECHT, H., WOODALL, J. M., KONNERTH, K., and PETTIT, D. G., 1966, *Appl. Phys. Lett.*, **9**, 221–3.

Light emitting devices utilizing current filaments in semi-insulating GaAs[†]

A. M. BARNETT, H. A. JENSEN, V. F. MEIKLEHAM, and H. C. BOWERS

General Electric Company, Electronics Laboratory, Syracuse, N.Y., U.S.A.

Abstract. The injection of electrons and holes from opposite junctions into a semi-insulator can cause a current-controlled negative resistance. The observation of current filaments in the high conductance region has been reported for GaAs. This paper will report some experimental and analytical studies of current filament formation in GaAs and an application for the basic filamentary device.

This double-injection device switches from a high-voltage low-conductance state to a low-voltage high-conductance state with the onset of double injection. The high-conductance state is marked by the formation of a current filament. Studies have been made of the current distribution of this filament by observing the recombination radiation. These experimental studies have been compared to an analytical model of current filaments.

The semi-insulator has been used to provide electrical isolation between devices, permitting an integrated array of light emitting switches (LES), to be developed for an optical display with inherent memory. In order to minimize the number of external lead connections, a configuration for matrix (row and column) address has been developed.

Monolithic integrated circuits have been fabricated with LES devices and appropriate load resistors on 1·27 mm centres. Circuits measuring 12·7 mm × 12·7 mm have been successfully developed.

1. Introduction

This current-controlled negative-resistance device performs both a light-emitting and a switching function. It is postulated that this device will be able to perform the memory function on the surface of an optical display provided that its negative-resistance characteristics can be satisfactorily reproduced.

In this paper recent development, in the analysis of the basic device will be presented. The application of these results has facilitated the development of a monolithic integrated circuit employing this device, as the basic light-emitting and switching element. The development of this circuit will be discussed.

2. Space charge limited negative resistance

The injection of electrons and holes into a semi-insulator in which the hole lifetime is much greater than the electron lifetime, $T_p \gg T_n$, produces a device exhibiting current-controlled negative resistance (Lampert 1962). Such a device is fabricated by putting a hole-injecting contact and an electron-injecting contact on opposite sides of a semi-insulating substrate.

When the hole-injecting contact is biased positive with respect to the electron-injecting contact, the current–voltage characteristic consists of a voltage rise at low currents until a threshold voltage, V_{TH} is reached. The onset of negative resistance occurs at this threshold voltage and is marked by a voltage drop from a high impedance pre-breakdown region to a low impedance post-breakdown region. The post-breakdown region is characterized by a current rise at a constant voltage, V_M, followed by a high current $I \propto V^{1.5 \text{ to } 2}$ relation. The characteristic curve of these devices is shown in figure 1.

A similar current-voltage characteristic has been observed in semi-insulating germanium, silicon, GaAs, and other materials (Holonyak 1962; Ashley and Milnes 1964; Ing, Jr. et al.

[†] This work was partially supported by the U.S. Army Electronics Command, Fort Monmouth, New Jersey.

1964; Gerhard and Jensen 1967). The pre-breakdown region has been analysed in great detail (Lampert 1962; Ashley and Milnes 1964).

Ridley (1963) predicted that devices exhibiting current-controlled negative resistance, similar to the type observed in these double injection studies, should conduct current in a filamentary mode in the high conductance state. Such a filament was first observed in double-injection devices in semi-insulating silicon at 77°K (Barnett and Milnes 1966). This work described the high-conductance region as filamentary over a specific range of semi-insulator widths. The filaments observed were in the high current $I \propto V^{1 \cdot 5 \text{ to } 2}$ region. Filaments have also been observed in the post-breakdown region of semi-insulating GaAs at room temperature (Barnett and Jensen 1968). These filaments were photographed in the low-current region where the current increases at a nearly constant voltage.

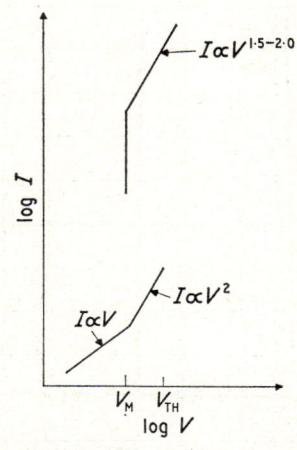

Figure 1. Theoretical voltage–current characteristics for double-injection device with the junction voltage drop subtracted.

3. Study of the filament

Previous studies of the filament (1966, 1968) have taken great care to develop planar injecting contacts. These devices had an n⁺ junction on one face of the semi-insulator. This face was alloyed to a metal substrate. A thin, one-micron, p⁺ junction was diffused into the opposite face. The semi-insulator was stepped in thickness enabling the wire to be applied to the top surface over a thicker section. This enables breakdown to occur at a point removed from the metal contacts.

The filament was studied by photographing the recombination radiation through the p⁺ junction. The stepped device has proved to be unsatisfactory for total device studies because the threshold voltage is strongly influenced by the amount of surface leakage current. (The effect of surface leakage will be discussed in greater detail at the end of this section.)

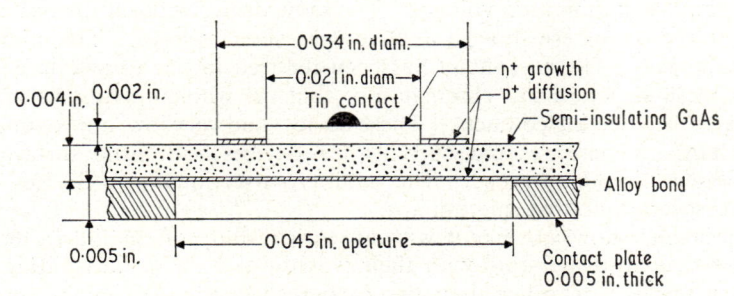

Figure 2. Drawing of device LES-J2.

The double injection devices for this study were fabricated on a 10^7 ohm-cm oxygen-doped GaAs substrate. The n+ junction was formed by liquid-phase epitaxial growth. This injecting contact was 50 μm thick and 0·79 mm in diameter. A p-type guard ring was placed around the n+ junction to minimize surface leakage current.

A one-micron p+ junction was diffused into the opposite face. The semi-insulator is 100 μm thick.

The p+ junction is alloyed to a gold-plated molybdenum contact with a 1·14 mm diameter aperture.

Figure 2 shows a drawing of the device developed for the present study. Figure 3 shows the voltage-current characteristic for this experimental device. Photographs of filaments taken through the p+ junction at several currents are shown in figure 4.

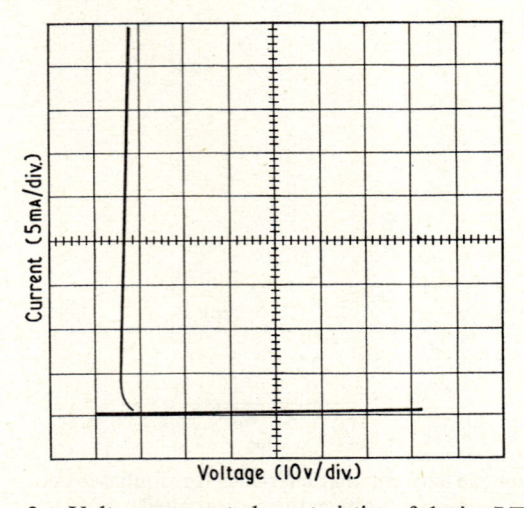

Figure 3. Voltage–current characteristics of device LES-J2.

A previous study developed a model of the filament based on a radial diffusion of carriers from a central core (Barnett and Milnes 1966). We have now developed a more complete theoretical model for the formation of filaments under double-injection conditions in semi-insulating GaAs. The analytical one-dimensional solutions presented by Lampert (1962) were extended. The solutions in this one-dimensional work were used to approximate some of the terms in the complete three-dimensional drift and diffusion current equation; the equation was then solved numerically. This equation predicted filaments in the order of 50–100 μm in diameter, for the semi-insulating GaAs of this study, which qualitatively agree with the present experimental observations.

The present experimental study has permitted several observations of the effects of surface leakage current. The guard-ring device for the filament study exhibited a threshold voltage of 70 volts and a 6-volt minimum voltage. The same device without the guard ring around the n+ solution regrown mesa showed a 30-volt threshold voltage. The minimum voltage remained unchanged. The n+ contact has been reduced in size to a square, 127 μm on a side, and the mesa structure was eliminated so that the whole device had the same cross-sectional area as the injecting contacts. These devices had very low, approximately 10 volts, threshold voltages. The threshold currents for devices with little surface leakage are usually in the one micro-amp range. The small cross-sectional area devices show leakage currents up to several hundred micro-amps.

These experiments show agreement with an earlier study of small-area devices (Weiser and Levitt 1964). Corresponding with their work was a theory attributing the negative resistance to a mechanism other than two-carrier space charge limited current (Dumke 1964).

Individual devices have been successfully operated at a 2 MHz rate. The average room-temperature quantum efficiency of the light output is 0·2% at 10 mA, and the light-emitting area is a circle approximately 75 μm in diameter.

4. Utilization of the double-injection device as a matrix addressed display element

In order to minimize lead complexity for multi-element semiconductor optical displays, matrix address is commonly employed. Figure 5 shows the circuit scheme for the matrix-address electronics. V_C is a steady-state voltage, V_A the column pulse voltage, and V_E the row pulse voltage. The display can be successfully operated when V_C biases all devices in a bistable state and $V_C + V_A$ (or V_E) is less than the threshold voltage for any device and $V_C + V_A + V_E$ is greater than the threshold voltage for all devices. A similar scheme can be used to turn off individual elements.

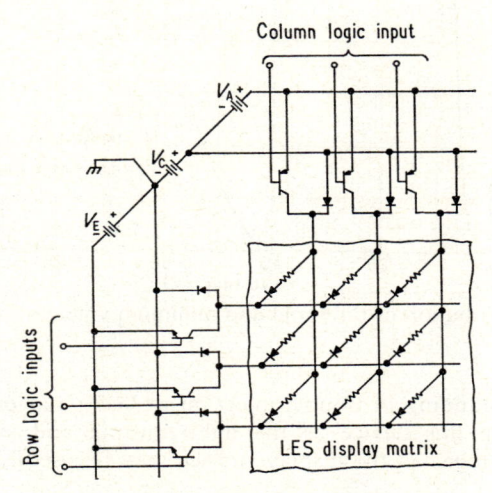

Figure 5. Matrix address electronics.

The display is built on opposite faces of a semi-insulating GaAs substrate. Silicon dioxide is deposited on the gallium face of a wafer cut in the [111] direction. Holes of 0·25 mm diameter are etched into this oxide and n⁺ islands are grown on the substrate by liquid-phase epitaxial growth. The wafer is then zinc diffused. This p-type diffusion is used for both the hole-injecting contact (on the arsenic face) and the diffused resistor which is connected to the n⁺ island by metallization. The load resistor is formed with an etched mesa structure. The resistor diffusion is also used as a substrate for the column metallization. Figure 6 shows the resistor n⁺ junction side of a 9×9 (81 element light-emitting switch integrated circuit). Figure 7 shows a close-up of a resistor-conductor pattern.

The hole-injecting contacts are isolated into row stripes which are perpendicular to the column metallization connecting the resistors and devices on the n⁺ face. The integrated circuit is then alloyed to a metallized pattern on an alumina substrate. A metallization is then applied from the substrate to the hole-injecting contact (top face). This metallization extends along the length of the contact to reduce the series resistance. The light is observed through the p⁺ junction. A photograph of a completed 9×9 circuit is shown in figure 8.

As each element on the integrated circuit is to be addressed with two-terminal threshold logic it is necessary that the threshold voltages be quite uniform. In addition the voltage drop across the resistor and the device in the high-conductance region must be reasonably uniform. Figure 9 shows the distribution of threshold and minimum voltages for an 81-element light-emitting switch integrated circuit.

This circuit can be operated with a 12 v steady-state bias, V_C, and 32 v pulses on the rows and columns, V_A and V_E. Seventy-seven of the devices can be successfully operated under these matrix-address conditions. The remaining four devices will not remain in the high-conductance state after the pulse voltage is removed.

These high-minimum-voltage devices are the result of the formation of an extraneous breakdown at a point removed from the n$^+$ junction, and its associated guard ring. Lower voltage devices, improved device structures, and more uniform processing should improve this condition.

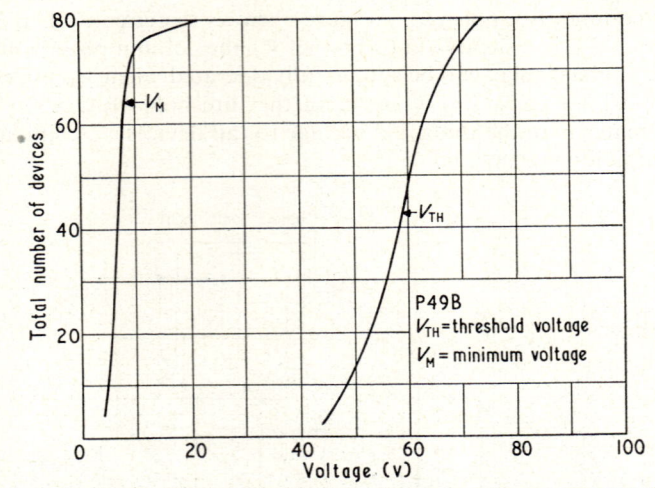

Figure 9. Distribution of threshold and minimum voltages for LES-IC P49B.

5. Conclusions

An improved understanding of double injection in GaAs has enabled us to develop an integrated circuit using this device as the light-emitting and switching element. This concept of performing logic on the display surface may be useful in future semiconductor displays.

It is postulated that this concept can be useful for desk-top message displays with memory using a larger energy gap material, a solid state image converter, or a direct conversion phosphor.

Acknowledgments

The authors would like to thank H. Spoor, R. Glusick and A. Simone for technical assistance. Valuable discussions were held with V. Russell, W. Willis, J. Dietz, J. Jones and D. Osborn.

References

ASHLEY, K. L., and MILNES, A. G., 1964, *J. Appl. Phys.*, **35**, 369.
BARNETT, A. M., and JENSEN, H. A., 1968, *Appl. Phys. Lett.*, **12**, 341.
BARNETT, A. M., and MILNES, A. G., 1966, *J. Appl. Phys.*, **37**, 4215.
DUMKE, W. P., 1964, *Proc. of 7th International Conference, Physics of Semiconductors*, 611, Dunod, Paris.
GERHARD, G. C., and JENSEN, H. A., 1967, *Appl. Phys. Lett.*, **10**, 333.
ING, JR., S. W., JENSEN, H. A., and STERN, B., 1964, *Appl. Phys. Lett.*, **4**, 164.
LAMPERT, M. A., 1962, *Phys. Rev.*, **125**, 126.
RIDLEY, B. K., 1963, *Proc. Phys. Soc.* (Lond.), **82**, 954.
WEISER, K., and LEVITT, R. S., 1964, *J. Appl. Phys.*, **35**, 2431.

Radiative tunnelling in GaAs: a comparison of theoretical and experimental properties

H. C. CASEY, JR. and DONALD J. SILVERSMITH[†]

Bell Telephone Laboratories, Incorporated, Murray Hill, New Jersey, U.S.A.

Abstract. The experimental properties of shifting-peak spectra for abrupt-asymmetrical junctions have been compared with Morgan's theory of photon-assisted (radiative) tunnelling. For this comparison the free-electron concentration on the more lightly doped side of the junction was varied from 3×10^{17} cm^{-3} to 8×10^{18} cm^{-3}, the voltage range was $\sim 1 \cdot 0$ to $\sim 1 \cdot 5$ v, and the temperature was varied from 12° to 255°K. The calculated and experimental spectra are in detailed agreement. No adjustable parameters are needed to calculate the spectral shape for any given temperature, applied voltage, and free-electron concentration. These results demonstrate that shifting-peak spectra often attributed to band filling may be described in terms of photon-assisted tunnelling.

1. Introduction

For germanium p–n junctions, which were the first semiconductor junctions considered in detail, the simple diffusion current theory (Shockley 1949) was sufficient to explain most of the observed behaviour. Silicon junctions required description in terms of the space-charge recombination-generation current (Sah, Noyce and Shockley 1957). From previously published studies of the I–V characteristics of GaAs junctions by Dumin and Pearson (1965) and the results presented here, it appears that explanation of GaAs junctions requires consideration of various tunnelling mechanisms. Several tunnelling mechanisms are illustrated by the comparison of the theoretical and experimental properties of radiative tunnelling in GaAs. Radiative tunnelling is concerned with spectra whose peak energy approximately equals the applied voltage and shifts in both position and intensity with the applied voltage. In this paper, experimental spectra for GaAs abrupt-asymmetrical junctions have been compared with the results of Morgan's photon-assisted (radiative) tunnelling theory (1966).

Initially, the shifting-peak spectra observed by Pankove (1962) were described as photon-assisted tunnelling. Similar spectra were obtained by Nelson *et al.* (1963) and were explained in terms of a band-filling model. It has not been clear whether two distinct injection luminescence processes occurred or whether there were two explanations for the same process. Because Morgan's results correctly predict the shifting-peak spectral shape for a given temperature, applied voltage, and free-electron concentration on the more lightly doped side of the junction, the shifting-peak spectra presented here have been designated as radiative tunnelling.

2. Experimental procedure

The samples used in this study were single-crystal, $\langle 100 \rangle$-oriented, Te-doped wafers obtained from crystals grown by the floating-zone technique. The free-electron concentrations, 0·3 to $8 \cdot 0 \times 10^{18}$ cm^{-3}, were determined by Hall measurements on each wafer before diffusion. For this concentration range, metallic-impurity conduction dominates, and the free-electron concentration is temperature-independent. For diffusion, a single wafer was sealed in a fused-silica ampoule along with the diffusion source which had a composition of 5, 50, and 45 atomic percent Ga, As, and Zn. This diffusion source was used because it provided an abrupt impurity profile and also gave a significant improvement

[†] Present address: Department of Metallurgy, Massachusetts Institute of Technology, Cambridge, Massachusetts.

in the reproducibility and planarity of the Zn diffusion as compared with other Zn diffusion sources (Casey and Panish 1968). Diffusion at 650°C for 10 h resulted in a junction depth of 3·6 μm. The Zn surface concentration was $1·7 \times 10^{20}$ cm^{-3} and the Zn concentration decreased abruptly below concentrations of about 10^{19} cm^{-3}. By standard masking and etching techniques, 4-mil diameter mesa diodes were formed.

Spectra described at 77°K were obtained by mounting diodes in a Dewar containing liquid nitrogen. At other temperatures the diodes were mounted on a cold finger in a variable-temperature cryogenic system. The spectral shape was measured with a Perkin-Elmer model 99 single-prism monochromator (resolution $\simeq$ 3 mev). The infrared radiation, chopped at 37·5 Hz, was detected with a dry-ice cooled Amperex 150 CVP (S-1 response) photomultiplier, and the resulting signal amplified by a Princeton Applied Research lock-in amplifier system which drives a Leeds and Northrup recorder. The resulting spectra were corrected for the wavelength dependence of the photomultiplier and the monochromator transmission.

Although measurements presented here are for a particular device, they have been verified as representative on a significant number of other devices. Several hundred diodes were investigated.

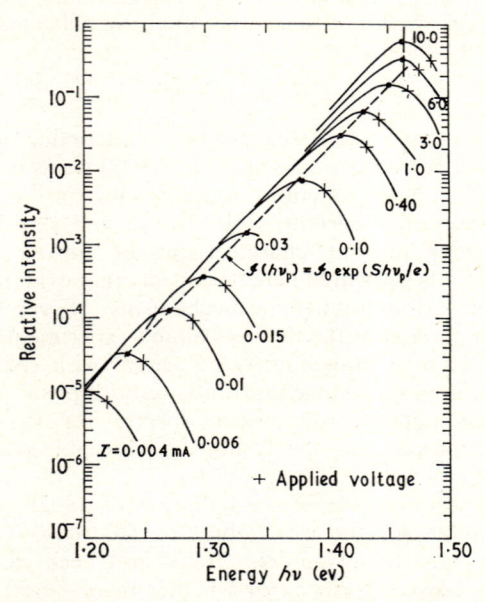

Figure 1. The 77°K shifting-peak spectra as a function of applied voltage for an abrupt-asymmetrical junction with $n = 1·74 \times 10^{18}$ electrons cm^{-3}.

3. Comparison of experimental spectra with Morgan's theory

3.1. *Experimental spectra*

The basic characteristics of the shifting-peak spectra at 77°K for abrupt-asymmetrical junctions are demonstrated in figure 1 for a diode with $n = 1·74 \times 10^{18}$ cm^{-3} on the more lightly doped side of the junction. The applied voltage is indicated by a cross +, and the current in mA is given for each spectrum. The emission on the low-energy side of the peak emission almost saturates, and the emission shifts to higher energy as the bias is increased. The energy at the maximum intensity for each spectrum has been marked with a solid circle. These solid circles have been joined by the dashed line to show that the relative peak intensity varies with peak energy $h\nu_p$ (or applied voltage) as

$$\mathcal{I}(h\nu_p) = \mathcal{I}_0 \exp(Sh\nu_p/e), \qquad (1)$$

where e is the charge and S is the logarithmic slope constant in V^{-1}, which has a value of

39 v⁻¹ for this diode. It is also useful to note that the peak emission energy occurs slightly below the applied voltage:
$$h\nu_p = e(V - V_0), \qquad (2)$$
where V_0 is about 0·020 v. At high bias the emission peak becomes stationary in energy while increasing in intensity. This stationary peak is band-edge emission due to thermal injection.

Shifting-peak spectra for linear-graded junctions are characterized by one value of logarithmic slope constant S_A at low voltage and another value $S_B \simeq 2S_A$ at higher bias (Archer et al. 1963). A comparison of shifting-peak spectra for linear-graded junctions with Morgan's theory (1966) has been given in the complete analysis of shifting-peak spectra (Casey and Silversmith, in press). For conciseness and clarity, the discussion presented here will consider only abrupt-asymmetrical junctions; however, the same concepts apply to graded junctions.

3.2. *Numerical evaluation of Morgan's theory and comparison with experimental spectra*

A brief description of Morgan's analysis (1966) will be given to provide physical insight into the process of radiative tunnelling. The essential feature of this analysis for an abrupt-asymmetrical junction is the determination of the electron and hole wave functions in the depletion region for the case of a parabolic potential barrier, rather than the usual approximation of a linear potential. The expression obtained for the radiative recombination rate will then be compared with the experimental results.

The radiative recombination rate $R(h\nu)$ is given by equation (2.8a) of Morgan (1966) as
$$R(h\nu) = 8\pi^2 B \int dE_n f(E_n - E_{fn}) f(E_{fp} - E_n + h\nu) \int 2\pi k_{\perp 1} \, dk_{\perp 1} M^2, \qquad (3)$$
where B contains the average interband matrix element that is assumed independent of the initial and final states, E_n and E_{fn} are the total electron energy and Fermi energy, $E_n - E_p = h\nu$ is the photon energy with E_p the total hole energy, the $f(E)$'s are the Fermi functions for electrons and holes, and $\hbar k_{\perp 1}$ is the momentum perpendicular to the field direction. The overlap integral M is
$$M = \int dx \, \psi_{1x} \psi_{2x}, \qquad (4)$$
where ψ_{1x} and ψ_{2x} are the electron and hole wave functions. The calculation of the spectral shape becomes a problem in finding ψ_{1x} and ψ_{2x} and the evaluation of equation (4) and finally equation (3).

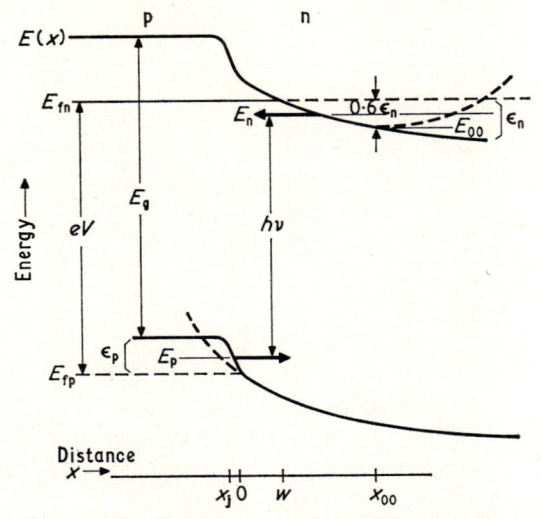

Figure 2. Energy E vs. distance x diagram for an abrupt diode more heavily doped on the p side, and degenerate though less heavily doped on the n side. (See text for description of quantities given in figure.)

Figure 2 shows the potential energy vs. distance diagram with the significant energies and distances labelled. This diagram illustrates the basic model for band-to-band radiative tunnelling: an electron of energy E_n and a hole of energy E_p tunnel into the depletion region and recombine with the emission of a photon of energy $E_n - E_p = h\nu$. The location of the Fermi levels with respect to the band edges at 0°K for parabolic bands is given by ϵ_n and ϵ_p. The energy gap is E_g, and the quasi-Fermi levels E_{fn} and E_{fp} are separated by the applied voltage V. The distance between the emergence of E_{fp} from the valence band and the emergence of E_{fn} from the conduction band is w, and x_j is the metallurgical junction. From the solution of Poisson's equation for both the space-charge region w and the band bending for $x > w$, a parabolic expression for the energy of the band edge was obtained. This parabolic function is utilized in the one-dimensional Schrödinger equation to obtain the electron and hole wave functions. As a consequence of the heavy doping on the p-side of the junction, the bias cancels in the potential function for the Schrödinger equation, and the resulting electron and hole wave functions are therefore independent of bias. If these wave functions are used to determine the recombination rate, the exponential character of all emission spectra for $h\nu < eV$ will be independent of bias, and will result in a constant emission rate at a given energy as the voltage is increased. The Fermi factors give the recombination formula bias dependence, but this becomes important only for $h\nu > eV$.

Integration of equations (4) and (3) with the wave functions obtained from the Schrödinger equation permits the radiative recombination rate to be written as

$$R(h\nu) = R_0 I_R \exp(\alpha h\nu/e), \qquad (5)$$

where R_0 represents prefactors that depend on physical constants, while I_R is essentially constant until $h\nu \simeq eV$, and then it decreases rapidly. The exponential factor α is equivalent to the logarithmic slope constant S and is given by

$$\alpha(\text{v}^{-1}) = (4\kappa m_n/n\hbar^2)^{1/2}\phi_0, \qquad (6$$

where κ is the dielectric constant, m_n is the effective electron mass, n is the free-electron concentration on the more lightly doped side of the junction, and ϕ_0 is a factor that is approximately 0·95 for these diodes. Equation (5) contains no adjustable parameters. The complete expressions for both R_0 and I_R are given by Casey and Silversmith (1969). (In order to retain use of the units of volts, amperes, and farads, the rationalized MKS units are used throughout this paper, while the equations by Morgan (1966) were originally

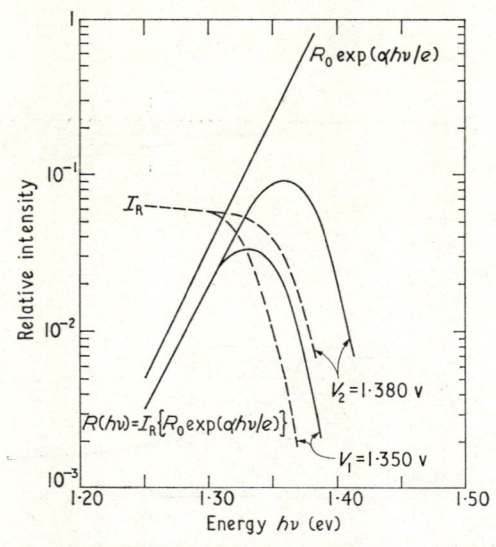

Figure 3. Calculation of the radiative recombination rate $R(h\nu)$ from equation (5) for $n = 1\cdot74 \times 10^{18}$ electrons cm^{-3} at 77°K. The quantity I_R is an integral given by Casey and Silversmith (1969) and has been evaluated by numerical integration.

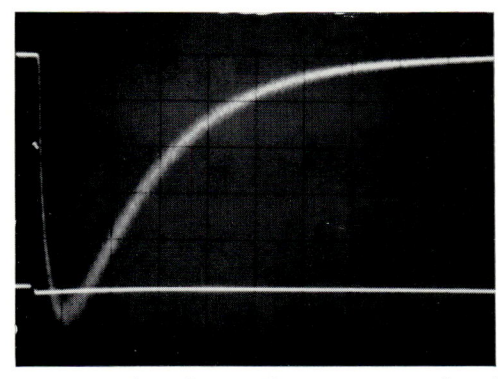

Figure 5. Visible light output vs. time for a 50 µs current pulse through the diode. Each horizontal division represents 500 µs. Current pulse appears as elongated dot at beginning of trace, 3 divisions up.

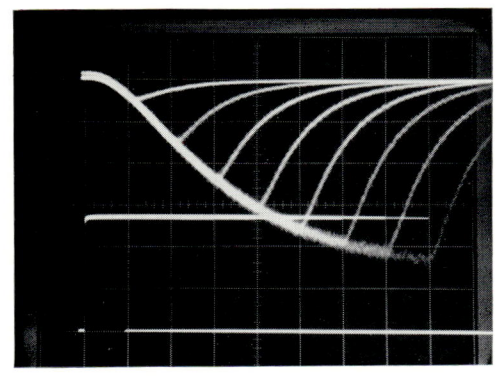

Figure 6. Visible light output vs. time for a series of current pulses from 1 ms long to 8 ms long, increasing by 1 ms steps. Each horizontal division represents 1 ms. Horizontal line near middle of picture represents the current pulses.

PLATE XVIII—A. M. Barnett et al.: paper 22

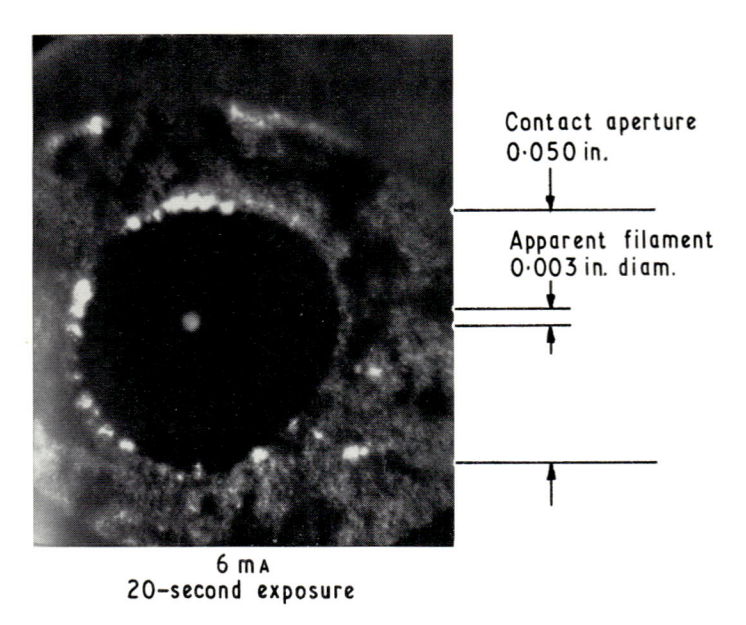

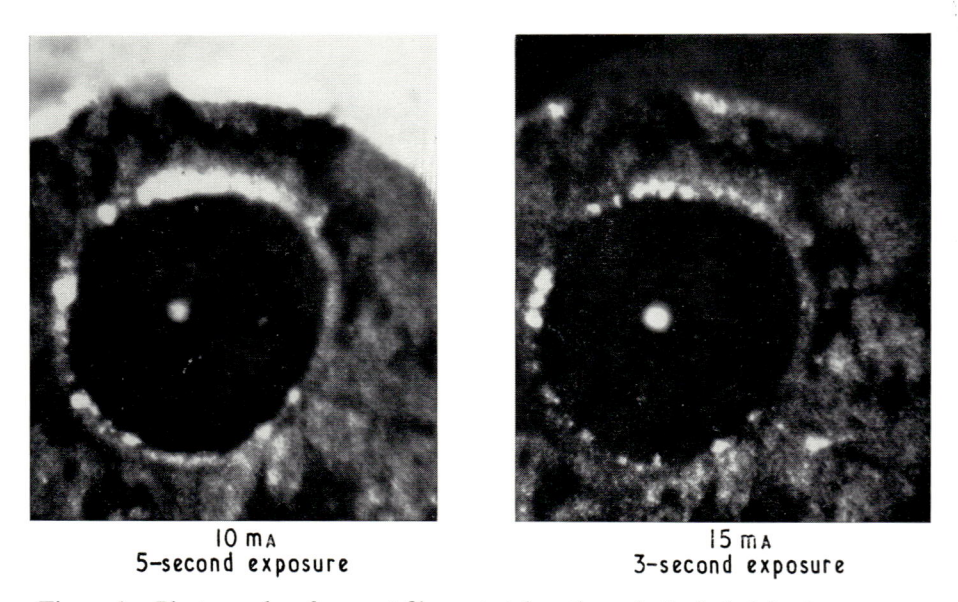

6 mA
20-second exposure

Contact aperture
0·050 in.

Apparent filament
0·003 in. diam.

10 mA
5-second exposure

15 mA
3-second exposure

Figure 4. Photographs of current filaments taken through the hole-injecting contact.

PLATE XIX—*A. M. Barnett et al.: paper* 22

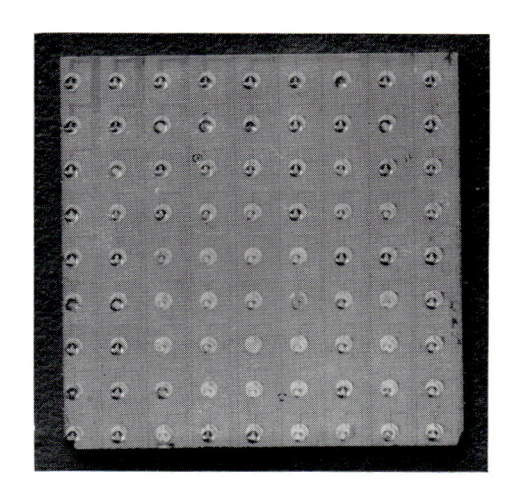

Figure 6. 9 × 9 element on 50 milli-inch centres integrated circuit, resistor—n$^+$ junction side.

Figure 7. Close-up of resistor—n$^+$ junction side.

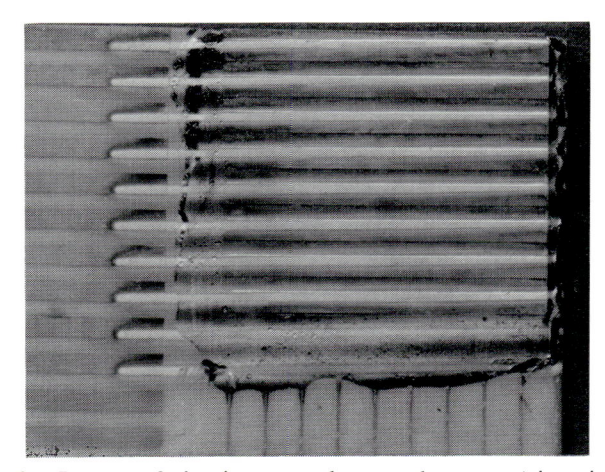

Figure 8. Integrated circuit mounted on a substrate, p$^+$ junction side.

written in Gaussian cgs units. However, distances and volumes are given in whatever units are most suitable.)

Figure 3 illustrates how the temperature, voltage, and free-electron concentration on the more lightly doped side of the junction determined the spectral shape. Equation (5) has been represented by two parts: the first part is $R_0 \exp(\alpha h\nu/e)$ which influences the spectral shape through the $(1/n)^{1/2}$ dependence of α, and the second part is I_R which contains the temperature and voltage dependence. In figure 3, $R(h\nu)$ has been calculated for $n = 1.74 \times 10^{18}$ cm^{-3}, $T = 77°$K, and $V_1 = 1.350$ v, and $V_2 = 1.380$ v. As the voltage goes from V_1 to V_2, the rapidly decreasing part of I_R shifts a corresponding amount and results in a change in position of the peak emission intensity by $(V_2 - V_1)$. The value of I_R remains constant at the energy $h\nu_p$ of the peak emission intensity $\mathscr{I}(h\nu_p)$, and $\mathscr{I}(h\nu_p)$ simply varies as $R_0 \exp(\alpha V)$. Therefore, α is equivalent to S of equation (1). In terms of the model represented in figure 3, the quantity $R_0 \exp(\alpha h\nu/e)$ represents the tunnelling probability, while I_R represents the quasi-Fermi level separation and the temperature distribution of carriers within each band. A change in temperature is then contained in the decreasing part of I_R.

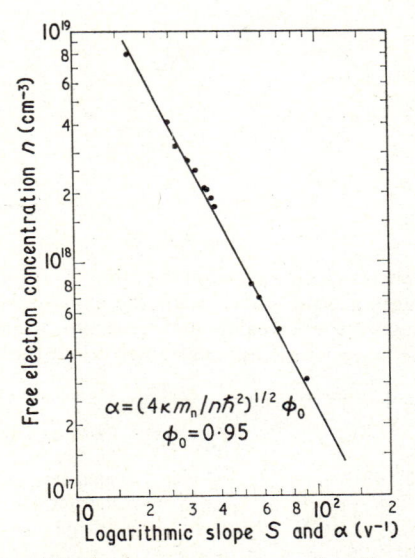

Figure 4. Variation of the logarithmic slope constant as a function of the free-electron concentration. The calculated values α from Morgan's theory are represented by the solid line, and α has been written in MKS units.

To compare the values of α given by equation (6) with the experimental logarithmic slope constant S, the shifting-peak spectra were measured for devices with n from 0.3 to 8.0×10^{18} cm^{-3}. The values of S for these devices are summarized in figure 4 by the solid circles. The line marked α is the logarithmic slope constant calculated from equation (6) with $m_n = 0.072\, m_0$ (Ehrenreich 1960) and $\kappa = 1.1 \times 10^{-12}$ coulomb^2/J-cm for $\kappa/\kappa_0 = 12.5$ (Hambleton et al. 1961). The calculated line is almost the best line that can be drawn through the experimental points. Values of the free-electron concentration contain an uncertainty from the assignment of the Hall-to-drift mobility ratio as unity; i.e., n was calculated from $1/(eR_H)$ where R_H is the Hall coefficient. Also, determination of S from the experimental data can have an uncertainty of perhaps $\pm 10\%$. Allowance for the maximum uncertainties still leads to extremely good agreement, and the $(1/n)^{1/2}$ dependence of the logarithmic slope constant predicted by equation (6) is certainly verified by the experimental data.

The spectra calculated from equation (5) for $n = 1.74 \times 10^{18}$ cm^{-3} and $T = 77°$K are shown in figure 5. The energy gap was taken as 1.515 ev (Sturge 1962). These spectra were calculated for the same applied voltages as given for the experimental spectra of figure 1 (for clarity some of the experimental spectra were not included in figure 1). For each voltage, the energy of the experimental peak emission $h\nu_p$ is shown by an arrow in order to demonstrate how well the variation of emission with voltage agrees with the experimental result. An overlay of figure 1 on figure 5 shows that the calculated and experimental spectra are virtually indistinguishable up to $V = 1.42$.

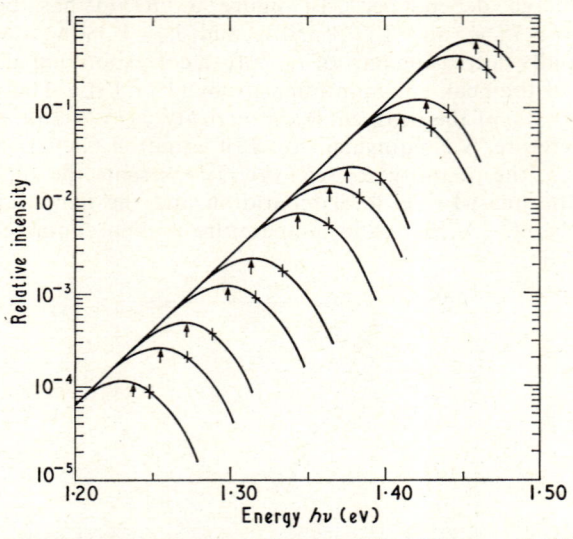

Figure 5. The shifting-peak spectra calculated from equation (5) for $n = 1.74 \times 10^{18}$ electrons cm^{-3} at 77°K, and the applied voltages are the same as for the experimental spectra in figure 1. The cross + gives the applied voltage V for the calculated spectrum, while the arrow ↑ designates the energy of the experimental peak intensity at the same V.

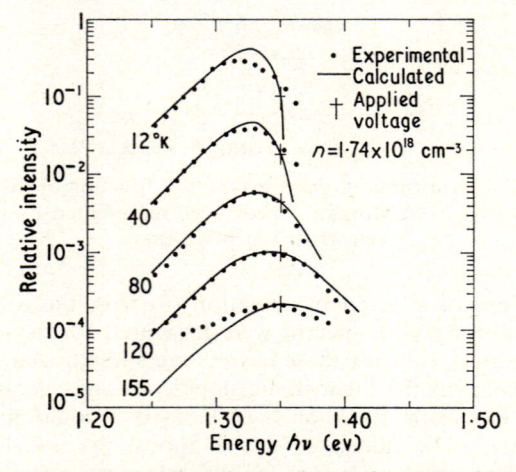

Figure 6. Comparison of the calculated and experimental spectra for a constant free-electron concentration and applied voltage, but variable temperature.

In figure 6 the experimental and calculated spectral shapes for a fixed voltage of 1·350 v and free-electron concentration of 1.74×10^{18} cm^{-3} are given as a function of temperature. At the lowest temperature, the low-energy emission agrees very well. However, at 12° and 40°K the experimental emission does not decrease as rapidly on the high-energy side of the peak as predicted by equation (5). For 80° and 120°K the experimental spectra are adequately described by equation (5). The 155°K experimental spectrum has additional emission in the low-energy tail. At lower bias the excess low energy emission is less and the experimental and calculated spectra agree more closely. At higher temperatures, lower biases are necessary in order to observe the shifting peak spectra. The comparison (not shown here) at 215° and 255°K is difficult because the experimental emission due to thermal injection becomes a significant portion of the total emission. It is clear, however, that the radiative tunnelling persists to high temperatures and is observed as a very broad low-energy tail on the band-edge emission.

The relation between the energy of the peak emission and the applied voltage, as represented by V_0 in equation (2), has often been considered in discussions of shifting-peak spectra. In figure 6 the variations of V_0 with temperature for the experimental and calculated spectra are seen to be in close agreement. These variations of V_0 result from the temperature dependence of the carrier distribution within the parabolic density of states. At higher temperatures more carriers are present at energies greater than the quasi-Fermi level separation, and thus the peak emission occurs at energies in excess of the applied voltage. In addition, V_0 at a given temperature depends on the free-electron concentration and is correctly predicted by equation (5). The dependence of V_0 on n is representative of the motion of E_{fn} into the band with increasing concentration.

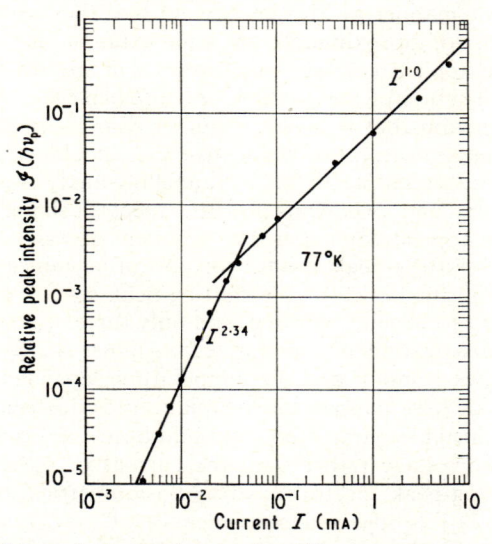

Figure 7. Dependence of the relative peak intensity $\mathscr{I}(h\nu_p)$ on the total junction current I for the diode shown in figure 1.

The 77°K emission spectra at 1·350 V for two diodes from the same wafer with identical areas were compared. It was found that the emission peak occurs at the same energy for each diode and has the same spectral shape. However, for one diode the current is 0·03 mA; for the other diode it is 1·1 mA. This comparison shows that current or current density is an irrelevant quantity when considering shifting-peak spectra and that the independent parameter in terms of excitation is the junction bias.

Another characteristic of the shifting-peak spectra is demonstrated in figure 7 where the peak intensity $\mathscr{I}(h\nu_p)$ is plotted as a function of current I. At currents between 4 and 40 μA, $\mathscr{I}(h\nu_p)$ varies as $I^{2\cdot34}$, so that in terms of equations (1) and (2), the current in this region may be written as

$$I = I_0 \exp(SV/2\cdot34) = I_0 \exp(16\cdot7\ V). \tag{7}$$

It should be emphasized that the approximately current-squared dependence was found in all cases only for very low currents, and results from non-radiative tunnelling through deep-lying recombination centres within the space-charge region (Morgan 1966). Morgan's theory predicts that $\mathscr{I}(h\nu_\mathrm{p})$ varies as $I^{2\cdot 20}$. The current exponent of 2·20 is in good agreement with the experimental value of 2·34, and thus the I–V behaviour expressed by equation (7) can be described by tunnelling to deep traps. For the more efficient devices at high current, $\mathscr{I}(h\nu_\mathrm{p})$ was linear with current as shown in figure 7, but no general dependence in the high-current region was obtained. For devices with $\mathscr{I}(h\nu_\mathrm{p}) \propto I^{1\cdot 0}$, the dominant injection mechanism is band-to-band radiative tunnelling and the internal quantum efficiency is 100%.

4. Summary and conclusions

In parts 1 and 2 of section 3, the detailed experimental and theoretical properties of shifting-peak spectra for abrupt-asymmetrical junctions have been presented. The unique step in Morgan's analysis (1966) is the use of a parabolic potential from the solution of Poisson's equation for both the space-charge region (the region between the point where the quasi-Fermi level crosses the conduction-band edge and the point where it crosses the valence-band edge) and the region of band bending on the more lightly, although degenerately doped n-side. Generally, the potential has been approximated by using a linear potential and therefore a constant field. Yunovich and Ormont (1966) calculated the spectral shape in the same manner as Morgan, except that they assumed a linear potential. The linear potential leads to approximately the same expressions for spectral shape except for a dependence on the applied voltage, which results in spectra that do not exhibit low energy tail saturation. The use of the potential as obtained from Poisson's equation leads to a potential energy function that is invariant under changes in the applied voltage, and therefore low-energy emission saturation results for the calculated spectra. The parabolic potential greatly increases the complexity of the tunnelling analysis with only a small change in the resulting prefactors and exponents for the spectral shape. However, this small change, the invariance to applied voltage, is the ingredient necessary for a theoretical model to describe correctly the shifting-peak spectra for abrupt-asymmetrical junctions.

Both experimentally and theoretically, the shifting-peak spectra for abrupt-asymmetrical junctions have been found to depend on three, and only three, quantities. The dependence on the free-electron concentration is summarized in figure 4, the voltage dependence is given by comparing figures 1 and 5, and the temperature dependence is shown in figure 6. Except for the very low or high temperatures, equation (5) describes in detail the shifting-peak spectra with no adjustable parameters. In addition, the external junction excitation has been shown to be the voltage rather than the current or current density as generally used. Because the shifting-peak spectra for abrupt-asymmetrical junctions have a narrow active region width $w \leqslant 600$ Å as observed experimentally (Casey et al. 1966), and the spectral properties are consistent with Morgan's analysis, strong justification exists for the designation of these spectra as radiative tunnelling.

Acknowledgments

The authors wish to acknowledge the many contributions made during this study. The emission spectra were measured by R. H. Kaiser. I. Hayashi provided the apparatus for the variable temperature measurements. J. R. Brews provided answers to many questions regarding radiative recombination. R. A. Logan and Mrs. H. V. Carlson made suggestions useful in preparing the manuscript. We would especially like to thank T. N. Morgan for discussions and correspondence on the interpretation of his tunnelling analysis. One of us (D.J.S.) would like to thank M. DiDomenico for the opportunity to work on this problem during a summer at Bell Telephone Laboratories.

References

ARCHER, R. J., LEITE, R. C. C., YARIV, A., PORTO, S. P. S., and WHELAN, J. M., 1963, *Phys. Rev. Lett.*, **10**, 483–5.
CASEY, H. C. JR., ARCHER, R. J., KAISER, R. H., and SARACE, J. C., 1966, *J. Appl. Phys.*, **37**, 893–8.
CASEY, H. C. JR., and M. B. PANISH, 1968, *Trans. Met. Soc. AIME*, **242**, 406–12.

Casey, H. C. Jr., and Silversmith, D. J., 1969, *J. Appl. Phys.*, **40**, 241–56.
Dumin, D. J., and Pearson, G. L., 1965, *J. Appl. Phys.*, **36**, 3418–26.
Ehrenreich, H., 1960, *Phys. Rev.*, **120**, 1951–63.
Hambleton, K. G., Hilsum, C., and Holeman, B. R., 1961, *Proc. Phys. Soc.* (Lond.), **77**, 1147–8.
Morgan, T. N., 1966, *Phys. Rev.*, **148**, 890–903.
Nelson, D. F., Gershenzon, M., Ashkin, A., D'Asaro, L. A., and Sarace, J. C., 1963, *Appl. Phys. Lett.*, **2**, 182–4.
Pankove, J. I., 1962, *Phys. Rev. Lett.*, **9**, 283–5.
Sah, C. T., Noyce, R. N., and Shockley, W., 1957, *Proc. IRE*, **45**, 1228–43.
Shockley, W., 1949, *Bell Sys. Tech. J.*, **28**, 435–89.
Sturge, M. D., 1962, *Phys. Rev.*, **127**, 768–73.
Yunovich, A. E., and Ormont, A. B., 1966, *JETP* (USSR), **51**, 1292–1305, translation: 1967, *Sov. Phys.–JETP*, **24**, 869–77.

CHAPTER 5

Microwave devices

1968 SYMP. ON GaAs   PAPER 24

Epitaxial GaAs Gunn effect oscillators: influence of material properties on device performance

L. COHEN, F. DRAGO, B. SHORTT, R. SOCCI and M. URBAN

The Bayside Laboratory, research centre of General Telephone and Electronics Laboratories Incorporated, Bayside, New York, U.S.A.

Abstract. This paper reports on a unified approach to cw Gunn device development which undertakes to relate performance characteristics such as power, efficiency, frequency, noise, and operating life with specific properties of the GaAs epitaxial material. The material was characterized by measurement of impurity profile and by microwave measurement of current density versus electric field in addition to the nominally specified values of carrier concentration, low-field mobility, and epitaxial thickness. Current density–electric field data taken for different regions of a single epitaxial n layer demonstrate that variations in device efficiency, operating life and current–voltage characteristics can be associated with material inhomogeneities exclusive of contact and circuit effects. The frequency dependence on bias voltage of cw Gunn oscillators is shown to be related to the impurity profile within the n-type epitaxial layer. The FM noise properties of these oscillators appear to be dependent on the degree of imperfection in the epitaxial gallium arsenide layer.

1. Introduction

It is well known that Gunn diodes fabricated from different epitaxial GaAs layers, as well as from a single epitaxial layer, can exhibit wide variations in device performance. The distributions of efficiency, frequency, and output power reported by Bott, *et al.* (1967) for epitaxial Gunn diodes are typical of the type of variations one can obtain. However, the different technologies used by various workers in the field to construct diodes, the varied methods of material fabrication, and the dependence of device performance on microwave circuit techniques make it difficult to relate performance variations to inherent differences within the epitaxial GaAs. This inability to separate cause and effect was the primary motivation for the work reported here.

The Gunn diodes discussed in this paper are sandwich type structures fabricated from n/n^+ epitaxial material. It is common practice when growing the n-type layer on the n^+ substrate to include a semi-insulating substrate in the reactor. The n growth on the semi-insulating substrate constitutes a control wafer on which Hall effect and resistivity measurements may be made. The values of Hall mobility, resistivity, and carrier concentration so obtained are then assumed likewise to characterize the n layer grown on the n^+ substrate. In the work reported here, we have further characterized the n-on-semi-insulating (n/SI) wafer by performing microwave measurements of current density versus electric field (J vs E) up to and beyond the Gunn threshold for different regions of the wafer. Measurements of the impurity profile for n/n^+ and n/SI wafers grown simultaneously have also been performed and it has been found that the two n layers have essentially the same type profiles. Finally, Gunn diodes were fabricated using the n/n^+ wafers in an attempt to relate the measured material parameters to device performance. As will be shown, the device's frequency versus bias performance, noise properties, efficiency, and operating life generally can be related to the impurity profile and the J vs E characteristics of the epitaxial material.

In some cases, however, some of the observed variations of frequency with bias could not be explained in terms of the measured material characteristics, indicating that more refined material characterization may be required.

153

2. Discussion
2.1. *Material*

Two wafers (2175 and 2218) showing different material characteristics were selected as test vehicles to correlate microwave device performance to material properties. All devices were constructed and tested in the same manner. The epitaxial layers used in the investigation were made at the Bayside Laboratory by vapour deposition in a gallium–arsenic trichloride system as described by Reid and Robinson (1968). An n/SI GaAs control wafer was grown simultaneously with each of the n/n+ wafers used for device fabrication. The Gunn diodes were constructed by the formation of ohmic contacts to the n and n+ surfaces by the vacuum deposition of a Au–Ge–In composition. The contacts were subsequently alloyed at a temperature of 425°C for thirty minutes in a forming gas (80% N_2–20% H_2) atmosphere. The ohmic contact to the n layer served as the cathode for the device in all cases.

Table 1 summarizes the low-field parameters for the epitaxial wafers selected for this discussion. The average values of these material parameters in table 1 give no indication that a particular epitaxial layer may be non-uniform in its properties, or would produce devices with significant differences in their performance. For this reason the material was further characterized by measurement of the impurity profile and microwave evaluation of J vs E.

Table 1. Material parameters of epitaxial wafer pairs

Wafer No.	Type	Substrate dopant	μ (cm^2 v^{-1} s^{-1})	n (cm^{-3})	ρ (Ω-cm)	t (microns)
2175	n/SI	Cr	6500	8×10^{14}	1·2	13
2175	n/n+	Si				18
2218	n/SI	Cr	7000	$1·9 \times 10^{14}$	4·7	14
2218	n/n+	Si				8

2.2. *J vs E characteristics*

The J vs E characteristic of the n-on-semi-insulating material was obtained by a high-power microwave technique similar to that reported by Breslau (1967). The measurements were made at 24 GHz and a magnetron was used to produce the electric fields required for characterization of the material up to and beyond the Gunn threshold. The n-on-semi-insulating measurement sample was in the form of a thin rectangular bar and was designed so that the impressed electric field in the sample was essentially uniform. The physical dimensions of all the samples were the same, a width of 0·28 mm, a length of 8·89 mm, and an overall thickness (epi-layer plus substrate) of 0·23 mm. The sample was centrally positioned across a reduced-height waveguide and was of such length that the required ohmic contact at each end of the sample extended beyond the broad walls of the waveguide. Thus, only the central portion of the sample was exposed to the microwave field and the measurement provided a direct measure of the quality of this central section of the sample independent of contact considerations.

Four samples from each of the n-on-semi-insulating wafers were measured. The current density, normalized to the low-field conductivity, σ_0, for the four samples from wafer No. 2218 is shown in figure 1. This figure shows that the peak normalized current density, threshold field, peak-to-valley current ratio, and behaviour above threshold for these samples is reasonably uniform. The average value of threshold field and peak-to-valley current ratio was 3250 v cm^{-1} and 1·2 respectively. The differential conductivity and the duration of its negative portion with respect to electric field was also found to be reasonably uniform for this wafer. As will be shown, devices fabricated from the 2218 n/n+ wafer exhibited a uniformity of performance consistent with the uniformity implied from the J vs E data.

The J vs E characteristics of the four samples taken from wafer 2175 are shown in figure 2. In comparison with the J vs E data on the previous samples (2218), these samples exhibited

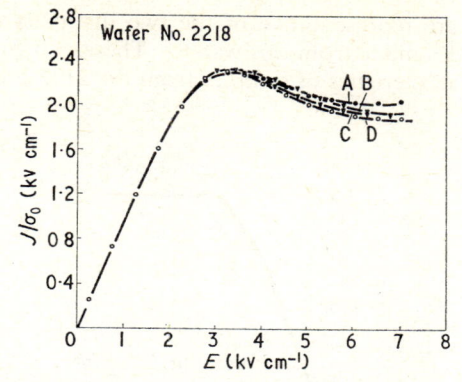

Figure 1. Normalized current density versus electric field characteristic measured at 24 GHz on four samples from a GT & E epitaxial n/i GaAs wafer. $n = 1.9 \times 10^{14}$ cm^{-3}; $\rho = 4.7$ ohm-cm; $\mu = 7000$ cm^2 v^{-1} s^{-1}.

Sample	E_{TH}	J_{peak}/J_{valley}
A	3350	1·13
B	3350	1·21
C	3150	1·21
D	3150	1·22

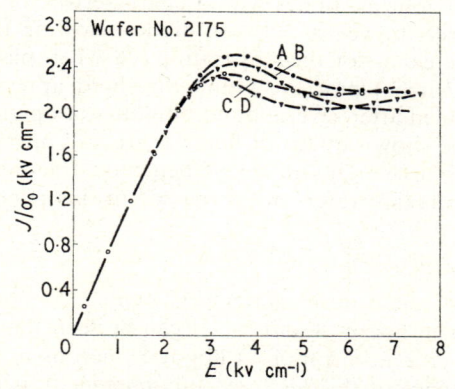

Figure 2. Normalized current density versus electric field characteristic measured at 24 GHz on four samples from a GT & E epitaxial n/i GaAs wafer. $n = 8.06 \times 10^{14}$ cm^{-3}; $\rho = 1.2$ ohm-cm; $\mu = 6500$ cm^2 v^{-1} s^{-1}.

Sample	E_{TH}	J_{peak}/J_{valley}
A	3500	1·16
B	3500	1·21
C	3300	1·08
D	3150	1·13

a lower average peak-to-valley current ratio, a larger spread in peak values of current density, and significant differences in behaviour above threshold. The non-uniformity of sample characteristics was also evident from the normalized differential conductivity versus electric field characteristics determined for this wafer. A region of positive differential conductivity for fields above 5 kv cm^{-1} was exhibited by samples C and D; this condition corresponds to the upswing observed in the J vs E characteristic for these samples. Gunn diodes made from the 2175 n/n$^+$ wafer were found to be less efficient and exhibited wider performance variations than those obtained with the relatively uniform 2218 wafer. The d.c. current–voltage curves for two diodes constructed from the n/n$^+$ 2175 wafer are shown in figure 3. These curves are representative of the two distinctly different types of characteristics obtained for diodes made from this wafer. These two curves are seen to be similar in shape to the J vs E characteristics of samples from the 2175 n-on-semi-insulating control wafer.

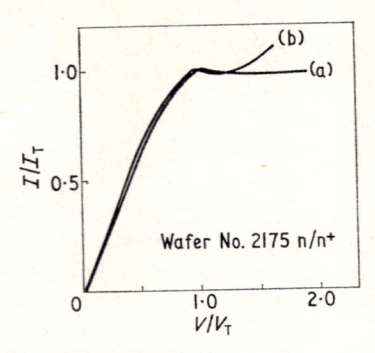

Figure 3. DC current–voltage characteristics for two Gunn diodes fabricated from the same wafer.

2.3. *Operational lifetime*

The diodes which exhibited an increase in current beyond the Gunn threshold as shown by curve (b) of figure 3 generally exhibited the same low-field resistance, output power, frequency, and noise behaviour as diodes with a characteristic of the type shown by curve (a). One distinct difference, however, between diodes showing the two types of current–voltage behaviour was the extent of their operating life when placed on life test. Diodes exhibiting an increase in current past the Gunn threshold appear prone to early failure. In life tests these diodes failed after several hundred hours of operation whereas diodes with a characteristic of the type shown by (a) of figure 3 are still operating after approximately 4000 hours of continuous test. This mode of behaviour has also been found in diodes constructed from other epitaxial wafers not discussed in this paper.

2.4. *Frequency vs bias characteristics*

The impurity profiles for the n on n⁺ wafers 2175 and 2218 obtained for sample regions of these wafers are shown in figures 4 and 5. It can be seen that wafer No. 2175 shows a localized decrease in carrier concentration at about 5 microns in from the surface of the n epitaxial layer. The presence of this localized inhomogeneity is particularly interesting in the light of the frequency vs bias characteristics for devices made from this wafer as shown in figure 6. The sudden transition in frequency from 11·8 GHz to 9·5 GHz for the pulsed

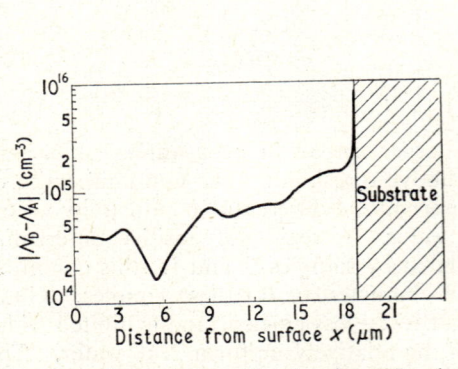

Figure 4. Impurity profile of wafer 2175 n/n⁺. 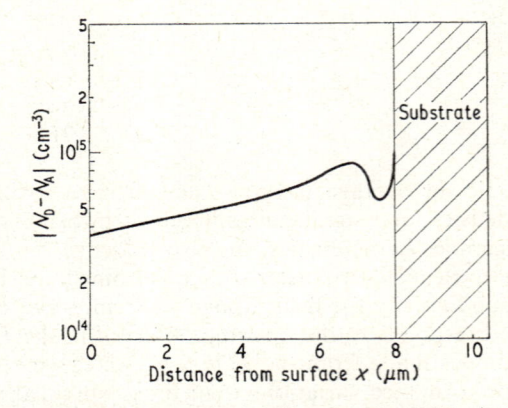   Figure 5. Impurity profile of wafer 2218 n/n⁺.

case, and from 8·8 GHz to 5·8 GHz for cw operation was observed to be typical for all diodes made from this wafer. The presence of this localized inhomogeneity is felt to be responsible for the frequency switching and, in fact, one is led to expect this type of behaviour by consideration of the results reported by Yamashita and Nii (1966) for inhomogeneous bulk samples. In contrast, the frequency vs bias behaviour for most of the diodes fabricated from wafer 2218 was found to be relatively independent of voltage with no transitions in frequency. This behaviour could be expected on the basis of the relatively flat impurity profile obtained from a sample region of this wafer.

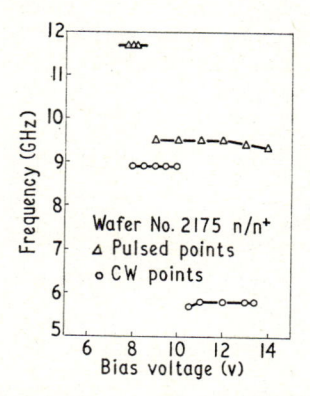

Figure 6. Typical cw and pulsed frequency vs bias for Gunn oscillators made from n/n+ wafer No. 2175.

The difference in frequency for the pulsed and cw modes of operations for these diodes as shown in figure 4 can be attributed to thermal effects. In the cw case elevated temperatures cause a decrease in carrier mobility and oscillation frequency. The diode frequency measured at a fixed pulsed voltage bias point was found to depend on temperature, as expected, and ultimately approached the cw value as the temperature of the diode was increased by means of an external heating element.

Although there is an apparent correlation between the frequency–bias voltage behaviour and impurity profile of the epitaxial layer, some diodes constructed from different regions of wafer 2218 exhibited the behaviour shown in figure 7. In this case, three distinct modes were observed with a sharp transition point between each one. For these diodes which exhibited multi-mode behaviour, no apparent differences in either threshold voltage, low-field resistance, or current–voltage characteristic were observed when compared to the devices which did not show this behaviour. The basis for this difference in performance is not clear. However, the consistency of our testing and fabrication methods and the observation of

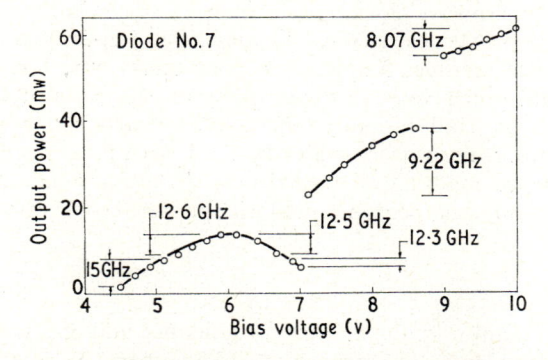

Figure 7. cw output power versus bias voltage for diode fabricated from wafer 2218.

this effect on both packaged and unpackaged diode chips lead us to believe that the effect is related to localized variations in material properties not identified by our characterization.

2.5. *FM Noise and efficiency*

The FM noise performance for diodes constructed from the two n/n^+ epitaxial wafers was measured at the frequency and bias point at which maximum output power was observed. Figure 8 shows typical data for diodes constructed from wafer 2218. Diodes constructed from wafer 2175 consistently exhibited higher noise characteristics. At 5 kHz off the carrier the r.m.s. frequency deviation was typically twice that exhibited by wafer 2218. The lower noise properties of diodes fabricated from wafer 2218 appear to reflect the more uniform impurity profile of this particular wafer.

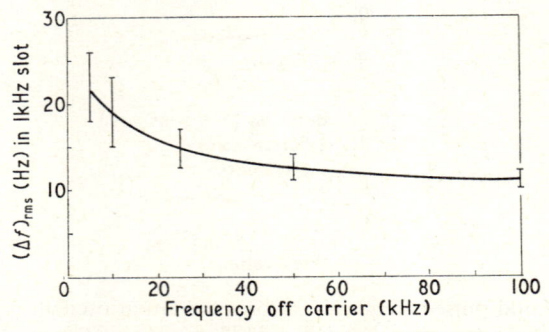

Figure 8. Typical r.m.s. frequency deviation measured in a 1 KHz slot as a function of frequency separation from carrier for Gunn diodes fabricated from wafer No. 2218.

It was also observed that the d.c. to r.f. efficiencies for diodes constructed from wafer 2218 were typically higher than those obtained from diodes fabricated from wafer 2175. Specifically, diodes constructed from the 2218 wafer exhibited an average efficiency of 1·45% for cw operation while those constructed from the 2175 wafer exhibited an average efficiency of 1·1%. In addition, diodes from the 2218 wafer could be interchanged in a fixed tuned circuit to give comparable optimum microwave performance. In contrast, diodes from wafer 2175 required individual circuit tuning to obtain optimum performance. The superior performance of the diodes constructed from the 2218 wafer relative to those fabricated from the 2175 sample are attributed to the greater uniformity and homogeneity of the epitaxial layer.

3. Conclusions

Two pairs of epitaxial wafers 2175 and 2218 have been examined in terms of their material parameters and their microwave device performance. Correlations were shown to exist between such material properties as the epitaxial layer impurity profile and the J vs E characteristic and device lifetime, frequency vs bias behaviour, FM noise and efficiency. Such correlations could not have been made solely on the basis of the standard material characterizations such as Hall mobility and average carrier concentration. However, certain diodes fabricated from n on n^+ wafer No. 2218 exhibited results not anticipated on the basis of the measured material characteristics. These anomalous results are thought not to be circuit effects but may be due to material properties not revealed by our measurements.

Acknowledgment

The authors wish to thank Mr. D. A. Fleri for guidance and encouragement, C. R. Smith for providing the life test data, J. Figueroa, A. Guthenberg, V. Lanzisera, and F. Monte for technical assistance.

References
BOTT, I. B., HILSUM, C., and SMITH, K. C. H., 1967, *Solid State Electronics*, **10**, 137–44.
BRESLAU, N., 1967, *Phys. Lett.*, **24A**, 531–3.
REID, F. J., and ROBINSON, L. B., 1968, *Preparation of Epitaxial Gallium Arsenide for Microwave Applications.* This Conference, paper No. 10.
YAMASHITA, A., and NII, R., 1966, *IEEE Trans. on Electron Devices*, ED-**13**, 196–7.

Gallium arsenide p–i–n diode as a microwave device

G. R. ANTELL

Standard Telecommunication Laboratories Ltd, Harlow, Essex, England

Abstract. Low capacitance switching diodes have been made in gallium arsenide. Many of the properties of the diodes depend on the type of GaAs used, e.g., bulk, vapour, or liquid epitaxial material.

Measurements have been made mainly on diodes made from epitaxial GaAs. Reverse breakdown voltages up to 350 v have been measured. The diodes have shown insertion losses of 0·15 dB at 4 GHz with isolations of 24–26 dB when measured in coax or in microstrip line.

The diodes can detect 4 GHz microwave energy with a tangential sensitivity of −22 dB, and this restricts their use as power switching diodes to frequencies above about 6 GHz. A relaxation effect has been found in the junction capacitance with a time constant of about 0·1 ms.

1. Introduction

P–I–N diodes have been made in silicon for a number of years now and have found uses as attenuators and phase shifters in microwave circuits. Silicon, by virtue of its large minority carrier lifetimes can have intrinsic regions that are several mils thick. This enables devices to be made which are capable of handling high power levels. The minority carrier lifetimes in GaAs on the other hand restrict the maximum intrinsic region to the order of about 20 μm. Thus GaAs diodes can be expected to be used to handle medium powers and the low minority lifetime should make the diode faster than an equivalent silicon device. Since there is always a materials aspect to any device made from GaAs this is covered in the first part of the paper.

2. Preparation of diodes

P–I–N diodes have been produced by coating an n-type slice of GaAs with either a pure or a zinc-doped silica film and diffusing the slice at 900°C for from 5 to 16 hours. The diffusion is carried out in either an open tube under a flow of wet hydrogen or in a sealed synthetic quartz capsule. The basic material requirements to produce p–i–n diodes are that the electron concentration should be in the region of 10^{16} cm^{-3} and the mobility should be in excess of 5000 cm^2 v^{-1} s^{-1}. The GaAs material can be pulled or boat-grown crystals, or epitaxial layers grown by the halide process. GaAs which has been grown by liquid epitaxy has not as yet given p–i–n diodes when diffused in the presence of water vapour. It is probable that gallium vacancies which are suppressed in growth by liquid epitaxy play some role in the production of p–i–n diodes. There is of course as yet no direct evidence that oxygen is responsible for the appearance of p–i–n diodes in GaAs but the indirect evidence has already been given (Antell 1968). There is some evidence to suggest that a silica film on the surface of the slice is necessary for the production of p–i–n diodes, possibly by ensuring that oxygen enters the GaAs. If there is no silica film present then oxygen will react with the surface and may volatilize as oxides of arsenic and gallium without entering the lattice. After diffusion the SiO$_2$ film is removed, contacts are alloyed in and mesas about 30–40 μm deep are etched out.

3. Characteristics of the diodes

3.1. Current–voltage characteristics

Diodes having mesa diameters of 2 mm or more and having reverse breakdown voltages of 200–250 v have been produced in bulk GaAs. In diodes made from vapour epitaxy the area of the junction has to be reduced to about 5×10^{-4} cm^2 to give a reasonable yield of diodes that have breakdown voltages in excess of 250 v.

The forward currents of the diodes are given by:

$$I = I_0 \left(\exp \frac{qV}{nkT} - 1 \right) \quad (1)$$

where n has a value of about 2·1. This relationship is followed over six orders of magnitude of current for diodes made from epitaxial material.

The diodes that have been produced using epitaxial GaAs have lower forward resistances than those made in bulk material. The shorter lifetime of minority carriers in bulk GaAs may contribute to this result but also the n$^+$ region in the epitaxial diode gives lower series resistance and higher injection efficiency than the more lightly doped region in bulk material.

A snap back in the reverse I–V characteristic occurs frequently in diodes produced in bulk GaAs, figure 1, but very rarely in epitaxial material and then only if there is no zinc in the SiO$_2$ film.

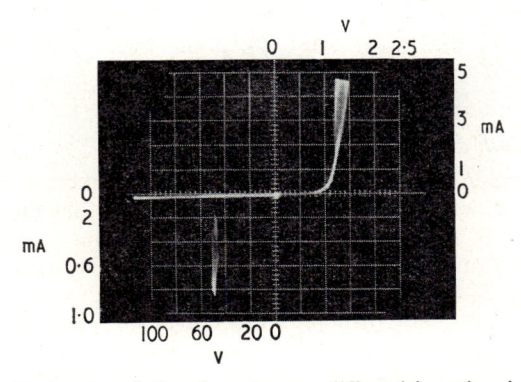

Figure 1. I–V characteristic of an oxygen-diffused junction in bulk GaAs.

Two distinct types of diode can be made in what appears to be nominally identical epitaxial GaAs. In one case diodes having reverse characteristics with a well-defined breakdown in the region of 200–350 v are produced, whereas in the other a looped reverse characteristic with a breakdown of about 20 v is obtained. These diodes show a delayed breakdown effect under pulsed conditions (Antell and White 1966) and there is little difference between the reverse breakdown voltages of both the hard and soft diodes when measured by pulses that are a few microseconds wide. That this is a genuine materials effect has been confirmed by simultaneously processing two epitaxial slices, whose electrical characteristics were about identical and which showed no significant differences when examined by x-rays and by the scanning electron microscope but which yet gave two entirely different types of diode.

3.2. Activation energies

The activation energy of generation centres in the depletion layers of diodes produced in bulk, vapour, and liquid epitaxial GaAs have been obtained from the variation with temperature of the reverse leakage current of the diodes (Sah, Noyce and Shockley 1957). The activation energies fall in groups according to the origin of the GaAs. Typical results are shown in figure 2 where it can be seen that in diodes made from bulk GaAs the generation centres are close to the middle of the energy gap. No difference was found between the

G

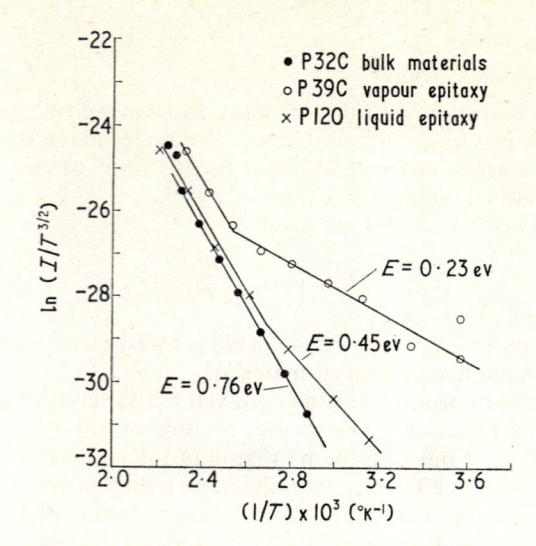

Figure 2. Activation energies of centres in depletion layer of oxygen-diffused junctions.

activation energies of hard and soft diodes which had been made from vapour grown epitaxial GaAs.

3.3. *Depletion layer width determinations*

Measurements on oxygen-diffused diodes show that the junction capacitance varies little with reverse voltages. It is this effect which has suggested that these diodes have a p–i–n type of structure. This is certainly the case with diodes that have been made from bulk GaAs but measurements on diodes from epitaxial material suggest that the diodes have a $p^+n^-n^+$ type of structure.

The capacitance of diodes made from epitaxial GaAs when measured at 1 MHz decreases as the temperature is lowered. The junction capacitance at room temperature also decreases as the measuring frequency increases and becomes constant above about 100 MHz. These two values of capacitance are about equal. These measurements support the idea that at low temperature the resistance of the n^- region becomes so great that it is by-passed by the reactance of the junction between the p^+ and n^+ regions. At very high frequencies, at room temperature, the resistance of the n^- region again becomes greater than the reactance of the shunt capacitance and the measured capacitance is approximately that between the n^+ and p^+ region. A more direct measurement of depletion layer width is given by the scanning electron microscope. Measurements are made on samples which have been cleaved perpendicular to the junction plane and contacts are made to the n^+ and p^+ regions. The magnitude of the beam-induced current that flows in the external circuit depends on the magnitude of the field in the junction and the lifetime of the electron-hole pairs produced by the beam.

Figure 3 shows the variation of the beam-induced current, recorded on an *X–Y* plotter as the electron beam passes across the junction in a bulk sample of GaAs. If the recombination rate of the carriers produced in the depletion layer is assumed to be constant then the field is constant. This result is what would be expected for a p–i–n diode. The depletion layer width as given by figure 3 agrees with that derived from junction capacitance measurements.

A similar trace through a junction in epitaxial material is shown in figure 4. Here the field distribution is more like that observed in p–n junctions.

Figure 5 shows the beam-induced currents in a junction in an epitaxial layer 30 μm thick. Although there is an apparently large variation in the depletion layer width the junction

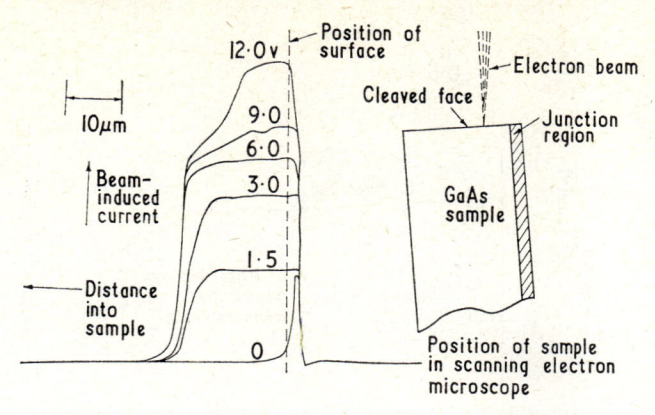

Figure 3. Scanning electron microscope beam-induced trace of oxygen-diffused junction in bulk GaAs.

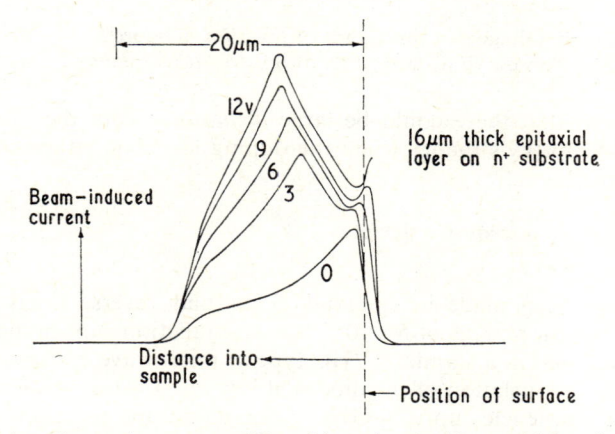

Figure 4. Beam-induced trace in oxygen-diffused junction in 16 μm thick epitaxial layer.

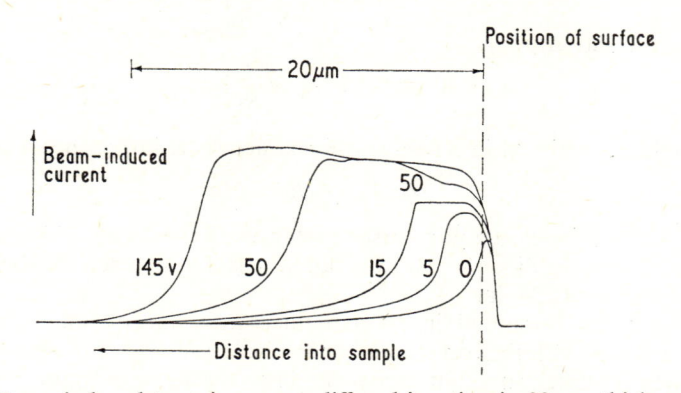

Figure 5. Beam-induced trace in oxygen-diffused junction in 30 μm thick epitaxial layer.

capacity changed little with increasing reverse voltage. Figure 6 shows the room-temperature capacitance of the junction in figure 5 as a function of reverse voltage as measured at 1 MHz together with the capacitance that has been computed from the depletion layer widths derived from figure 5. It can be seen that the capacitance variation with voltage as seen by the scanning electron microscope is that which would be expected for an abrupt type of junction whereas the measured capacitance changes much less rapidly with voltage.

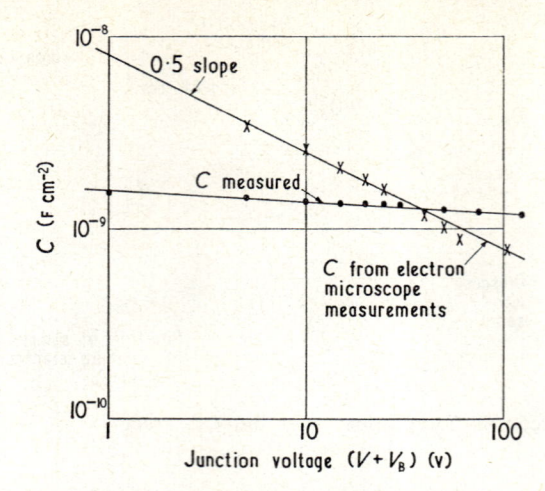

Figure 6. Variation with voltage of capacitance of junction in figure 5. Lower curve is measured. Top curve is derived from S.E.M. measurements.

Chang (1966) showed that there could be large deviations from the junction capacitance as calculated from Schottky's theory when the doping levels on either side of the junction were very asymmetric.

4. Behaviour of diodes as microwave devices

4.1. *Diode as a low-level switch*

Diodes which had been made in epitaxial GaAs had reverse breakdown voltages of from 200 to 300 v, junction areas of 5×10^{-4} cm^2 and junction capacitance of about 0·3 pF. The diodes were mounted in a standard VHC type of microwave encapsulation.

The insertion loss and isolation of the diodes at low microwave powers were measured in a coaxial system at frequencies up to 4 GHz. The minimum insertion loss is given by:

$$L_1 = 20 \log_{10} \left(1 + \frac{1}{2}\sqrt{\frac{R_1}{R_2}}\right) \text{ dB} \qquad (2)$$

and the maximum isolation by:

$$L_2 = 20 \log_{10} \left(1 + \frac{1}{2}\sqrt{\frac{R_2}{R_1}}\right) \text{ dB} \qquad (3)$$

These values would be achieved in a line whose characteristic impedance is given by

$$Z_0 = \sqrt{(R_1 R_2)} \qquad (4)$$

where R_1 and R_2 are the forward and reverse resistances of the diode. The insertion loss at 4 GHz with zero bias on the diode was 0·15 dB and the isolation when the diode was passing 50 mA forward current was 24 dB.

Diode chips were also mounted in 50 ohm microstrip line on an alumina substrate. When mounted in series with the top conductor the isolation was only about 14 dB without matching but when mounted in shunt across the line, as shown in figure 7 an isolation of 26 dB was measured.

4.2. *Diode as a detector*

When there was zero bias on the diode, a microwave power level of 10–20 mw at 4 GHz caused a d.c. voltage to appear across the diode. If a forward current of about one microamp was passed through the diode it acted as a detector with a tangential sensitivity of −22 dBm. When it is realized that the apparent depletion layer width is about 16 μm this result shows that the structure of the diode is not that of a simple p–i–n diode.

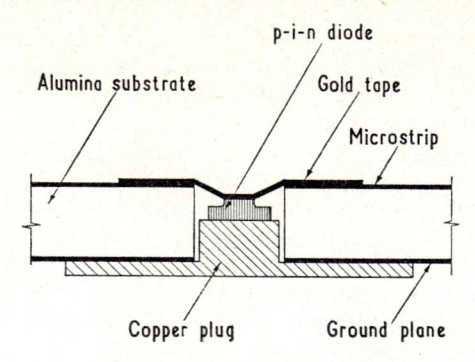

Figure 7. Section of diode mounted in microstrip.

This ability of the diode to rectify microwave power at 4 GHz limits its usefulness as a phase shifter. For example, if the diode is biased at a current of 50 mA it will effectively reflect low levels of microwave power but as the power level rises there will come a time when the microwave power can switch the diode current off and thus in part of the cycle microwave power is reflected by the diode in a low impedance state and part when in a high impedance state. Under these conditions VSWR measurements are not very meaningful. A diode was set up to terminate a 50-ohm line and was biased at a forward current of 50 mA. Measurement of the VSWR at low power levels showed that the resistance of the diode was 1·7 ohms. The VSWR was then measured when power levels of 25 mw and 1 w respectively were applied to the diode. The data obtained are shown in table 1.

Table 1. Variation in VSWR obtained from a forward biased p–i–n diode

Microwave power	VSWR at frequency noted		
	1 GHz	2·8 GHz	6 GHz
25 mw	19	19	29
1 w	9·5	10	32

It can be seen from the table that at frequencies of 1 and 2·8 GHz the VSWR decreases as the power level increases. At 6 GHz the diode fails to rectify the microwaves and the VSWR is relatively independent of the incident power level and agrees reasonably well with the value of VSWR measured at low power levels.

Thus the diodes are useful as switches only at frequencies in the region of 6 GHz upwards.

4.3. *Diode as a fast switch*

It had been shown that a diode passing a forward current of 50 mA could be switched into a non-conducting state in about 3–5 ns by applying a reverse biasing pulse.

To measure the performance of the diode as a microwave switch it was mounted in waveguide in such a manner that it connected the cross-bar member of the waveguide to coaxial transition to the centre conductor of the coaxial cable. Fast, balanced detectors mounted at the end of an attenuating section of waveguide were used to measure the microwave power. The waveguide acted as a high-pass filter which prevented the pulses that were applied to the diode reaching the detectors.

The diode was set up to pass a steady current of 50 mA and the parasitics of the diode were tuned out by means of a wave trap in the coaxial cable behind the diode. Pulses, 20 ns wide were applied to the diode through the coaxial cable and the microwave power at a frequency of 4 GHz was switched off in less than 5 ns.

When longer pulses were applied to the diode it was found that there was a relaxation effect present with a time constant in the region of 0·2 ms. This effect is mainly a change in capacitance of the junction with time after the diode has been switched off. Since the parasitics of the diode have been tuned out for a particular value of junction capacitance this relaxation effect detunes the diode during part of the time that it is switched off. That the effect is mainly a change in capacitance is shown by the fact that the diode can be tuned up to give a high isolation at any point during the reverse pulse. This relaxation effect is shown in figure 8 which is made up of three superimposed traces. The top trace is the zero power reference level with power increasing downwards. The third trace at the top is a measure of the microwave power that leaks past the diode when it is permanently switched off. The centre trace is obtained when the diode is switched off by a long pulse and shows the relaxation effect in which the isolation at the front end of the pulse is much higher than at the rear. This effect is considered to be due to a change in the capacitance as deep trapping levels empty when the junction is reverse biased. The carriers coming out of these deep levels give rise to a resistive term in the diode impedance and this cannot be tuned out.

The diode can be tuned up to give good isolation at either the front or back end of the pulse or an intermediate value of capacitance can be tuned out so that all the three traces in figure 8 can be merged into a single trace.

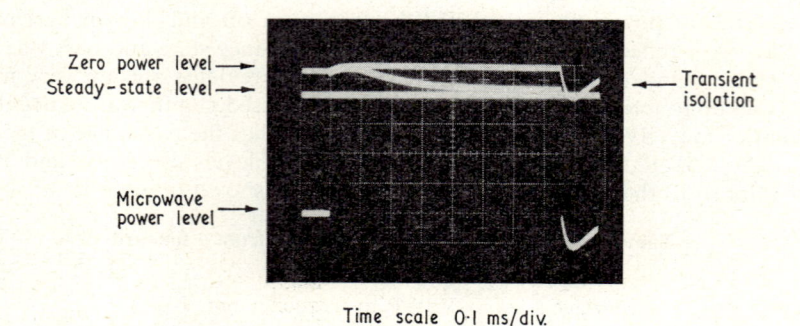

Figure 8. Demonstration of relaxation effect in reverse biased junctions.

5. Conclusion

It has been shown that a GaAs p–i–n diode can perform at microwave frequencies but that its ability to rectify microwave power up to about 4 GHz limits its use as a phase-shifting element to frequencies of 6 GHz upwards.

The relaxation effect in the junction capacitance during reverse switching restricts the use of the diode to fast switching circuits when it would not remain in any one state for more than a few microseconds. If good isolation is required together with fast switching with the diodes being reverse biased for a considerable time when two diodes would be required. One would be tuned to give high isolation at the front end of the pulse and the second tuned to give good isolation at the rear end of the pulse.

Acknowledgments

The author would like to thank M. Driver for his work on the scanning electron microscope and S. Allen for some of the microwave measurements and for useful discussions.

This work was done under a CVD contract and is published by permission of The Ministry of Defence (Navy Department).

References

ANTELL, G. R., 1968, *Brit. J. Appl. Phys.* (*J. Phys.* D) ser. 2, **1**, 113–4.
ANTELL, G. R., and WHITE, A. P., 1966, *Proc. Int. Symp. on GaAs.* Inst. Phys. and Phys. Soc. Conf. ser. No. 3, pp. 201–5.
CHANG, Y. F., 1966, *J. Appl. Phys.*, **37**, 2337.
SAH, S., NOYCE, R. N., and SHOCKLEY, W., 1957, *Proc. IRE.*, **45**, 1228–43.

Bulk GaAs travelling-wave amplifier

J. KOYAMA, S. OHARA, S. KAWAZURA, and K. KUMABE

Electrical Communication Laboratory, Nippon Telegraph and Telephone Public Corporation, Musashino-shi, Tokyo, Japan

Abstract. The possibility of suppressing the domain formation in n-type GaAs diodes with large nL product to make the diodes suitable for d.c.-stable microwave amplifiers was explored analytically and experimentally. Experiments show that decreasing the transverse thickness T of a planar-type GaAs diode and attaching high dielectric material to it tend to suppress the domain mode even if the nL product of the diode is larger than the critical value. We show analytically that in thin planar diodes the oscillation criterion is determined by nT product as $\epsilon_1 nT/\epsilon_2 > 1\cdot3 \times 10^{11}$ cm^{-2}, where n is the carrier density, ϵ_1 and ϵ_2 are the permittivity of diode and outside medium respectively. In stabilized bulk GaAs biasing above the threshold field, a stable excitation of travelling-space charge wave mode is possible. With this mode we made an experiment on a bulk GaAs travelling-wave amplifier. Typically, samples 0·5 mm long, 2 mm wide and 100 μm thick, with donor densities in the range of 9×10^{12} to 8×10^{13} cm^{-3}, mobility 5000 cm^2 v^{-1} s^{-1}, were prepared with evaporated Au–Sn ohmic cathode and anode contacts. The contacts were connected with striplines of external circuits directly and the specimen was biased using a 1 μs long pulse with a repetition rate of 100 ps^{-1}. In the course of this work we have been able so far to obtain 17 dB net gain with 3 dBm saturation power at the frequency of 1200 MHz, and to show that the device is unilateral. The gain of reverse direction was less than -25 dB.

1. Introduction

Microwave amplification in n-type bulk GaAs has been observed by Thim *et al.* (1966). This effect is due to the negative conductance which appears when an electric field in the GaAs diode exceeds several thousand volts per cm. Most of the experiments, however, were on reflection-type amplification, in which a signal field was applied uniformly between the electrodes of the diode and the amplified reflection wave was obtained.

As the impedance of a diode becomes negative resistance only at the frequency which is determined by the transit time of the carrier in the diode, the bandwidth of reflection-type amplifiers is essentially narrow. Moreover, this type of amplifier is apt to be unstable if there is slight mismatching in circuits. The negative resistance in GaAs, however, is caused by a bulk effect, so that a distributed amplification is possible. In order to confirm the possibility, we carried out a small signal analysis in the bulk of negative conductance and found a growing space charge wave mode which travels with the same velocity of the carrier flow.

Using this mode we made an experiment on travelling-wave amplification with a GaAs diode in which a signal is injected at the cathode to excite the space charge wave. This wave grows along the length of diode and an amplified signal is obtained at the anode. In order to achieve a high gain travelling-wave amplifier, it is obvious that the GaAs diode must have a sufficient length and the current density should be large. These conditions, however, make the nL product of the diode so large that the generation of Gunn oscillation in it is inevitable. To suppress the onset of instabilities in the diode of large nL product, we have found two effective methods. The first is to reduce the transverse thickness T of a planar type GaAs diode. The second method is to enclose the diode within a material of high dielectric constant such as barium titanate.

The effect of these measures is a reduction of the strength of domain field by leaking the field out of the bulk. We analysed these phenomena theoretically and concluded that when the thickness T of the planar type Gunn diode is very thin, the criterion for the onset of Gunn oscillation is determined by nT product rather than nL product. Utilizing this effect, we made d.c.-stable diodes and succeeded in obtaining net gain in travelling-wave type amplification.

Independently of our experiments, Robson, Kino, and Fay (1967) at Stanford University have succeeded in obtaining the same type of amplification.

2. Growing wave in bulk GaAs

2.1. *Propagation constant of travelling space charge wave*

We consider a thin planar structure of n-type GaAs, in which the x dimension is infinite and the electron current flows in the z direction. The behaviour of carriers in bulk is described by the phenomenological equation of motion together with the Maxwell equations. The mobility μ of the electron in a GaAs bulk changes considerably with the field strength E. To take account of the effect we express μ as follows:

$$\mu = \mu_0 + \frac{\partial v}{\partial E} E \qquad (1)$$

where μ_0 is the constant mobility.

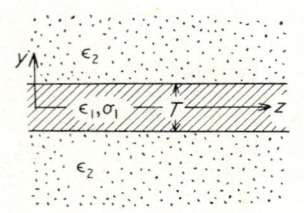

Figure 1. Planar structure of the semiconductor.

Linearizing the equations with respect to the a.c. parts, which have a factor $\exp[j\omega t - \Gamma z]$ and considering the boundary condition illustrated in figure 1, we obtain a propagation constant of travelling space charge wave, that is

$$\Gamma = j\frac{\omega}{v_0} + \frac{\alpha(\omega_d/v_0)}{1 + \epsilon_2 v_0/\epsilon_1 \omega T} \qquad (2)$$

where v_0 is average velocity of carrier flow, ω_d is dielectric relaxation frequency and $\alpha = \mu/\mu_0$. In deriving equation (2) we make the following assumptions:

(1) The phase velocity of the space charge wave is nearly equal to the carrier velocity v_0 which is about 10^7 cm s^{-1}.

(2) The thickness of the planar bulk is small compared to the wavelength in it.

(3) The diffusion effect is negligible.

The value of α in equation (2) becomes negative when the field strength in bulk exceeds the threshold, which is about $3\cdot 5$ kv cm^{-1}, and this effect gives rise to a growing wave mode. In the case of n-type GaAs with carrier density 10^{13} cm^{-3} and mobility 6000 cm^2 v^{-1} s^{-1}, using the data of velocity-field curve from Butcher and Fawcett which gives the value of α as $-0\cdot 3$ at the steepest slope in the curve, the magnitude of gain is about 200 dB mm^{-1}.

2.2. *Suppression of Gunn oscillation by two dimensional effect*

When a thin planar Gunn diode is operated under constant-voltage conditions, a small fluctuation may grow spontaneously into large-signal instability or Gunn oscillation. The criterion for the onset of the instability is determined by zero impedance of the diode. According to the analysis of McCumber and Chynoweth (1966) the impedance becomes zero when $\exp(-\Gamma L) + \Gamma L - 1 = 0$, where L is the length of the diode. This transcendental equation has a denumerable number of zeros. The root corresponding to the Gunn mode is written as $\Gamma L = \xi_1 \pm j\eta_1$, which is a quadratic equation with respect to the frequency ω. If the imaginary part of the root of this equation has negative sign, the current instability grows with time. The condition of the onset of instabilities is derived as

$$\frac{d}{dE} \ln v(E) < -\frac{(1+2\epsilon_2 L/\eta_1 \epsilon_1 T)}{4\pi e n L} \xi_1 \epsilon_1 \tag{3}$$

When the thickness of the planar diode is small compared to its length, the inequality (3) becomes

$$\frac{\epsilon_1}{\epsilon_2} nT > 1\cdot 3 \times 10^{11} \text{ cm}^{-2} \tag{4}$$

This criterion indicates that if we make the thickness of Gunn diode sufficiently small and enclose it within a material of high dielectric constant, the generation of instabilities can be suppressed even if the nL product of the diode is larger than the critical value of 5×10^{11} cm^{-2}.

3. Experimental

3.1. Sample preparation

Single crystals of n-type GaAs used in the experiments were Monsanto oxygen-doped material. The donor density was in the range 9×10^{12} to 6×10^{13} cm^{-3}, mobility 6000 cm^2 v^{-1} s^{-1}. Typically, samples 0·5 mm long, 2 mm wide, and 0·1 mm thick were prepared with evaporated Au–Sn ohmic cathode and anode contacts on the same side of the sample. Holding the diode on a plane surface, we lapped the other side of it with carborundum until instabilities or oscillations no longer appeared at any biasing voltage. As a matter of fact, the critical thickness T_0 below which the diode becomes d.c.-stable is not always as small as indicated in the criterion (4). This is partially due to the damage layer which is produced by lapping the surface of the sample, and this layer decreases the effective thickness of the active region in a planar diode.

Table 1. Typical characteristics of samples

Sample	Length (mm)	Width (mm)	Thickness (μm)	n (cm^{-3})	nL (cm^{-2})	nT (cm^{-2})
No. 1	0·6	1·0	50	9×10^{12}	$5\cdot 4 \times 10^{11}$	$4\cdot 5 \times 10^{10}$
No. 2	0·6	2·0	100	1×10^{13}	$6\cdot 0 \times 10^{11}$	$1\cdot 0 \times 10^{11}$
No. 3	0·6	1·0	20	6×10^{13}	$3\cdot 6 \times 10^{12}$	$1\cdot 2 \times 10^{11}$

The dielectric constant ϵ_2 of the surrounding medium is another factor which affects the suppression of instabilities. Sheets of barium titanate of various dielectric constants were pressed tightly to both surfaces of the sample using mechanical pressure. This proved to be effective in suppressing the onset of instabilities to a certain extent, but not so efficient as expected from criterion (4). The effective value of ϵ_2, however, depends not only on the outer medium but also on the inactive layer of the surface in the planar diode; therefore the surrounding material of high dielectric constant is considered to have little influence on the suppression of instabilities. For this reason we did not use high dielectric material in the experiments on amplification.

3.2. Coupling circuit

For coupling the signal wave into a diode various types of couplers were tried. For instance, a slow-wave circuit of meander lines was made by photo-etching and combined with the diode, but the coupling was proved to be very weak and no gain was obtained. The most efficient coupler was the ohmic contact of the diode In figure 2 the detailed structure of the coupler is illustrated. The signal is injected to a diode through a stripline which is connected directly with the ohmic contacts of the sample. In this construction, the ohmic contacts of the diode are used in common for signal couplers and d.c. biasing electrodes. In figure 2, the ground plate which is close to the sample is aluminium foil and was insulated from the sample by a polyethylene sheet 20 μm thick.

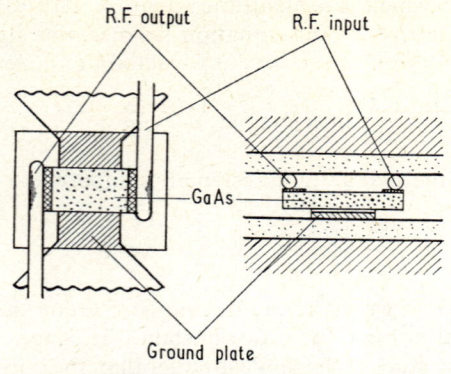

Figure 2. Structure of the coupler.

The function of this plate is firstly to focus the signal field around the coupler and secondly to prevent the direct coupling of the signal between the input and output electrode. These effects of the plate become remarkable with decreasing thickness of the planar diode. If this contact plate is removed from the sample, the signal cannot be injected into the diode through the ohmic contact coupler.

The diode was biased using pulses several micro-seconds long with a repetition rate of 100 ps^{-1}. The signal frequencies used in the experiments were 1 to 2 GHz. In order to prevent the signal power from leaking toward the pulse source, the length of the pulse transmission line was set at $\lambda/4$ for the frequency of 1·2 GHz.

4. Experimental results

4.1. *Current–voltage characteristics*

Figure 3 is the typical current–voltage relationship of a d.c.-stabilized diode. The current increases at first linearly with the applied voltage. Above V_0, which corresponds to the threshold field of an ordinary Gunn diode, the curve bends slightly downward, but the linear relationship between the current and voltage still holds. In this sample No. 3 no instabilities break out at any biasing voltage, but in ordinary samples noise-type oscillations occur at a certain voltage V_T which is much higher than V_0. In any case there is a region above V_0 in which current increases with applied voltage without any oscillations. In this region a diode is in differential negative conductance and amplification is possible in it.

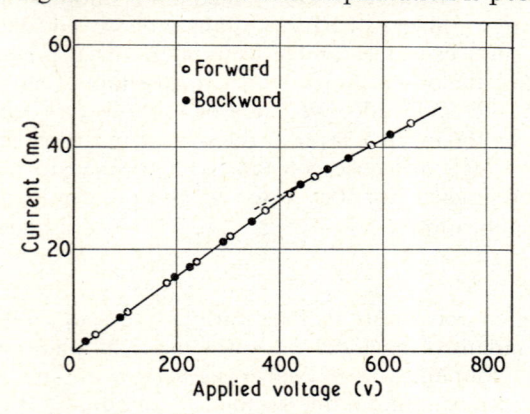

Figure 3. Current versus voltage.

4.2. *Signal input-output characteristics*

Figure 4 shows typical input and output characteristics of sample No. 2, measured at a frequency of 1·2 GHz. The parameter in the figure is the applied voltage. The upper curve

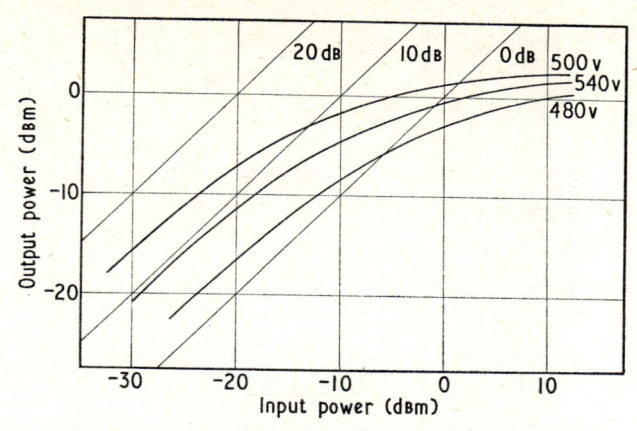

Figure 4. Output power versus input power at 1·2 GHz.

is for an applied voltage of 500 v which is the voltage near the onset of oscillation. These data were taken by adjusting the stab tuner so as to obtain the maximum output power at each measuring point. It will be seen in this curve that the net gain or available gain in the small-signal region is 14 dB and the saturation power is about 3 mw. In the course of this experiment we have been able, so far, to obtain a 17 dB net gain at the frequency of 1·25 GHz. Figure 5 shows gain versus frequency in the small-signal region measured on sample No. 2. At fixed bias voltage, the magnitude of gain first increases with frequency, reaches maximum and then decreases again. The optimum frequency which gives the maximum gain is 1·25 GHz for all bias voltages. In this experiment the length of all samples is fixed at 0·6 mm which corresponds to a 200 MHz Gunn oscillator, so there is no particular reason why the sample should respond to a signal frequency of 1·25 GHz. Therefore the frequency characteristics in figure 5 are mainly caused by the couplers and circuit properties. The $\lambda/4$ pulse line is one factor which is responsible.

Figure 6 shows gain versus current at fixed bias voltage. The parameter is the frequency. These data were measured by changing the temperature of the diode. If constant mobility is assumed then each temperature corresponds to a certain carrier concentration. Therefore figure 6 is considered to show the gain versus carrier concentration characteristics. The theoretically predicted gain is proportional to the carrier density n as seen in equation (2), and the qualitative agreement with the experimental results is good.

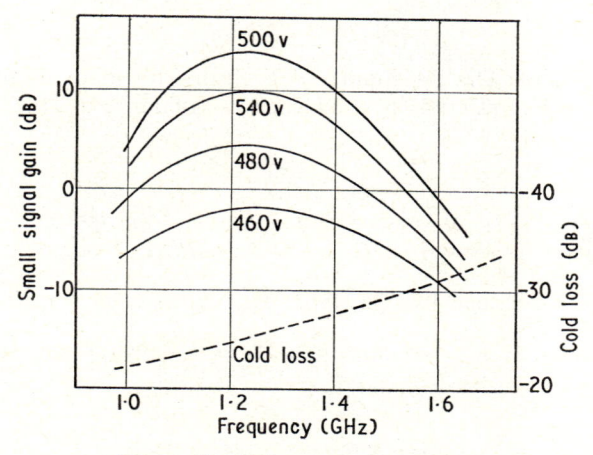

Figure 5. Gain versus frequency.

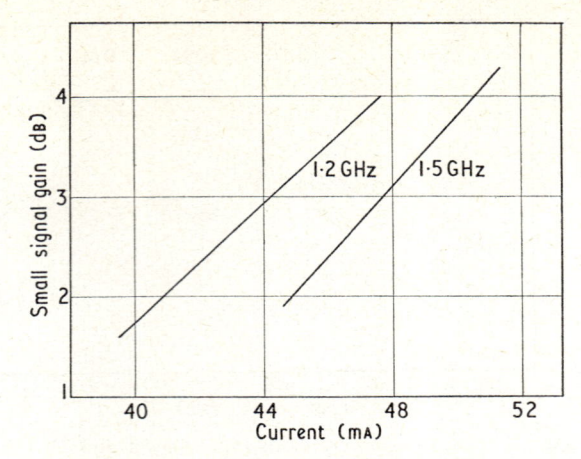

Figure 6. Gain versus bias current.

5. Discussion

It has been shown analytically that a travelling space charge wave mode whose amplitude grows with current flow is possible in bulk material of negative conductance, and this growing wave has been demonstrated experimentally. The gain obtained in experiments, however, is less than that predicted by analysis. The first reason for this discrepancy is the non-uniformity of the field strength in the diode. The static voltage distribution along the diode was measured using a probe, and it was found that the region in which the field strength exceeds the threshold is narrow compared to the whole length of the diode. Therefore the active region which contributes to the amplification of the signal is only a small fraction of the diode.

The second reason is the large loss of signal power at the input and output couplers. The total coupling loss is estimated to reach -12 dB at a frequency of 1·2 GHz and -13 dB at 1·5 GHz. Then the maximum net gain of 17 dB indicates that the electronic gain in the bulk is of the order of 30 dB.

It should be noted that in these experiments the cold loss between two contacts was more than -25 dB when the biasing field was below the threshold, and the reverse gain in the opposite direction was of the same magnitude as that of cold loss. So we can conclude that the device is unilateral.

We would expect that in future with improvement of the heat sink and using suitable material with a positive resistance-temperature coefficient, cw operation can be performed.

Acknowledgment

We should like to thank Drs. M. Sumi and T. Suzuki for many useful discussion on the theoretical analysis. We gratefully acknowledge the guidance of Dr. T. Miwa and Dr. B. Oguch throughout this study.

References

BUTCHER, P. N., and FAWCETT, W., 1967, *Brit. J. Appl. Phys.*, **18**, 755–9.
MCCUMBER, D. E., and CHYNOWETH, A. G., 1966, *Proc. IEEE Trans. Electron Devices*, **ED-13**, 4–21.
ROBSON, P. N., KINO, G. S., and FAY, B., 1967, *Proc. IEEE Trans. Electron Devices*, **ED-14**, 612–15.
THIM, H. W., and BARBER, M. R., 1966, *Proc. IEEE Trans. Electron Devices*, **ED-13**, 110–16.

Design calculations for cw millimetre wave L.S.A. oscillators

T. J. RILEY

Bell Telephone Laboratories Incorporated, Murray Hill, New Jersey, U.S.A.

Abstract. In order to fabricate small diameter, reproducible oscillators by a batch fabrication process, an etched mesa approach has been adopted. The maximum active length for a given temperature rise within the device has been calculated as a function of carrier concentration, for inverted bonding to a copper heat sink. The n+ GaAs substrate introduces parasitic series resistance, due to a millimetre wave skin depth of typically several microns, and distributed shunt capacitance to the heat sink. The power dissipated by this parasitic R–C low pass filter and the power delivered to a matched load have been computed. The improvement obtainable by metallizing the substrate to a thickness of several skin depths, typically several thousand ångströms, has been computed. The procedures involved in constructing such an oscillator are briefly described. The limit on the number of separate mesas that can be operated in parallel is determined by impedance matching and the requirement that the lateral dimensions of the array in the direction of wave propagation shall not exceed $\lambda/4$. For the various parameters assumed, the impedance limit ranges from 17 mesas at 50 GHz to 3 at 150 GHz with continuous power outputs of the order of 3 w and 0·5 w respectively. However, due to mechanical bonding considerations, the 3 mesa oscillator is considered to be the basic unit from which larger arrays may be constructed if desired. Since there is, at the present time, no definite knowledge of the intervalley scattering time this effect has not been included in the calculations.

1. Introduction

There are several restraints imposed upon the design of cw bulk effect oscillators, especially those operating at millimetre wavelengths in the L.S.A. mode. The long path for heat flow through the semiconductor, for sandwich structure oscillators, restricts the active length. The total active area is limited by impedance matching considerations. Finally, the diameter of a circular device is limited by the thermal spreading resistance of the heat sink and by standing wave effects within the active region.

The use of a stripe geometry can improve both the skin effect and thermal resistance restrictions. However, a number of small circular devices arranged in an array further improves the heat sinking, since the heat flux pattern is three dimensional, instead of two dimensional with stripe geometry. Design calculations can now be based upon a single etched mesa, with the power output and load impedance obtained by scaling to the appropriate number of mesas. The maximum number of mesas is determined here by

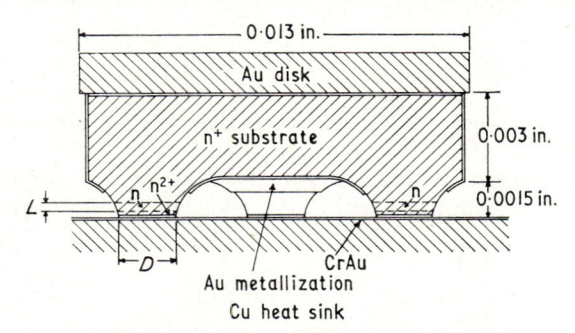

Figure 2. Cross-sectional view of three-mesa bulk effect oscillator.

rather arbitrary impedance matching limits since actual limits are circuit dependent. Although the size of the array must be somewhat less than $\lambda/4$ in the direction of wave propagation, it is possible to design circuits which enable this to be exceeded in the direction perpendicular to the direction of propagation. For this reason the meander line geometry of figure 1 has been adopted.

However, the most mechanically stable unit is the three-mesa or tripod structure, shown in figure 2, which can be considered a basic unit from which larger arrays can be constructed, if desired, thus reducing the stringent requirements upon surface planarity. Such a unit can also be cooled effectively with a simple copper heat sink and thus serves as a good vehicle for initial experimentation.

2. Device structure

For these calculations, a degenerately doped n^{++} GaAs layer of 2 μm thickness has been assumed to make ohmic contact to the active region, with a minimum of defects introduced into the active region such as may occur with metallic alloys (Cox and Hasty 1968). This goal is routinely achieved with the liquid tin epitaxial growth process.

A metallic alloy based on silver or gold of about 1 μm thickness has been assumed to be adequate for thermocompression bonding the degenerate region to a gold-plated copper heat sink. The thermal resistance of such a layer is negligible compared to the other contributions. The circular contact area may be defined either by photoresist techniques or by an evaporation mask.

The mesa structure of the active region is defined by photoresist and an anisotropic etch such as methanol–bromine. Subsequent to etching, the photoresist may serve as an evaporation mask for gold metallization between mesas, if desired, as shown in figure 2. The purpose is to reduce the parasitic series resistance of the n^+ GaAs substrate due to a skin depth of

$$\delta = (\rho/\pi\mu f)^{1/2} \qquad (1)$$

where ρ is the substrate resistivity, μ is the permeability, and f is the frequency. It is typically only several microns at millimetre wavelengths. The sheet resistance is

$$R_{\text{sh}} = \rho/\delta = (\pi\mu\rho f)^{1/2} \; \Omega/\text{square} \qquad (2)$$

which therefore increases with frequency, but decreases with decreasing resistivity.

A gold film of several thousand ångströms, which is about twice the skin depth, will thus reduce the sheet resistance by a factor of 30. It will not be possible to obtain quite so large a reduction in practice due to the impossibility of metallizing exactly to the edge of the active region, but this has been neglected in the following calculations. Similarly, the edges of the chip may be metallized, after cleaving, by an oblique evaporation during which the active region is protected by the overhanging n^+ substrate.

3. Heat extraction

The maximum temperature rise within an n^{++}–n–n^+ GaAs sandwich structure oscillator, with a single heat sink, is given by Knight (1967) as

$$\Delta T_{\max} = \left[T_\infty + \frac{(QL)D}{2K_s} \right] \exp\left[QL\left(\frac{L}{2K_1} + \frac{W}{K_2}\right)\right] - T_\infty \qquad (3)$$

The ambient temperature is $T_\infty\,^\circ\text{K}$, Q is the power density in w cm^{-3}, L is the active length, D is the diameter of a circular mesa, W is the thickness of the n^{++} layer (all in cm), and K_s, K_n, K_{n++} are the thermal conductivities of the heat sink, n, and n^{++} regions. K_1 and K_2 are thermal constants of the active region and n^{++} region respectively, given by $K_n = K_1/T$ and $K_{n++} = K_2/T$.

The power density is given by

$$Q = E_\text{B} J = E_\text{B} e v_\text{s}(n/f) f \qquad (4)$$

where E_B is the bias field of typically 10^4 v cm^{-1}, e is the electronic charge, v_s is the satura-

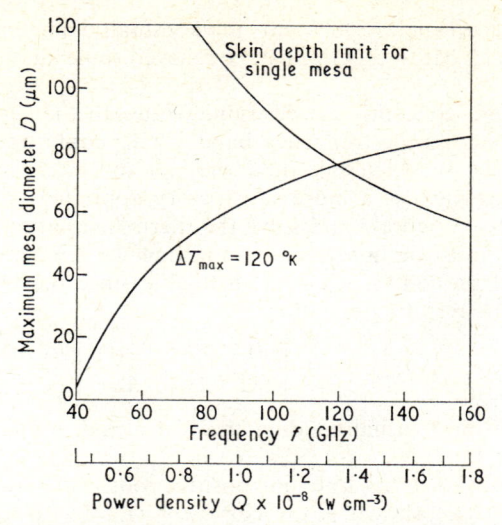

Figure 3. Thermal and skin depth limits upon maximum mesa diameter for constant fL product. $n/f = 7 \times 10^4$ s cm^{-4}; $fL = 5 \times 10^7$ cm s^{-1}; $QL = 5 \cdot 6 \times 10^4$ w cm^{-2}.

tion velocity of electrons (i.e., 10^7 cm s^{-1}), n is the carrier concentration and f is the frequency. For L.S.A. operation the ratio (n/f) should be a constant of about 7×10^4 s cm^{-3} for optimum efficiency. Similarly, it is advantageous for design across a frequency band to keep the product fL, which is proportional to the number of transit lengths, constant and equal to about $5v_s$. Thus the heat flux, QL, incident on the n^{++}–n interface is constant and equal to $5 \cdot 6 \times 10^4$ w cm^{-2}.

Equation (3) may now be re-arranged and solved for maximum mesa diameter as a function of frequency with $\Delta T_{\max} = 120°$K, $T_\infty = 300°$K, $W = 2$ μm and $K_s = 3 \cdot 88$ w cm^{-1} °K^{-1} (copper). This function is plotted in figure 3 together with the skin depth limit for the diameter of a single mesa of $(9 \times 10^8$ cm Hz$)/f$ given by Copeland (1967b). It is apparent that the long path for heat flow in the active region, represented by the L^2 term in equation

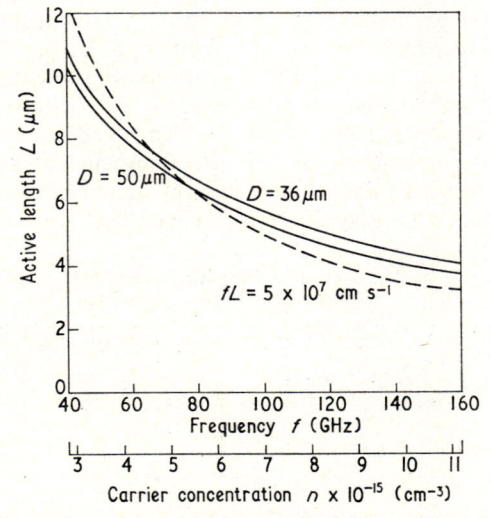

Figure 4. Thermal limit upon active length for constant mesa diameter. $n/f = 7 \times 10^4$ s cm^{-3}; $\Delta T_{\max} = 120°$K.

(3), is dominant at low frequencies and results in very small diameters. This rapid decrease of diameter for frequencies below 50 GHz can be relieved somewhat by using slightly smaller values of fL product.

With the multiple mesa concept, it is desirable to design the single mesas for relatively constant power output across the frequency band. This could be achieved by a constant diameter and constant fL product, but these are not optimum conditions for minimum temperature rise. Consequently, a more satisfactory approach is to choose an optimum diameter, e.g., 50 μm, from figure 3 and solve the thermal equation for L. To accomplish this easily, equation (3) must be linearized by omitting the temperature dependence of the thermal conductivity of the GaAs, which is justifiable since the temperature rise is to be restricted to 120°K. The resulting equation is

$$\Delta T_{max} = QL\left[\frac{D}{2K_s} + \frac{L}{2K_n} + \frac{W}{K_{n++}}\right] \tag{5}$$

where K_n and K_{n++} are the thermal conductivities of n and n^{++} GaAs at the appropriate temperatures (Maycock 1967).

This quadratic equation for L has only one positive root, which is plotted as a function of frequency in figure 4. Also shown is the curve for $fL = 5 \times 10^7$ cm s^{-1}, which serves to indicate that the departure from this value is not large. It might be emphasized that for optimum design of cw L.S.A. oscillators the active length must be carefully controlled. For practical application, the epitaxial layer is grown thicker than required, the carrier concentration is measured, the design frequency is thus determined, and the active length tailored by etching.

4. Equivalent circuit

The active region of the device has been represented by a negative resistance equal to 12 R_A, as calculated by Copeland (1967b) where R_A is the active low field resistance, in parallel with the geometrical capacitance, C. The necessary peak R.F. electric field of 8×10^3 v cm^{-1} for L.S.A. operation has been assumed to exist across the active region. The inactive cathode region where electrons are accelerated has been assumed to be constant at 0·4 μm for a uniform d.c. electric field of 10^4 v cm^{-1}.

The spreading resistance, R_s, arising from the surface of the substrate has been approximated by a curvilinear squares analysis for the geometry of figure 1 and then lumped at the base of the mesa where the largest contribution is located. The equivalent circuit is shown in figure 5. The parallel capacitance, C_p, between the substrate and the heat sink has been approximated by two parallel plates with fringing fields neglected.

This network is in reality a distributed R–C transmission line acting as a low-pass filter, but the lumped element approximation is considered adequate for the present purposes. An additional series resistance, R_c, arises from the cleaved edges of the chip. This has been calculated for the meander-line geometry, i.e., assuming current flow on two faces only.

The impedance of both active and passive regions has been computed and the matched load impedance obtained. The magnitude of the voltage, V_L, across the real part of the

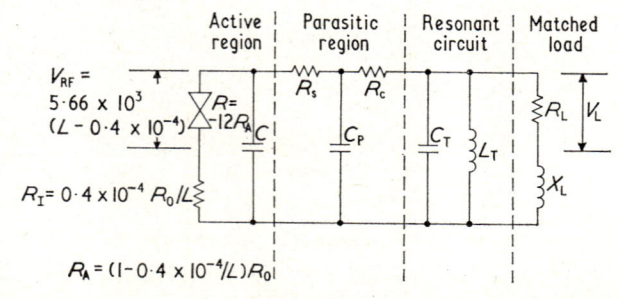

Figure 5. Lumped element approximation of equivalent circuit.

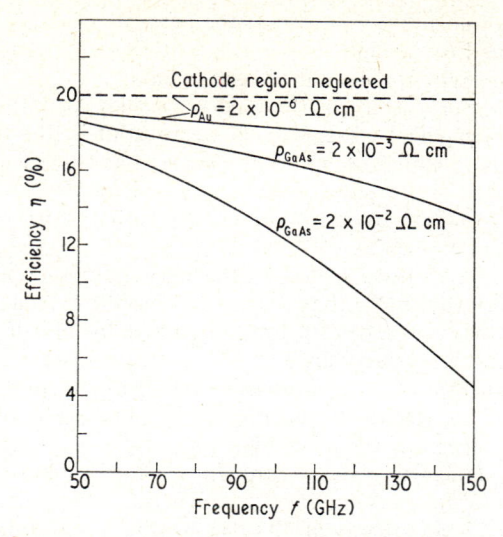

Figure 6. Oscillator efficiency, with gold coated and n⁺ GaAs substrates, when optimized at each frequency. $D=50\ \mu\mathrm{m}$; $n/f = 7 \times 10^4\ \mathrm{s\ cm^{-3}}$; $\Delta T_{max} = 120\,°\mathrm{K}$; L calculated from thermal equation.

load was computed and the R.F. power output found. The efficiency was derived by assuming a d.c. bias field of $10^4\ \mathrm{v\ cm^{-1}}$ and an operating current corresponding to an average electron velocity of $10^7\ \mathrm{cm\ s^{-1}}$ to obtain the d.c. input power. The design curve of efficiency versus frequency is plotted in figure 6, where the active length L is calculated from the thermal equation.

For a completely metallized substrate, the change in efficiency with frequency is negligible, when such factors as loss due to a finite intervalley scattering time, waveguide losses, and loss due to the inert cathode region where electrons are accelerated, are neglected. The inclusion of the inert cathode layer results in an unavoidable loss which increases with frequency due to a decreasing active layer thickness. In addition, the series resistance of an unmetallized n⁺ GaAs substrate causes a significant decrease in efficiency with increasing frequency. The intervalley scattering time effect has not been included in these calculations

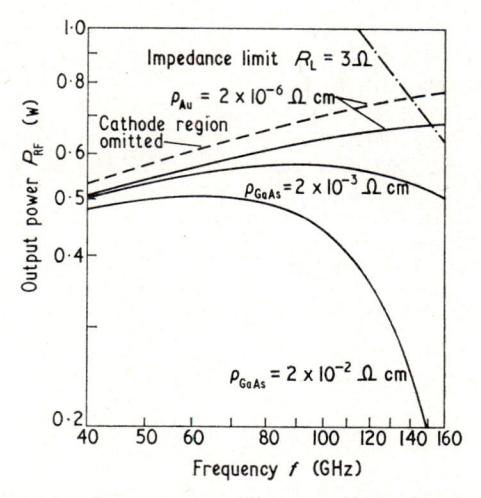

Figure 7. R.F. output power for three-mesa oscillators optimized at each frequency. $D=50\ \mu\mathrm{m}$; $\Delta T_{max} = 120\,°\mathrm{K}$; $n/f = 7 \times 10^4\ \mathrm{s\ cm^{-3}}$; L calculated from thermal equation.

since its magnitude is not accurately known at the present time. In fact, knowledge of the magnitude of these other loss mechanisms may assist in an experimental determination of the scattering time from oscillator output measurements.

The maximum R.F. power output for these 20% efficient oscillators is plotted in figure 7 as a function of frequency. The increase in power output is due to the slight increase in fL product with frequency, which compensates somewhat for the losses described above. Although each oscillator is optimized at only a single frequency, it can in practice operate over about an octave bandwidth by letting (n/f) vary from 5×10^4 s cm^{-3} to 1×10^5 s cm^{-3} without drastic changes in efficiency.

The output power has also been scaled to the basic 3 mesa unit discussed previously, from which it may be seen that more than $\frac{1}{2}$ w cw is possible even at the higher frequencies. Also included is the somewhat arbitrary limit obtained by plotting the output power at which the real part of the load impedance is 3 Ω, a figure which is hopefully fairly representative of the state-of-the-art of impedance matching at millimetre wavelength. It should be noted that the reduction of load impedance due to the presence of the parasitic and geometrical capacitances has been included.

The slope of the impedance limit is not constant since it depends upon the magnitude of the parasitic elements in a complex manner. Numerical computation shows that it might be possible to operate a 17 mesa array at 50 GHz, with a continuous power output of 3 w.

Since the original calculations of Copeland (1967b) of a maximum efficiency of 18·5% for GaAs L.S.A. diodes, experimental determinations of the velocity vs. field characteristic by Ruch (1967) indicate that even higher efficiencies should be possible. Efficiencies of n–GaAs oscillators operated at room temperature by Berson (1967), using short pulses, have approached 30% and single-chip cw devices have reached from 4 to 6% (Mackensie 1968). For cw efficiency to approach pulsed efficiency, the effective heat sinking provided by the multiple-mesa approach is essential.

5. Conclusions

The heat extraction from an etched mesa L.S.A. oscillator has been considered and a design curve relating active length of carrier concentration derived. The equivalent circuit including parasitic elements has been analysed and the decrease in efficiency at millimetre wavelengths has been calculated. An arbitrary impedance limit of 3 Ω has been used for the computations which indicate that roughly 3 w of continuous power should be obtainable at 50 GHz and 0·5 w at 150 GHz with 20% efficient oscillators. The effect of the intervalley scattering time has not been included as its magnitude is still uncertain.

References

BERSON, B. E., 1967, *NEREM Record*, **9**, 24.
COPELAND, J. A., 1967a, *Bell Syst. Tech. J.*, **46**, 284.
COPELAND, J. A., 1967b, *J. Appl. Phys.*, **38**, 3096.
COX, R. H., and HASTY, T. E., 1968, *Electrochem. Soc. Fall Meeting, Extended Abstracts*, **17**, 454.
KNIGHT, S., 1967, *Proc. IEEE*, **55**, 112.
MACKENSIE, L. A., 1968, Private Communication.
MAYCOCK, P. D., 1967, *Solid State Electronics*, **10**, 161.
RUCH, J. G., 1967, *Appl. Phys. Lett.*, **10**, 40.

CHAPTER 6

Other devices

A GaAs pn-junction FET and gate-controlled Gunn effect device

R. ZULEEG

Solid State Electronics Branch, McDonnell Douglas Astronautics Company–Western Division, Santa Monica, California 90406, U.S.A.

Abstract. A single gate pn-junction GaAs field-effect transistor was developed, which can be designed to operate in the stable mode as a depletion type unipolar field-effect transistor and in the unstable mode as a gate-controlled negative resistance Gunn device.

The GaAs films are grown on a semi-insulating substrate of GaAs by vapour-phase epitaxy with impurity concentrations in the range from 1 to 5×10^{15} cm^{-3} and a mobility of 7000 cm^2 v^{-1} s^{-1}. The gate junction is formed by a closed-tube zinc diffusion technique. The diffusion is masked by a sputtered silica film. Gate, source, and drain contact areas are defined in the silica film by standard photo-lithographic resist-etch techniques.

Devices operating as normal field-effect transistors with a 25 μm long and 250 μm wide channel have resulted in transconductances of 2 to 3 mA v^{-1}, with a corresponding cut-off frequency of 1·5 to 2 GHz. Operation in a gated Gunn-mode was obtained in devices which did not saturate in drain current due to the pinch-off mechanism, but achieved saturation due to the limiting velocity of charge carriers. Above a critical field, voltage-controlled Gunn oscillations were produced from 0·6 to 1·2 GHz by varying the gate potential.

1. Introduction

With advances in epitaxial growth of gallium arsenide the fabrication of planar and mesa type pn-junction field-effect transistors was achieved (Turner, Winteler, Steinemann 1966). Combination of the semi-insulating properties of a gallium-arsenide substrate with the epitaxial growth of thin layers of gallium arsenide leads to a unipolar transistor structure with a single gate configuration. Such a structure was proposed employing a Schottky barrier gate (Mead 1966) and experimental results were reported (Hooper *et al.* 1967). The design of a single gate pn-junction field-effect transistor was reported previously (Turner 1966, Zuleeg 1968). This paper describes the fabrication technique and presents electrical characteristics of such devices in more detail.

The single gate configuration offers a low gate capacitance which is combined with the high electron mobility in n-channel devices to realize microwave frequency operation of unipolar transistors. Since the gain–bandwidth product of a unipolar transistor is proportional to the channel mobility and inversely proportional to the gate capacitance, a frequency improvement factor of 3 to 5 over silicon devices with comparable geometry, e.g., gate capacitance, is predicted when considering the mobility ratio only. The GaAs p-channel device has frequency response lower by an order of magnitude for the same geometry because of the lower hole mobility. It was the subject of a study of radiation effects, which indicated improvement in radiation tolerance to fast neutrons and in transient response to ionizing radiation, when compared with silicon junction field-effect transistors.

N-channel devices exhibit, above a critical electric field, microwave frequency oscillations which are due to the transit-time limited Gunn effect in n-type GaAs. This negative resistance behaviour immediately adjacent to the drain current saturation has already been reported. (Winteler, Steinemann 1966; Zuleeg 1968). Experimental results on the operation of n-channel GaAs junction field-effect transistors as gate-controlled Gunn devices will be presented here.

2. Device fabrication technology

A batch fabrication process was developed, which can be executed with techniques now standard for silicon planar device fabrication. In figure 1 the five principal steps for the fabrication of the epitaxial GaAs junction field-effect transistor are presented.

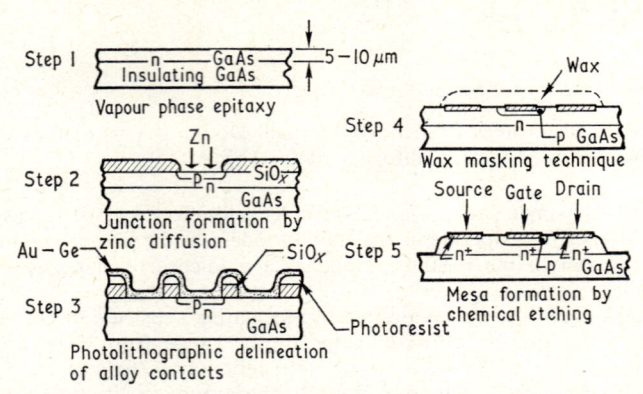

Figure 1. Gallium arsenide junction field-effect transistor fabrication technology.

Step 1

The starting material for the device is an epitaxially grown n-type GaAs film on top of a chemically prepared GaAs substrate of semi-insulating property, which is oriented in the (100) plane. A vapour phase epitaxial reactor is employed to grow films of carrier concentrations between 1×10^{15} and 5×10^{15} cm^{-3} with electron mobilities at room temperatures between 6000 and 7000 cm^2 v^{-1} s^{-1}. By timing the growth, film thicknesses of 2 to 20 μm were obtained.

Step 2

A silica film, which was deposited on the GaAs wafer from a low-sodium-content quartz source using RF sputtering techniques, served as a diffusion mask. For the diffusion mask against zinc a film thickness of about 5000 Å was sufficient. This diffusion was carried out in a closed system at 700°C using a ternary source of Ga–As–Zn in a composition of 10 : 110 : 80 by weight. For a diffusion depth of 2 μm or less the sideways diffusion at the silica surface interface (Turner 1966) was negligible, but became more pronounced in deeper diffusions which required a prolonged cycle. The openings in the silica film are cut by standard photolithographic resist-etch techniques. Gate widths as small as 2 to 5 μm are feasible for obtaining microwave transistor dimensions. The surface concentrations for a 700°C diffusion are around 1×10^{20} cm^{-3} and capacitance–voltage measurements on these pn-junctions indicated a close relation of $1/C^2 \propto V$ which reveals a rather abrupt junction profile. The diffusion of the pn-junction for the devices reported is confined to a line width of 12·5, 25, and 50 μm across the entire GaAs slice. This design eases the mesa formation and extends the pn-junction across the formed mesa surface thus ensuring complete pinch-off for proper device operation. An angle lapped and stained section of a p-diffusion into an epitaxial film of 5 μm thickness is shown in figure 2a.

Step 3

Photoresist is again applied to the wafer after the diffusion is completed and facilitates

(a) engraving and opening the source, drain, and gate contact areas, and
(b) breaking up and removal of the evaporated gold-germanium layer during alloying.

The alloying of metal contacts to source, drain, and gate region is performed at 550°C in a reducing atmosphere of hydrogen on a strip heater. The Au–Ge alloy produces n$^+$ contacts to the n-type GaAs and a non-rectifying contact to the degenerate p$^+$ gate region. During alloying the photoresist 'burns off' and removes the evaporated material outside the contact areas. Excess material is removed together with the silica film in hydrofluoric acid.

Step 4

The mesa area is defined by a wax evaporation technique with subsequent chemical etching. Wax is evaporated onto the GaAs surface at 10^{-4} torr pressure through a metal mask.

Step 5

The chemical etching through the n-type film gives complete dielectric isolation of the individual devices on the semi-insulating substrate. Figure 2b is a photograph of four isolated devices on a common substrate at this processing step. Thereafter the devices are diced and after cleaning are ready for mounting on headers or substrates using thermo-compression bonding for the electrical connections. Figure 3a gives a perspective view of a

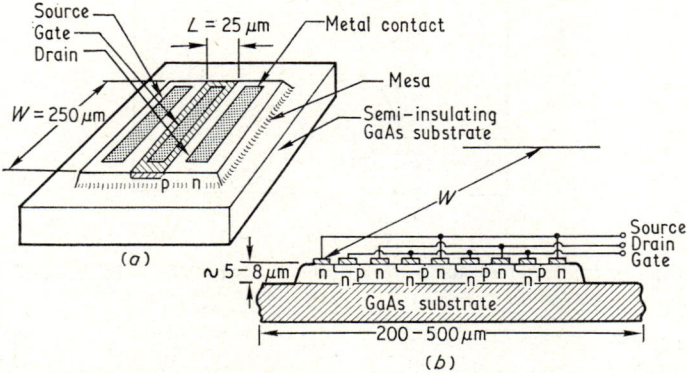

Figure 3. Gallium arsenide junction field-effect transistor geometry (a) Single gate mesa structure; (b) Interdigitated power structure.

single gate GaAs pn-junction field-effect transistor, which was used for the investigations. This low-power device can be expanded into a high-power, high-frequency device by interdigitated device configurations as shown in the cross-section, figure 3a. Since GaAs devices can be safely operated at 350 to 400°C, the prospects for a high-power field-effect transistor are excellent and should be investigated.

3. Electrical characteristics

3.1. Normal field-effect transistor operation

The drain voltage–current characteristics of a device with a W/L aspect ratio of 10 is shown in figure 6a. This device is stable and displays normal field-effect transistor characteristics which requires that the pinch-off voltage for reduction of the drain current to small values, is about the same as the drain voltage necessary for the drain current to saturate. Furthermore, the square law characteristic for drain current as a function of gate voltage in the saturation region is obeyed. Maximum transconductance of this device is approximately 1·5 mA V^{-1} and the pinch-off voltage is -20 V. To assess the high frequency properties of these devices the Y-parameters were measured as a function of frequency with the General Radio Admittance Bridge. Figure 4 gives the real and imaginary parts of these admittance parameters for one of the best devices. The imaginary parts show all the capacitive susceptance behaviour and from $Y_{12} = \omega C_{12}$, one obtains $C_{12} = 0.26$ pF. The theoretical maximum frequency of oscillation with $g_m = 2.3$ mA V^{-1} is therefore

$$f_{max}(\text{theor.}) = \frac{g_m}{2\pi C_{12}} \simeq 1.5 \text{ GHz}. \tag{1}$$

From the experimental results the actual maximum frequency of oscillation is 1·2 GHz, which can be also estimated from (Zuleeg 1967)

$$Y_{11(r)} \simeq g_m \left(\frac{\omega}{\omega_0}\right)^2. \qquad (2)$$

Degradation of the frequency response of the device structure is due to parasitic device elements. These elements are represented in figure 5, which gives the pi-section equivalent circuit of a field-effect transistor (Zuleeg 1967). A distributed RC-network acts as a low-pass filter in the input circuit and affects the frequency response of $Y_{21(r)}(\omega) = g_m(\omega)$.

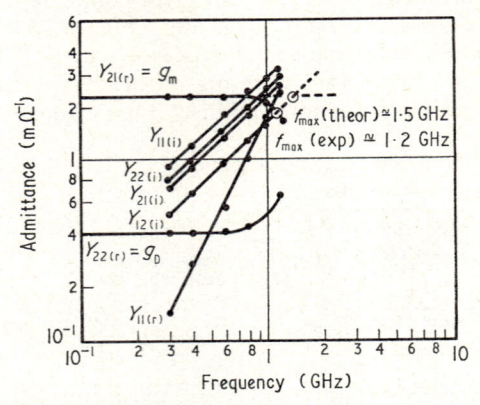

Figure 4. Admittance parameters as a function of frequency for single gate GaAs-JFET. $V_D = 20$ v; $V_G = -2$ v; $I_D = 10$ mA; $W = 250$ μm; $L = 25$ μm.

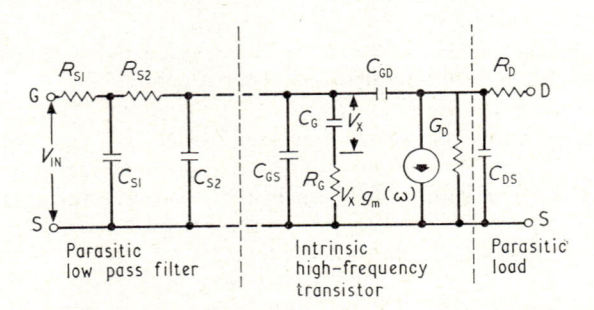

Figure 5. Pi-section equivalent circuit of FET.

With $R_s = 0$, the intrinsic angular cut-off frequency of the device is equal to $\omega_0 = (R_G C_G)^{-1}$, but with $R_s \neq 0$, the premature angular cut-off frequency is given by $\omega_{op} = (2C_{GS} C_G R_G R_s)^{-1/2}$ (Zuleeg 1967). In addition to the parasitic components of the input, the output has R_D and C_{DS} as contributing factors to the reduction of power gain. They become effective above 600 MHz, when the real part of Y_{22} starts to increase.

It is concluded from these measurements that microwave frequency operation of GaAs junction field-effect transistors is possible with a gate length of 5 to 10 μm, but structural refinements are imperative for the reduction of parasitic device elements which will ultimately determine the frequency response.

3.2. Unstable mode of field-effect transistor operation

Devices with channel heights of 3 to 5 μm and short channel lengths, e.g., less than 25 μm, exhibit current saturation at voltages much less than the pinch-off voltage and deviate in this respect from the normal mode of field-effect transistor operation. Limiting velocity of electrons in high electric fields is responsible for this behaviour. The device in figure 6b has for example a pinch-off voltage of 40 v, but saturation of drain current occurs at about 15 v. Exceeding electric fields of about 5×10^3 v cm^{-1} introduces a negative differential resistance and the device becomes unstable, thus generating high-frequency oscillatory conditions. These oscillations are associated with the transit-time limited Gunn effect in n-type GaAs. Negative resistance behaviour immediately adjacent to the saturation of drain current and accompanied by Gunn oscillations was previously reported for structures meeting the high field velocity saturation criterion (Winteler and Steinemann 1966). The three-terminal Gunn element gives the highest frequency of oscillation with the gate tied to the source contact, i.e., zero gate voltage. Increasing negative gate bias decreases the frequency of oscillation and ultimately suppresses oscillation at some finite value of drain current, permitting normal field-effect transistor operation. One explanation for the electronic frequency control of the Gunn mode of operation is the variation of transit-time and point of nucleation of the domains, since increasing negative gate voltage results in a widening of the pinch-off region in the channel. A tuning range of 2 : 1 can be achieved with this gate-controlled Gunn effect device. By mechanically adjusting the point of nucleation of the high field domain a tuning range of 5 : 1 was reported. (Haydl 1968.)

Figure 6b is a representative drain current–voltage characteristic of such an unstable device with gate voltage as variable. Figure 6c and d present the voltage–current characteristics of another device on a different scale. At $V_G = 0$, the typical negative resistance behaviour of the electronic transfer mechanism is displayed with subsequent high frequency oscillations. Applying -6 v to the gate reduces the current and shifts the onset of negative resistance oscillations to a higher drain voltage.

For high-frequency evaluation, the device was mounted on an alumina substrate which could be inserted into a 1N23 package. Figure 7a is a photograph of a mounted device on the substrate and figure 7b the 1N23 package assembly with the plastic or ceramic tubular support removed. In a two-terminal configuration, the gate is returned to source by bonding to the source tab. For the three-terminal configuration, the gate lead is fed through a hole in the tubular support. Experimental results of controlling the oscillations by varying the gate voltage with the device placed into a tunable cavity are shown in figure 8. A tuning range of 600 MHz to 1·2 GHz was available with the cavity and at zero and -12 v gate bias the maximum amplitude of oscillations corresponded to 1·1 GHz and 720 MHz respectively. Further investigations are in progress using a wideband coaxial structure to assess phase-locking properties and power capabilities of this three-terminal Gunn device and will be reported at a later date.

4. Conclusions

The fabrication technique for a single gate pn-junction GaAs field-effect transistor was presented and has been successfully employed to produce experimental devices for evaluation of their electrical characteristics. It was shown that this device can be designed to operate in a stable mode as a depletion type field-effect transistor with feasible applications as an amplifier and standard oscillator in the microwave frequency region. For optimized structures with respect to geometry and material, a frequency response to 5 GHz should be possible.

The feasibility of a gate controlled Gunn device was demonstrated. It requires field-effect transistor operation whereby the drain saturation current is obtained at a drain voltage less than the voltage necessary to pinch-off the channel. This gate control mechanism of the Gunn oscillations suggests a variety of applications in microwave functional or integrated electronics, such as frequency modulators, phase-locked and keyed, and electronically tuned oscillators.

Acknowledgment

The author is grateful to S. H. Watanabe and D. Dion for the material preparation and device fabrication and to U. Ranon and J. Golan for the high-frequency measurements.

References

HAYDL, W. H., 1968, *Appl. Phys. Lett.*, **12**, 357.
HOOPER, W. W., and HOWER, P. L., 1967, International Electronic Devices Meeting, Washington, D.C.
HOOPER, W. W., and LEHRER, W. I., 1967, *Proc. IEEE*, **55**, 1237.
MEAD, C. A., 1966, *Proc. IEEE*, **54**, 307.
TURNER, J. A., 1966, *Proc. Symp. on GaAs*, Institute of Physics and Physical Society Conf. Series No. 3, p. 213.
WINTELER, H. R., and STEINEMANN, H., 1966, *Proc. Symp. on GaAs*, Institute of Physics and Physical Society Conf. Series No. 3, p. 228.
ZULEEG, R., 1967, *Solid-State Electronics*, **10**, 559.
ZULEEG, R., 1968, *Proc. IEEE*, **56**, 879.

The Schottky barrier gallium arsenide field-effect transistor[†]

P. L. HOWER, W. W. HOOPER, D. A. TREMERE, W. LEHRER, and C. A. BITTMANN

Fairchild semiconductor research and development laboratory, Palo Alto, California 94304, U.S.A.

Abstract. Simplified analyses of the junction field-effect transistor show that high-frequency performance is directly related to the drift velocity of the carriers in the channel. Because of the high drift velocities attainable in n-type GaAs, this material is particularly attractive for use as F.E.T. channel material.

The advantages of an n-channel GaAs F.E.T. using a Schottky barrier gate in conjunction with a semi-insulating substrate have been previously described (Hooper and Hower 1967). Measurements of unilateral gain U on experimental devices show that this quantity decreases at 6 dB/octave, in agreement with simple theory. A typical value for U at 1 GHz is 10 dB, corresponding to a maximum frequency of oscillation of 3 GHz. These values are comparable to those now obtainable with bipolar transistors. In addition, the GaAs F.E.T. has the advantage of being operable over a wide temperature range.

Comparison with theory shows the measured U to be less than the value calculated from channel dimensions and impurity concentration. Possible explanations for this discrepancy are discussed. The behaviour of the characteristics with temperature and some interesting deviations from conventional F.E.T. behaviour are described.

1. Introduction

It has been shown (Hooper and Hower 1967) that GaAs field-effect transistors can be made with useful gain extending to microwave frequencies. This paper will outline the design considerations and will describe the observed electrical behaviour of this type of F.E.T., which in many respects is similar to Shockley's original junction F.E.T.

If an F.E.T. is to be used as a high-frequency amplifier, particular attention must be paid to the following characteristics: (*a*) transit time; (*b*) feedback capacitance; (*c*) lossy parasitic elements.

Small values of transit time can be attained by using a short channel length and by choosing a material for the channel in which majority carriers attain high drift velocities. Large values of feedback capacitance (C_{gd}) increase the loss contributed by the extrinsic resistance R_D in series with the drain terminal. In addition, it is desirable to keep C_{gd} small to avoid difficulties with neutralization. In saturation, the feedback capacitance of an ideal F.E.T. is quite small compared to the gate capacitance (C_{gs}). However, for conventional diffused junction F.E.T.s (which have two gates), C_{gd} is several times the ideal value due to the unavoidably large area of the second gate.

The ideal value of C_{gd} can be obtained by using a structure proposed by C. A. Mead (1966) in which the channel is n-type GaAs grown epitaxially on a semi-insulating substrate. In this F.E.T. there is only one gate, a reverse biased Schottky barrier, and source-drain isolation is achieved by means of the semi-insulating substrate. This type of substrate is preferable over a p–n junction to avoid losses due to the series R–C network introduced by the junction.

From transit time considerations, n-type GaAs is desirable because the electron drift mobility in GaAs is several times that in silicon or germanium. We expect this advantage

[†] This work sponsored by the Air Force Avionics Laboratory, Air Force Systems Command, Wright-Patterson Air Force Base, Ohio.

to be maintained even when channel fields become large enough to invalidate the use of low field mobilities in calculating high-frequency performance.

2. Design criteria

For a quantitative measure of high-frequency performance we shall use unilateral gain, U, defined by Mason (1954) and given by the formula

$$U = \frac{|y_{21} - y_{12}|^2}{4(g_{11}g_{22} - g_{12}g_{21})}. \tag{1}$$

Throughout this paper we shall refer to the common source y-parameters in the form: $y_{ij} = g_{ij} + jb_{ij}$.

A suitable circuit model for the F.E.T. is shown in figure 1. Here R_S and R_D represent unmodulated resistances in series with the channel. It is desirable to keep these resistances small, since their effect is to decrease U from the ideal value.

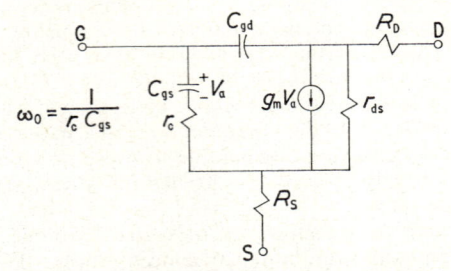

Figure 1. Small-signal circuit model of the GaAs F.E.T.

To give an idea of the capabilities of the GaAs F.E.T. we shall calculate U for the ideal case, $R_S = R_D = 0$. The unilateral gain is then given by

$$U = \frac{g_m^2}{4g_c g_{ds}} \left(\frac{\omega_0}{\omega}\right)^2. \tag{2}$$

As a figure of merit we shall use the maximum frequency of oscillation, f_{max}, which is defined as the frequency for which U has decreased to unity. From equation (2)

$$f_{max} = \frac{\omega_0}{4\pi} g_m / \sqrt{(g_c g_{ds})}. \tag{3}$$

The terms on the right-hand side of this equation are functions of bias, film dimensions, and film doping. In this section we derive an approximate formula for f_{max} as a function of these device variables.

The most important term in equation (3) is the frequency ω_0, which is defined in figure 1. For the ideal case we are considering, ω_0 is the radian frequency for which $|y_{21}|$ has decreased to 0·707 of its low-frequency value.

Physically, we would expect the transit time to be related to ω_0, and that decreasing the transit time τ would increase ω_0. Using the y-parameter analysis of Hauser (1965) it is possible to show that this is indeed the case and that, for an F.E.T. in saturation, ω_0 and τ are related by

$$\omega_0 \simeq 4/\tau. \tag{4}$$

The transit time is defined by

$$\tau = \int_0^L \frac{dx}{v_d(x)} \tag{5}$$

with drift velocity v_d and channel length L. For an F.E.T. in saturation and for small channel fields, the gradual channel solution (Shockley 1952) gives, for low field mobility

$$\tau = \frac{3}{2} \frac{L^2}{\mu_0 U_0} \frac{1+3\sqrt{\eta}}{(1-\sqrt{\eta})(1+2\sqrt{\eta})^2}. \tag{6}$$

The term U_0 is the total electrostatic potential necessary to deplete the film and is given by

$$U_0 = V_p + V_B = \tfrac{1}{2} N_D a^2 q/\epsilon$$

where N_D is the film concentration (assumed uniform), a is the film thickness, and V_B is the built-in voltage of the gate junction.

The quantity η is a normalized value of gate bias defined by

$$\eta = (V_B - V_{GS})/U_0.$$

As η varies from 0 to 1, the width of the channel at the source end of the gate decreases from the film thickness, a, to zero, and the coefficient of $L^2/\mu_0 U_0$ in equation (6) increases from 1·5 to infinity, the latter condition corresponding to a completely pinched-off channel.

As the average field in the channel, which is equal to $(V_p - |V_{GS}|)/L$, reaches the non-linear portion of the drift velocity characteristic, equation (6) gives optimistically small values of τ.

Since we are dealing with pinch-off voltages typically in the range 1 to 5 v and a channel length of approximately 5 μm, it is not difficult to exceed the field of 3×10^3 v cm^{-1}, where v_d goes through a maximum. If the average field exceeds this value it is desirable to have an alternate solution for τ.

At present there is no theory for calculating either low- or high-frequency F.E.T. behaviour which satisfactorily includes the negative mobility portion of the $v_d(E)$ characteristic of GaAs. For our purposes we shall approximate the true situation by using in place of equation (5), the equation

$$\tau = L/\bar{v} \tag{7}$$

where $\bar{v}$ is an average drift velocity which will be some fraction of the peak electron drift velocity ($\sim 2 \times 10^7$ cm s^{-1}).

The remaining terms to be calculated are the transconductance, g_m, the channel conductance, g_c, and the output conductance, g_{ds}. From the analysis of Hauser (1965) we find $g_c/g_m \simeq 2\cdot 5$ over the entire range of gate bias. The transconductance in saturation is given by $g_m = G_0(1-\sqrt{\eta})$ where the conductance G_0 is the unmodulated channel conductance, and equal to $q\mu_0 N_D a Z/L$, Z being the transverse dimension of the gate.

The output conductance g_{ds} is determined by the decrease in channel length with drain-gate bias. From the analysis of Hower (1967) we can write

$$g_{ds} = \frac{G_0}{3} \frac{a}{L} \frac{G(\eta)}{h(\theta)} \tag{8}$$

where

$$G(\eta) = 1 - 3\eta + 2\eta^{3/2} \tag{9}$$

and $h(\theta)$ is a dimensionless quantity that relates g_{ds} to the drain-gate bias through the following two equations:

$$h(\theta) = \left(1 - \frac{\theta}{L/a}\right)^2 \theta \ln(1+\theta^2)$$

$$(V_{DG} + V_B)/2U_0 = 2 - \theta^2 + (1+\theta^2)\ln(1+\theta^2). \tag{10}$$

For typical bias values, $h(\theta)$ lies in the range 0·2 to 2·0.

Substituting the relevant terms in the equation for f_{max} we find

$$f_{max} = 0\cdot 35 \frac{\bar{v}}{L} \sqrt{\left[\frac{1-\sqrt{\eta}}{G(\eta)} \frac{L}{a} h(\theta)\right]}. \tag{11}$$

Considering present technological limits, reasonable estimates for film thickness, channel length, and bias are as follows:

$$a = 1\ \mu m$$
$$L = 6\ \mu m$$
$$\eta = 0.25$$
$$(V_{DG} + V_B)/U_0 = 1.8\ \text{(corresponding to } h(\theta) = 1.0)$$

For $\bar{v}$ we shall use approximately one-half the peak value, or $\bar{v} = 1 \times 10^7$ cm s^{-1}. Using these numbers in equation (11), we get $f_{max} = 14$ GHz.

Present experimental devices, for which R_S and R_D are not negligible, have measured f_{max} values of approximately 3 GHz. In addition to R_S and R_D, which account for some of the reduction in f_{max}, there are other factors to be considered, as discussed in section 4.

3. Construction

The Schottky barrier gate GaAs F.E.T. is made entirely by thin film metal depositions. For this reason the GaAs is not subjected to the high temperature of diffusion ($\sim 750°$C) encountered in a junction F.E.T. which can cause deterioration of the electrical properties. An additional advantage is that the processing steps are reduced to a minimum.

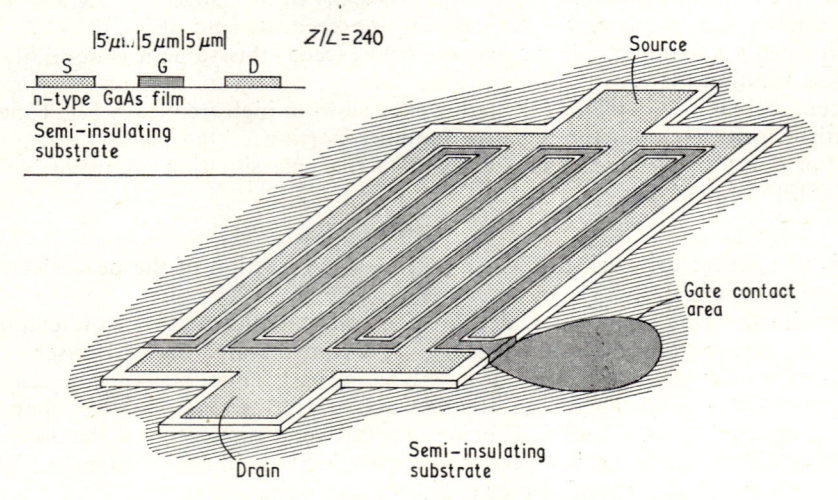

Figure 2. Contact spacing and construction.

The F.E.T. is constructed on an n-type GaAs epitaxial film, typically 1 μm thick with carrier concentration $N_D = 5 \times 10^{15}$ cm^{-3}, and electron mobility $\mu_n = 6000\text{--}7000$ cm^2 v^{-1} s^{-1}. The epitaxial film is grown by the vapour transport method (Knight et al., 1965) on a semi-insulating ($\sim 10^7$ Ω-cm) chromium-doped GaAs substrate. As noted in section 1, the use of a semi-insulating substrate results in a device with very low parasitic loss compared to a conventional p-substrate structure.

As the first step, the epitaxial film is masked to define the source and drain contact areas, as shown in figure 2, and an alloy of silver, indium, and germanium is evaporated at 500°C. The source-drain mask is then washed away, removing the unwanted metal outside the contact area, and the remaining contact metal alloyed at 600°C.

The wafer is again masked and the epitaxial film etched down to the high-resistivity substrate. This step isolates the source and drain outside the active gate area, and provides a region of low carrier concentration for location of the gate contact pad.

Finally, the wafer is masked to define the gate area, aluminium is evaporated at room temperature, and the gate mask is washed away to remove the unwanted metal. After completion of this step, the F.E.T. appears as shown in figure 2. The devices are then scribed apart and mounted using tin die-attach in TO-18 headers.

4. Electrical behaviour
4.1. D.C. characteristics

Devices assembled from a number of wafers and processed as described in the preceding section all tend to show similar electrical behaviour. A typical drain characteristic is shown in figure 3(a). Although this characteristic is similar to that of a conventional junction F.E.T., there are certain differences between the two types of F.E.T.s.

In an ideal F.E.T. the transconductance in saturation ($g_{m,\,\text{sat}}$) is equal to the source-drain conductance at $V_{DS}=0$, (g_{dso}), provided V_{GS} is fixed. The ratio $g_{m,\,\text{sat}}/g_{dso}$ can also be calculated for the case where fixed resistances R_S and R_D are in series with the channel. For $V_{GS}=0$ and the contact spacing shown in figure 2, this ratio is 1·1. However, the data from figure 3(a) give $g_{m,\,\text{sat}}/g_{dso}=0\cdot75$, and for some wafers we notice a further decrease in this ratio to $\sim 0\cdot5$. There are two explanations for this discrepancy. One, already mentioned, is that when the F.E.T. is in saturation, high fields in the channel reduce the apparent mobility and hence $g_{m,\,\text{sat}}$ is reduced from the expected value.

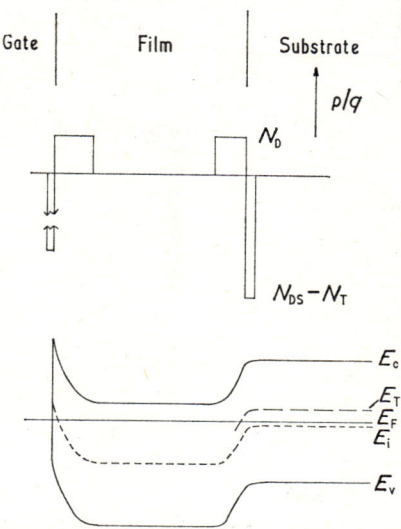

Figure 4. Energy band diagram and charge profile along a line normal to the gate.

A second possible cause of the discrepancy is the presence of a depletion layer at the film–substrate junction. As we note below, the presence of a large trap density in the substrate will result in the formation of a dipole layer at the film–substrate junction that behaves similarly to a p–n junction. Figure 3(b) shows experimental evidence of this depletion layer. By increasing the negative voltage applied to the substrate, the drain current can be drastically reduced.

The presence of a dipole layer can be explained using the deep acceptor (N_T), shallow donor (N_{DS}) model for Cr doped GaAs discussed by Cronin and Haisty (1964). If $N_T \gg N_{DS}$ the energy band diagram and charge densities will be as shown in figure 4. In the neutral part of the substrate only a small fraction (N_{DS}/N_T) of the acceptors are filled with electrons, while in the negatively charged layer, all the acceptors are filled, giving a net charge density there of $q(N_{DS}-N_T)$.

It is not unreasonable to expect N_T to be in excess of 5×10^{16} cm^{-3} for high resistivity substrates and thus the substrate-film junction can be expected to have many properties in common with a gate. One distinguishing feature, however, is that the substrate 'gate' is not readily accessible electrically due to the high resistivity of the substrate. Because of the large substrate resistance, a certain time (~ 0.5 ms) is required to charge the substrate gate, thus explaining the loops observed in the drain characteristic.

4.2. Behaviour with temperature

The drain characteristic for different ambient temperatures is shown in figure 5. In saturation, both the transconductance and the drain current are relatively insensitive to variations in temperature. However, the changes that do occur are in the same direction as the temperature variation of the film mobility.

4.3. High frequency behaviour

The measured y-parameters for a typical F.E.T. are shown in figure 6. The observed frequency dependence of the y-parameters agrees with that predicted by the circuit model of figure 1. Not shown in this model are the capacitances contributed by the package. There is approximately 0.7 pF stray capacitance from the G to S and D to S terminals, and 0.05 pF between the D and G terminals.

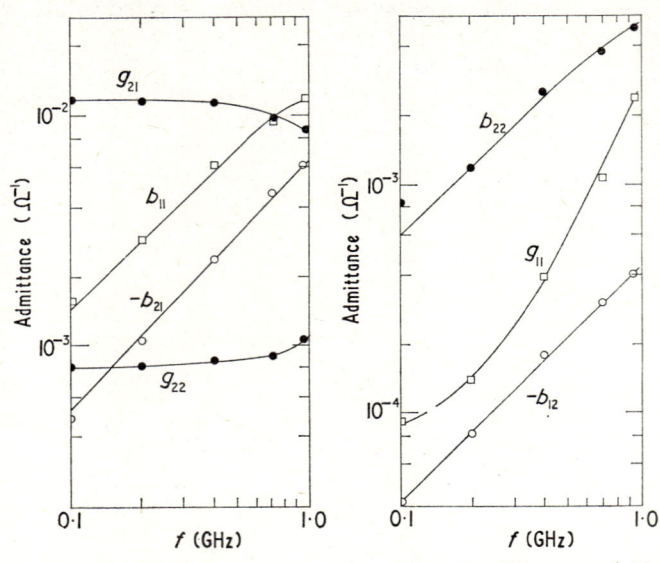

Figure 6. y-parameters vs. frequency. $V_{DS} = 6.0$ v, $I_D = 24$ mA, and $V_{GS} = -1.3$ v.

In figure 7 we show the unilateral gain computed from measured y-parameters using equation (1). As predicted, U shows a 6 dB/octave decrease with frequency. Extrapolating to 0 dB, we find $f_{max} = 3$ GHz. From an analysis similar to that of section 2, but one which assumes $R_S > 0$, we compute $f_{max} = 11$ GHz for the bias point noted in figure 6. The reason for this discrepancy is not presently understood; however, there are two factors that require further discussion.

First, the use of equations (4) and (7), together with $\bar{v} = 1 \times 10^7$ cm s^{-1} is an attempt to adapt the low field theory to the case where high channel fields are present. It is possible that an F.E.T. analysis which properly accounts for the negative mobility portion of the $v_d(E)$ characteristic will lead to a smaller $\bar{v}$ and hence smaller f_{max}.

Second, in the calculation of f_{max} it is assumed that the extrinsic source resistance R_S is independent of bias and is calculable from the contact spacing shown in figure 2. As we have noted previously, it is possible for the substrate to act as a gate; for this case a depletion layer will extend into the film from the substrate junction to a distance determined by the potential difference between substrate and film. This behaviour increases the apparent value of R_S and reduces f_{max} from the calculated value.

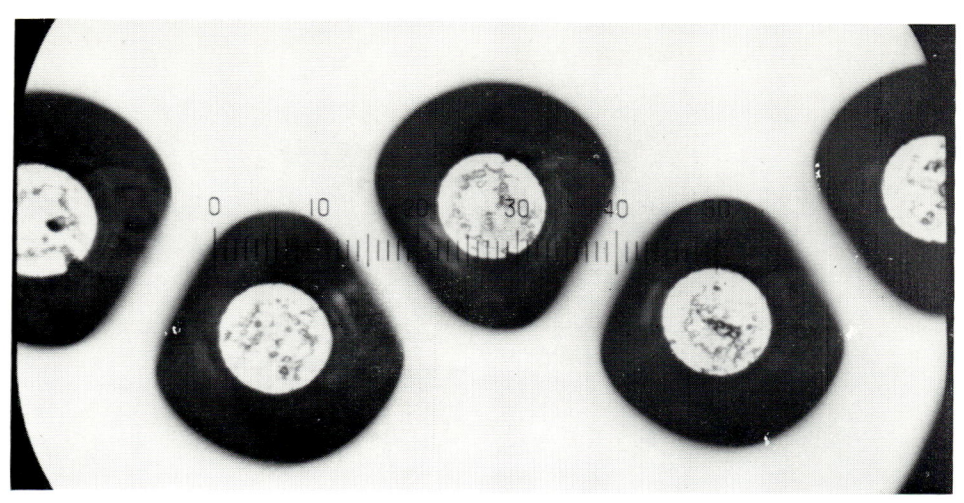

Figure 1. Meander line geometry for etched mesas. Scale, 5 μm per division.

PLATE XXI—R. *Zuleeg*: *paper* 28

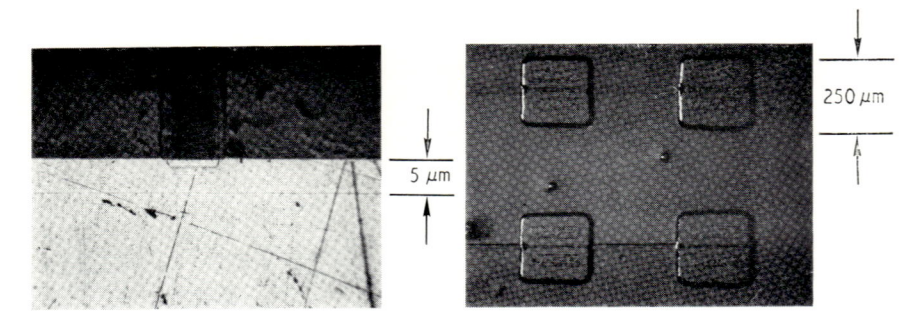

Figure 2. Gallium arsenide junction field-effect transistor design (*a*) Angle lapped and stained section; (*b*) Photograph of mesa-etched devices.

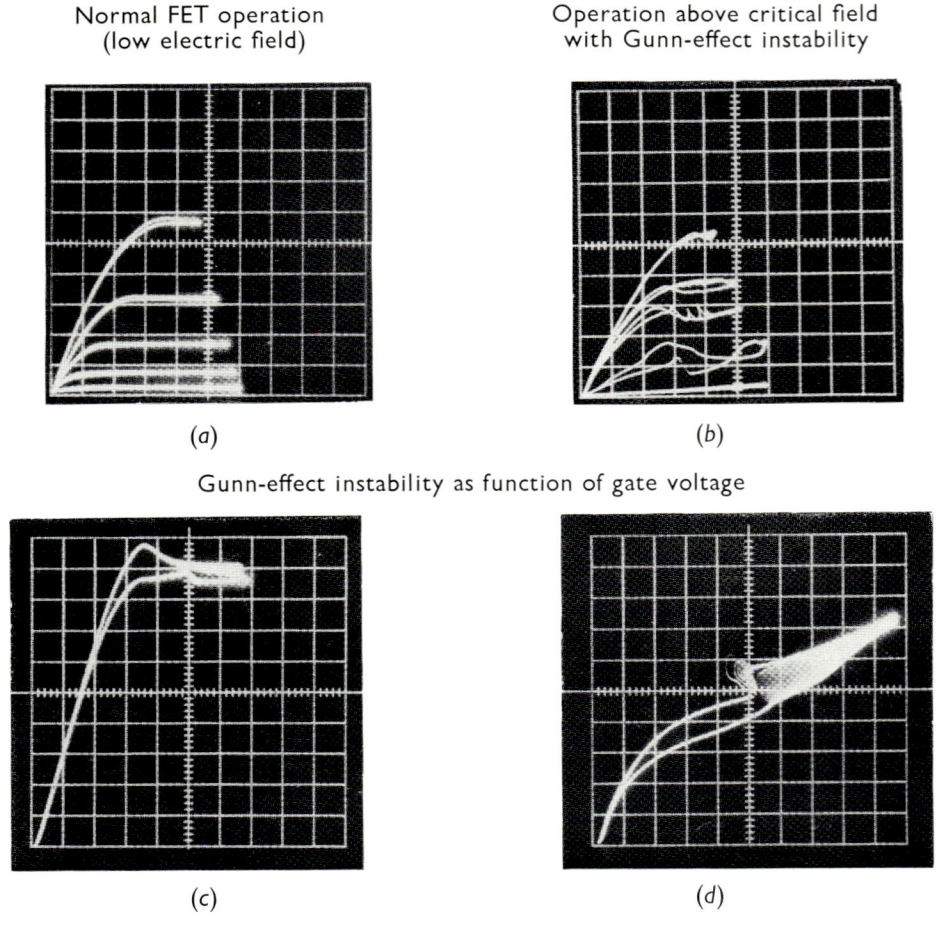

Normal FET operation
(low electric field)

Operation above critical field
with Gunn-effect instability

(*a*)　　　　　　　　　　(*b*)

Gunn-effect instability as function of gate voltage

(*c*)　　　　　　　　　　(*d*)

Figure 6. Voltage–current characteristics of experimental GaAs JFET's operating in the stable and unstable mode.

(*a*) Vertical: I_D in 2 mA/div
　　Horizontal: V_D in 5 v/div
　　Negative: V_G in 5 v/step
(*c*) Vertical: I_D in 2 mA/div
　　Horizontal: V_D in 10 v/div
　　$V_G = 0$ (gate tied to source)

(*b*) Vertical: I_D in 5 mA/div
　　Horizontal: V_D in 5 v/div
　　Negative: V_G in 10 v/step
(*d*) Vertical: I_D in 2 mA/div
　　Horizontal: V_D in 10 v/div
　　$V_G = -6$ v

PLATE XXII—*R. Zuleeg: paper 28*

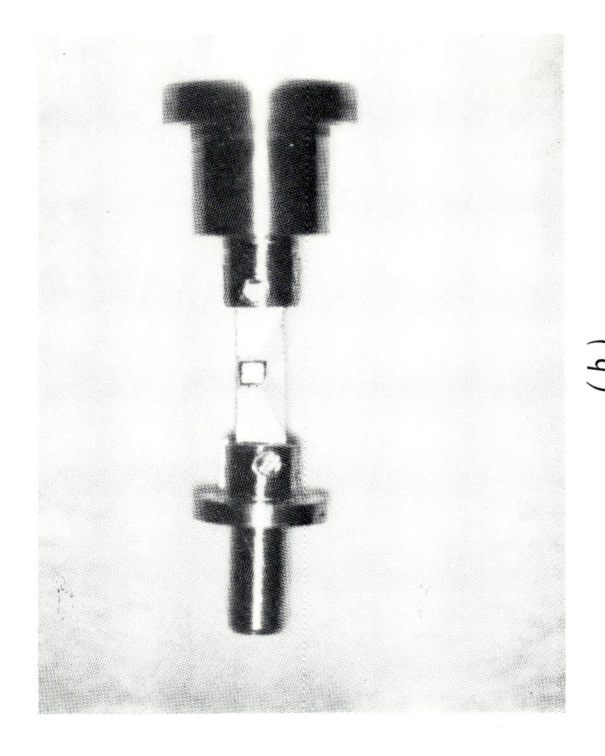

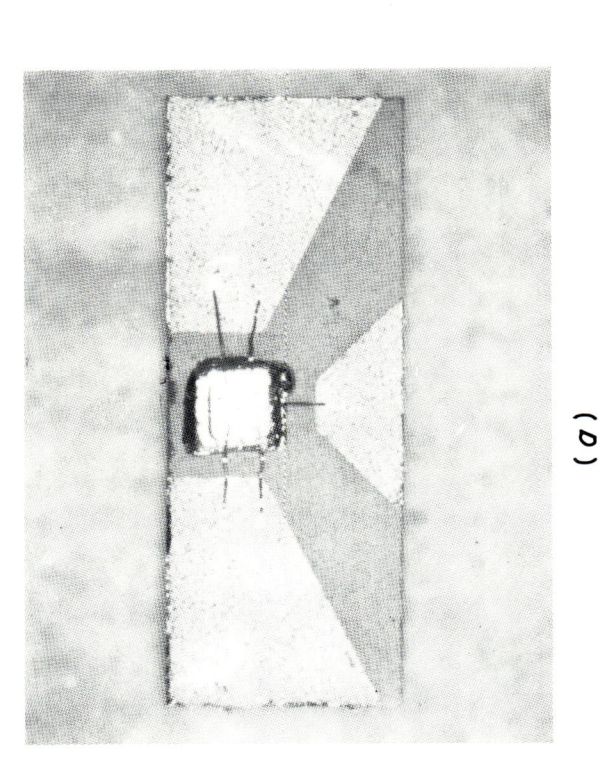

(*a*)

(*b*)

Figure 7. Microwave packaging of devices (*a*) Alumina substrate with mounted device; (*b*) 1N23 package assembly (tubular support removed).

PLATE XXIII—R. *Zuleeg: paper* 28

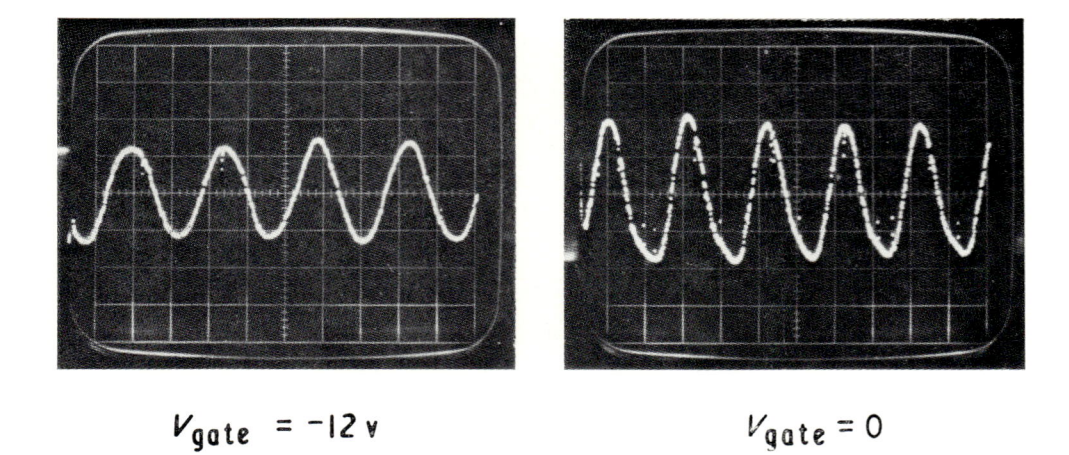

Figure 8. Gunn-oscillations with tunable cavity at $V_G = 0$ and $V_G = -12$ v vertical: 200 mv/div; horizontal: 0·5 ns/div; Bias: 60 v.

4.4. Comparison with silicon

Because it is possible to attain higher drift velocities in GaAs, we expect the performance of a GaAs F.E.T. to be superior to that of a silicon F.E.T. for the same channel length. As an experimental check of this idea, a silicon Schottky barrier gate F.E.T. was made using the same mask dimensions as for the GaAs F.E.T. For the silicon F.E.T., the Schottky barrier is aluminium, and the source-drain contacts are made by ion implanting antimony and using aluminium for the metallic contact. The substrate is high resistivity n-type (2×10^3 Ω-cm) and the film concentration is approximately 5×10^{15} cm^{-3}.

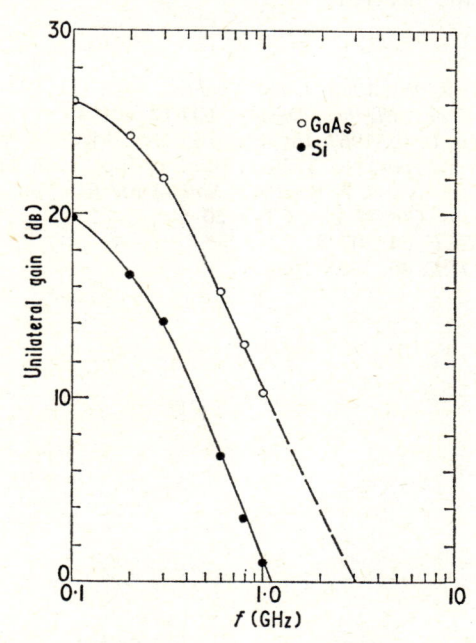

Figure 7. Unilateral gain vs. frequency for a GaAs and Si F.E.T. having identical contact geometries and approximately the same film thicknesses and concentrations.

For both devices, the film thickness is approximately 1 μm. Pinch-off voltages and transconductance for the two devices are given below.

	$g_{m, sat}$ (mm hos)	V_p (v)
GaAs	12	3·6
Si	4	3·4

The unilateral gain for the silicon F.E.T. is plotted in figure 7, and we find $f_{max} = 1 \cdot 1$ GHz. This reduction from the GaAs device is largely due to the difference in transit time, since the ratio $g_m^2/g_c g_{ds}$ is approximately the same for the two devices. At 1 GHz the noise figure (optimized) is typically 4 dB.

4.5. Conclusion

We have described the theory, fabrication, and performance of an n-channel Schottky barrier gate GaAs F.E.T. Theoretical estimates indicate that the device structure described is capable of achieving f_{max} values in excess of 10 GHz. Present experimental devices typically have an f_{max} of 3 GHz, with unilateral gain of 10 dB at 1 GHz. This performance exceeds that of a comparable silicon F.E.T. and is competitive with microwave bipolar transistors.

In addition to the superior high-frequency performance, the simple construction and natural isolation of the GaAs F.E.T. make it attractive for use in both hybrid and monolithic microwave integrated circuits. Although work remains to be done in understanding the details of device behaviour, we anticipate that the GaAs F.E.T. will make a substantial contribution to the field of microwave devices and systems.

Acknowledgment

We thank B. Cairns and R. Fairman, who grew the GaAs films, and also E. Yim, who supplied the material for the Si F.E.T.

References

CRONIN, G. R., and HAISTY, R. W., 1964, *J. Electrochem. Soc.*, **111**, 874–7.
HAUSER, J. R., 1965, *IEEE Trans. Electron Devices*, **ED-12**, 605–18.
HOOPER, W. W., and HOWER, P. L., 1967, Int. Electron Dev. Meeting, Washington, D.C.
HOWER, P. L., 1967, *Technical Report* No. 4726–1, Stanford Elec. Lab., Stanford, Calif.
KNIGHT, J. R., EFFER, D., and EVANS, P. R., 1965, *Solid-State Electron.*, **8**, 178.
MASON, S. J., 1954, *IRE Trans. Cir. Theory*, **CT-1**, 20–5.
MEAD, C. A., 1966, *Proc. IEEE*, **54**, 307–8.
SHOCKLEY, W., 1952, *Proc. IRE*, **40**, 1365–76.

Implications of carrier velocity saturation in a gallium arsenide field-effect transistor

J. A. TURNER and B. L. H. WILSON

Allen Clark Research Centre, The Plessey Company Limited, Caswell, Towcester, Northants, England

Abstract. The high mobility and peak velocity of electrons in GaAs favour its use in a field-effect transistor. In structures of short gate length L the electrons may attain their peak velocity at the drain end of the gate electrode before the channel is pinched off by the ohmic potential drop along it, that is while the elementary theory shows the F.E.T. to be unsaturated. If the drain voltage is increased beyond this point, further potential drop must take place at the drain end of the channel which can deepen only in so far as the field can diverge free of the gate equipotential: the drain current remains essentially constant in channels of an adequate ratio L/a of length to depth.

A model of this behaviour predicts how the drain pinch-off voltage $V_d = V_d{}'$ $[I_d = I_{d\,\text{max}}\, V_g = 0]$ become less than the cut off voltage $V_g = V_0$ $[I_d = 0]$. The transconductance is less and the input capacitance greater than the values predicted by the normal theory, so that the gain–bandwidth product finally becomes proportional to L^{-1} rather than L^{-2}. The transconductance also decreases less rapidly with gate voltage.

Practical structures have been fabricated in epitaxial layers on a semi-insulating substrate using a metal semiconductor barrier diode for the gate. The results from these structures have been compared with the theory and reasonable agreement is obtained. The advantages of using GaAs for this device are outlined.

1. Introduction

The high majority carrier mobility in GaAs makes it an obvious choice for a high frequency F.E.T. and the application of classical F.E.T. theory suggests that operation at say X-band can be obtained for gate lengths of a few microns. Capacitative strays can be much reduced by the use of semi-insulating GaAs. However, for structures with gate lengths L under ten microns the carriers approach their limiting velocity and transit-time-limited operation may be expected to give a frequency limitation tending to L^{-1} rather than L^{-2} expected from the classical treatment.

It might also be expected that Gunn oscillations or negative resistance would be found: these are seldom found in practice in the structures to be described and the gate electrode appears to inhibit beneath it the production of fields above that required to allow carriers to attain their maximum velocity. In so far as this is true a relatively simple theory allows analytical expressions to be developed for the transconductance, input capacitance, and gain–bandwidth product, at least for structures with a large ratio of channel length to depth. The theory describes the entry into velocity saturation in terms of a single parameter which can be expressed in terms of:

(1) The ratio of gate voltage required to pinch off the channel for zero drain current to that required to pinch off the drain current for zero gate voltage.

(2) The ratio of the field giving maximum velocity to the field across the fully depleted channel for zero drain current.

(3) The dimensions and conductivity of the channel.

2. Theory

Two mechanisms may be responsible for the saturation of current observed in field-effect transistors. In the case considered by Shockley (1952), the ohmic potential drop in the channel becomes comparable with the potential, V_0, necessary to cut off the channel in the absence of current. For a long channel it is possible to separate the saturated channel into two regions. In the 'gradual' region nearer the source the 'gradual approximation' holds; the edge of the depletion region is nearly parallel to the direction of current flow so that the voltage across the depletion region at any point is equal to the sum of the gate voltage and the ohmic potential drop down the appropriate portion of the channel. In the 'fully depleted' region nearer the drain the potential is determined by the electrodes, the fixed ionized impurities, and the condition of current continuity at its boundaries. The matching between these solutions has been considered in detail by Wu and Sah (1967). A third region of ohmic potential drop may also occur between the 'depleted' region and the drain contact, that is in the passive drain series resistance. The fully depleted region can only penetrate into the channel for a distance of the order of the channel width, so that for long channels the current remains nearly constant at a value which can be predicted from the behaviour in the gradual regions. For channels of shorter aspect ratio the effective channel length falls appreciably with increasing drain voltage and finite drain conductance dI_d/dV_d occurs even when the channel is pinched off. It may be shown that where the carriers have not reached velocity saturation when they enter the 'depleted' region the subsequent deepening (and narrowing) of the channel is such that velocity saturation is also not serious in the 'fully depleted' region.

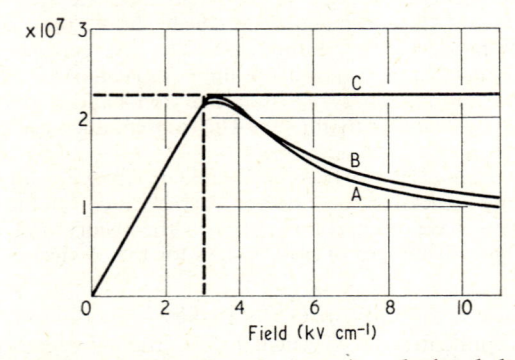

Figure 1. Velocity field characteristics for GaAs. A, calculated by Butcher and Fawcett; B, measured by Ruch and Kino; C, simplified form used in F.E.T. analysis.

The alternative possibility is that velocity saturation occurs while the gradual approximation still holds in the channel, as originally suggested by Grosvalet *et al.* (1963). We suppose in the first instance that the velocity field curve has the simple form shown in figure 1. The situation is illustrated in figure 2, for a one sided F.E.T. with a Schottky barrier gate. The source-drain voltage has increased to the point where velocity saturation has just been reached at the narrowest point of the channel PP'. Up to this point the channel has been in the unsaturated condition as described by normal F.E.T. theory and the Shockley gradual approximation holds, at least on the source side of the constriction. In a long narrow channel further increase of drain current is not possible. For suppose that on increasing the

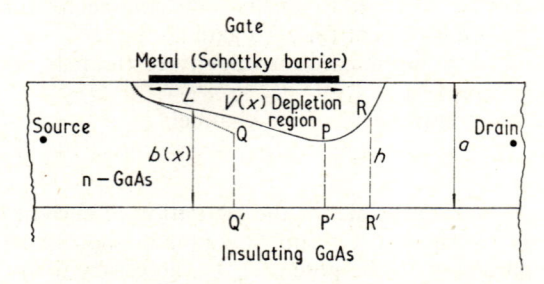

Figure 2. Cross-section of a Schottky barrier F.E.T.

voltage between source and drain the current does increase, the ohmic potential drop in the channel will be more rapid and velocity saturation must occur earlier say at QQ'. Further narrowing of the channel is impossible as the free carriers have already reached their saturation velocity, unless the channel is simultaneously deepened to allow current flow by additional free electrons. But this implies that the 'fully depleted' region has penetrated to the plane QQ', substantially under the gate electrode. This is only possible for distances of the order of the channel width and for a long narrow channel the effect is negligible. But narrowing of the channel is required if the ohmic potential drop a short distance to the right of Q is to be equated to the extra potential drop due to ionization of further fixed impurities between the gate electrode and the edge of the depletion region. Thus the constriction PP' cannot in fact move substantially towards the source end of the channel and for a long narrow channel the source-drain current remains constant for further increases in voltage. This voltage must be dropped to the right of PP' where the channel may deepen unconstrained by the gate electrode equipotential. The solution in this region appears to be a genuine two-dimensional problem which can for example be solved by relaxation methods as indicated by Magowan and Ryan (1968).

The equations determining the breakpoint on the normal current–voltage curve due to velocity saturation have been outlined for a uniformly doped channel by Hauser (1967). The channel width at the constriction at the onset of velocity saturation has been reduced from its value a to a value $a(1-u)$ where

$$u^2 = (V_d' - V_g)/V_0 \tag{1}$$

Here V_d' is the voltage between the end of the channel and the source, and will be equal to the drain voltage at the onset of velocity saturation, after subtracting the voltage in the passive drain series-resistance between the end of the channel and the drain electrode. V_g is the applied gate voltage, usually for an n-type channel. V_0 is the cut-off voltage, the gate-source voltage which cuts off the channel conductance for zero drain voltage.

The current flowing through the constriction is therefore

$$I_m = I_0(1-u) \tag{2}$$

where

$$I_0 = nqv_m Za. \tag{3}$$

Now as the current is just entering velocity saturation it can also be obtained from the normal theory of the unsaturated junction F.E.T. which gives

$$I_d = I_p(3u^2 - 2u^3 - 3t^2 + 2t^3) \tag{4}$$

where

$$I_p = V_0 nqu\, aZ/(3L), \tag{5}$$

the pinch-off current for zero gate bias given by normal junction F.E.T. theory

$$t^2 = (V_s - V_g)/V_0. \tag{6}$$

We shall usually suppose that the source voltage V_s is zero.

If we put $I_d = I_m$, we obtain the value of u for a given t (or the drain voltage for a given gate voltage), at which the break occurs in the junction F.E.T. characteristics owing to velocity saturation.

For example for $t=0$,

$$\frac{I_0}{I_p} = \frac{3u^2 - 2u^3}{1-u}. \tag{7}$$

The quantity I_0/I_p may be expressed in terms of the channel dimensions from (3) and (5), putting $V_0 = nqa^2/2\epsilon$

$$I_0/I_p = 6\epsilon v_m L/\sigma a^2 = 3E_m L/V_0. \tag{8}$$

The cubic (7) relates I_0/I_p to the quantity u which is directly observable. The reduced current at which saturation occurs may be calculated from equations (2) and (7) (cf. figure 11 of Hauser 1967).

Figure 3 shows the effect of velocity saturation for the general case for different values of the calculated quantity I_0/I_p. The full lines show the normal F.E.T. characteristic for a long channel. Owing to velocity saturation this characteristic is subject to a break at the point of intersection with the dotted lines. The break point varies according to the parameter I_0/I_p. At the point of intersection the current saturates as is shown explicitly for the

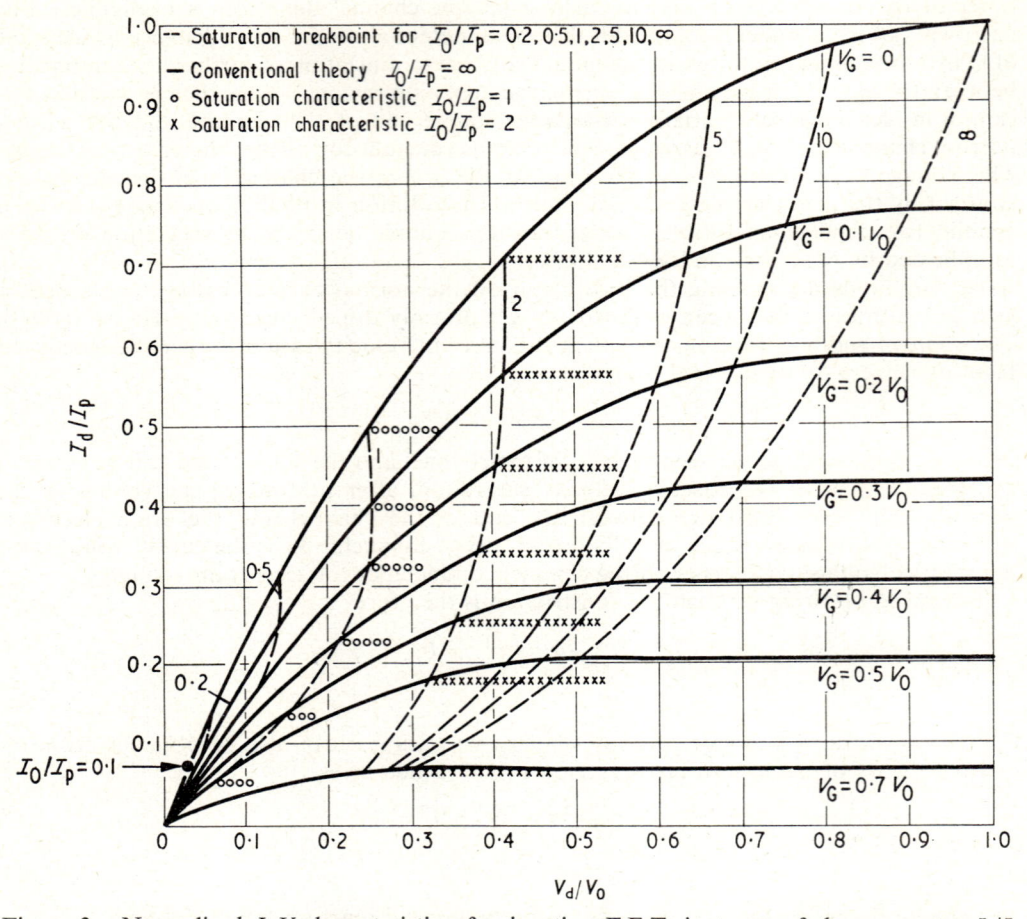

Figure 3. Normalized I–V characteristic of a junction F.E.T. in terms of the parameter I_0/I_p describing velocity saturation.

cases $I_0/I_p = 2$ and ∞. The sharp breakpoint is, of course, associated with the break in the velocity field characteristic and will be rounded in practice. For $I_0/I_p > 1$ the channel is more or less fully depleted before velocity saturation occurs and its effect will be less severe.

The effect of velocity saturation is seen to produce a number of marked effects on the characteristics, which can be used to diagnose its presence and estimate its effect:

(1) The pinch-off voltage V_d' in the presence of current is less than the cut off voltage V_0 for zero source-drain current.

(2) The pinch-off voltage V_d' depends weakly on gate voltage near zero gate voltage. V_d' may even increase with increasing $|V_g|$.

(3) The spacing of the characteristics becomes more uniform, i.e., the saturation transconductance becomes more uniform and not necessarily a monotonic decreasing function of gate voltage, $|V_g|$. The spacing of the characteristics is also changed in the normal F.E.T. theory by variation in carrier concentration in the depth of the channel.

(4) The critical field $(V_d' - V_g)/L$ for the onset of velocity saturation tends to E_m as $u \to 0$ or as $I_0/I_p \to 0$. Significant deviations from this value occur in structures at present feasible.

(5) As the temperature of the device is changed, large changes occur in the quantity I_0/I_p, as a consequence of the change in mobility.

(6) Velocity saturation is particularly easily observed in GaAs F.E.T.s, free from thermal effects and breakdown and at normal working voltages.

In practice the velocity-field characteristic does not have the ideal shape shown in figure 1. For $v < v_m$ the departures are apparently small in GaAs. The figure also shows the experimental curve obtained by Ruch and Kino (1967) and the theoretical curve computed by Butcher and Fawcett (1966). Once the peak velocity has been reached it is possible that the decreasing velocity field characteristic becomes important. This will, however, only exacerbate the situation at a plane QQ' within the 'gradual' ohmic channel, as a larger ohmic potential drop than E_m will be more difficult to take up in an increased potential drop across the walls of the channel. Thus it seems that the pinch-off PP' will still remain substantially fixed and determine the current–voltage characteristic. The occurrence of Gunn domains in the channel seems unlikely as they would tend to be nucleated at the constriction PP' where the field is highest, at least for voltages not too much in excess of the saturation voltage. Deviations from the present theory are most likely to occur for short wide channels, small gate voltage, and large drain voltages. They presumably occurred in some of the structures described by Winteler and Steinemann (1967). Negative resistance effects have only occasionally been observed in the devices described below under the conditions of operation.

3. Small signal parameters

3.1. *Transconductance*

The transconductance in velocity saturation is not equivalent to that just outside it, as the imposition of a change in gate voltage ΔV_g alters the point at which velocity saturation is reached.

Equations (1), (2), and (4) imply that

$$I_d' = I_d'(V_d', V_g) \quad \text{and} \quad I_m = I_m(V_d', V_g) \tag{9}$$

so that for a given V_g, the condition $I_d' = I_m$ determines V_d and hence I_d'. The transconductance is calculated by considering an excursion ΔI due to a gate voltage change ΔV_g, subject to the condition that

$$I_d' + \Delta I_d' = I_m + \Delta I_m \tag{10}$$

and then by proceeding to the limit of $\Delta I / \Delta V_g$. Thus

$$dI_d' = \frac{\partial I_d'}{\partial V_d'} dV_d' + \frac{\partial I_d'}{\partial V_g} dV_g = \frac{\partial I_m}{\partial V_d'} dV_d + \frac{\partial I_m}{\partial V_g} dV_g = dI_m \tag{11}$$

Eliminating dV_d,

$$\frac{dI_d'}{dV_g} = \left(\frac{\partial I_d'}{\partial V_d} \frac{\partial I_m}{\partial V_g} - \frac{\partial I_m}{\partial V_d} \frac{\partial I_d}{\partial V_g} \right) \Big/ \left(\frac{\partial I_d}{\partial V_d} - \frac{\partial I_m}{\partial V_d'} \right) \tag{12}$$

Substituting in equations (1), (4), and (6) gives the transconductance

$$g_m = \frac{dI_d'}{dV_g} = \frac{3I_p}{V_0} \frac{(1-t)(3u^2 - 2u^3 - 3t^2 + 2t^3)}{6u - 9u^2 + 4u^3 - 3t^2 + 2t^3}. \tag{13}$$

When $V_g = V_s = 0$, $t = 0$

$$g_m = \frac{3I_p}{V_0} \frac{(3-2u)u}{6-9u+4u^2} = \frac{g_{m0}(3-2u)u}{6-9u+4u^2}. \tag{14}$$

Equation (14) describes the reduction in transconductance, owing to velocity saturation, from the value $g_{m0} = 3I_p/V_0 = \sigma Za/L$ given by normal junction F.E.T. theory. The expression may also be written as

$$g_m = \frac{6(1-u)}{(6-9u+4u^2)u} v_m \epsilon \frac{Z}{a} = \frac{6(1-u)}{(6-9u+4u^2)u} g_{m1}. \tag{15}$$

The variation of g_m/g_{m1} is also shown in figure 4 and it will be seen that g_m/g_{m1} increases without limit as $u \to 0$:

$$g_m \to (v_m \epsilon \sigma/2L)^{1/2} Z \quad u \to 0 \tag{16}$$

The limit for a given gate width is set only by voltage breakdown and dissipation in high conductivity material.

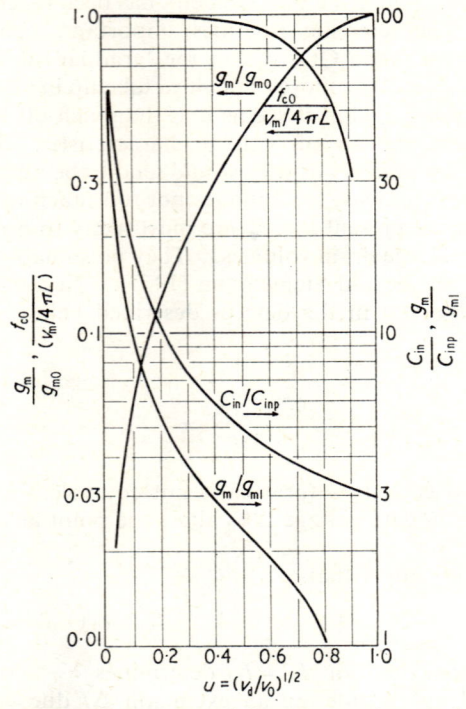

Figure 4. Variation of transconductance and input capacitance for zero gate bias with the parameter u.

3.2. Input capacitance

The procedure used for calculating the input capacitance is to determine the total charge associated with the depletion region and to determine its rate of change with gate voltage under the conditions of F.E.T. operation. The charge associated with the input is integrated over the entire gate length, making the usual assumption that the channel is adequately long so that pinch-off occurs at the end of the channel nearest the drain. Charge in the depletion region beyond the pinch-off point is associated with feedback capacitance. The case where velocity saturation does not occur has been treated by Richer (1963) and Das (1966).

Referring to figure 2,

Total charge

$$Q = \int_0^L nqZ[a - b(x)] \, dx \tag{17}$$

$$= \int_0^L nqZa v^{1/2} \, dx \quad \text{where} \quad v = (V - V_g)/V_0 \tag{18}$$

and V is the voltage in the channel at the point x.

The x-dependence is eliminated using the ohmic drop in channel

$$I\,\mathrm{d}x = Znq\mu b(x)\,\mathrm{d}V \tag{19}$$

$$= Znq\mu a(1-v^{1/2})\,\mathrm{d}V \tag{20}$$

integrating,

$$IL = Znq\mu a \int_{v_s}^{v_d} (1-v^{1/2})\,\mathrm{d}V \tag{21}$$

where $v_s = t^2$, $v_d = u^2$ as defined in (1) and (6)

Thus

$$\frac{\mathrm{d}x}{L} = \frac{(1-v^{1/2})\,\mathrm{d}V}{\int_{v_s}^{v_d}(1-v^{1/2})\,\mathrm{d}V}. \tag{22}$$

Hence from (20) and (24)

$$Q = nqZaL\,\frac{\left[4v^{3/2} - 3v^2\right]_{v_s}^{v_d}}{2\left[(3v - 2v^{3/2})\right]_{v_s}^{v_d}}$$

$$= \frac{V_0 ZL\epsilon}{a}\,\frac{4u^3 - 3u^4 - 4t^3 + 3t^4}{3u^2 - 2u^3 - 3t^2 + 2t^3}. \tag{23}$$

The desired quantity $\mathrm{d}Q/\mathrm{d}V_g$ is normally obtainable for an unsaturated F.E.T. through the explicit dependence of u and t on V_g. An additional term arises when velocity saturation has just occurred through the dependence of the voltage V_d' under which velocity saturation occurs on the gate voltage.

$$C_\mathrm{in} = \frac{\mathrm{d}Q}{\mathrm{d}V_g} = \frac{\partial Q}{\partial t}\frac{\partial t}{\partial V_g} + \frac{\partial Q}{\partial u}\frac{\partial u}{\partial V_g} + \frac{\partial Q}{\partial u}\frac{\partial u}{\partial V_d'}\frac{\mathrm{d}V_d'}{\mathrm{d}V_g} \tag{24}$$

$$= \frac{\partial Q}{\partial t}\frac{\partial t}{\partial V_g} + \frac{\partial Q}{\partial u}\frac{\partial u}{\partial V_g}\left(1 - \frac{\mathrm{d}V_d'}{\mathrm{d}V_g}\right). \tag{25}$$

The quantity $\mathrm{d}V_d'/\mathrm{d}V_g$ may be obtained from equation (11)

$$\frac{\mathrm{d}V_d'}{\mathrm{d}V_g} = -\left[\frac{\mathrm{d}I_d'}{\mathrm{d}V_g} - \frac{\mathrm{d}I_m}{\mathrm{d}V_g}\right] \Big/ \left[\frac{\mathrm{d}I_d'}{\mathrm{d}V_d'} - \frac{\mathrm{d}I_m}{\mathrm{d}V_d}\right]. \tag{26}$$

If $t=0$,

$$\frac{\mathrm{d}V_d'}{\mathrm{d}V_g} = \frac{4u-3}{6-9u+4u^2}. \tag{27}$$

It is interesting to note that although $\mathrm{d}V_d'/\mathrm{d}V_g = 1$ for $u=1$ as expected, it becomes negative for $u < \frac{3}{4}$, as indicated in figure 3.

Equations (25) and (26) give rise to a complex general solution for the input capacitance as an algebraic function of u and t.

If $t=0$

$$C_\mathrm{in} = \frac{\epsilon ZL}{a}\,\frac{3(36 - 84u + 67u^2 - 18u^3)}{u(3-2u)^2(6-9u+4u^2)}. \tag{28}$$

The u-dependence of C_in is shown in figure 4. The value for $u=1$, $C_\mathrm{inp} = 3\epsilon ZL/a$, is three times the input capacitance when the channel under the gate is fully depleted.

The increase in transconductance as u falls is then partially compensated by a rise in input capacitance.

The current-gain–bandwidth product is given by

$$f_{co} = \frac{g_m}{2\pi C_{in}} = \frac{v_m}{4\pi L} \frac{4(1-u)(3-2u)^2}{(36-84u+67u^2-18u^3)} \qquad (29)$$

from (15) and (28).

Figure 4 shows how f_{co} approaches $V_m/4\pi L$ as $u \to 0$. Thus f_{co} can approach within a factor of 2 of the naive transit time frequency. This factor arises from the use of a channel of finite thickness and can be reduced by applying gate voltage or by using a conducting channel whose conductivity increases with distance from the gate.

4. Limitations of present treatment

A number of further refinements would be required in a full treatment. The input capacitance is essentially a lumped element and at high frequencies the distributed nature of the R–C transmission line must be taken into account. Practical devices also incorporate a series resistance r_s in the source which reduces the transconductance to $g_m/(1+r_s g_m)$. In the structure illustrated in figure 5 where the source-to-gate spacing equals the gate length the series resistance is approximately $1/g_{mp}$ and the classical current gain bandwidth product is reduced to $g_m/2$ for $u=1$. Further resistance is associated with the contacts. As velocity saturation becomes important g_m/g_{mp} declines and the series resistance becomes less important. The effect favours designs for lower values of u. Some capacitive strays are to be expected at each terminal from the mounting. Stray capacitance from the bonding pads, etc., can, however, be much reduced by the use of semi-insulating GaAs. Practical devices also show an impurity concentration gradient in depth as the epitaxial interface with semi-insulating material is seldom sharp.

5. Experiment

The validity of the theory presented in the previous section was tested by comparing experimental and theoretical results of batches of GaAs F.E.T.s with gate lengths of 25 μm, 12 μm, and 4 μm. The devices were produced in n-type ($2-6 \times 10^{15}$ cm^{-3}) epitaxial layers of GaAs 1–2 μm thick grown on a semi-insulating GaAs substrate. The ohmic source and drain contact were produced by direct metallizing with a silver-tin alloy and the gate was an aluminium metal-semiconductor barrier diode.

The structure in figure 5 was used for the 12 μm and 4 μm gate devices. In this structure the gate contact area is removed from the active layer so as to minimize the capacity associated with it. This is achieved by mesa etching the layer and running the gate metallizing down the side of the mesa and on to the semi-insulating substrate. Figure 5 shows a completed 4 μm device.

The transconductance, input capacitance, and the effect of gate voltage on the transconductance were determined for these devices.

6. Correlation: practice with theory

To effects a correlation of experimental and theoretical results, many parasitic effect must be taken into account. These include the reduction of the transconductance by series resistance and the increase of the input capacitance introduced by the inter-electrode capacity of the header on which the device is mounted. Both these degrading effects, however, can be inferred from direct electrical measurements. The series resistance is the combination of the resistance between the end of the channel and the edge of the source contact with the contact resistance between the metal and GaAs. It can be estimated by comparing the resistance of two adjacent devices of different source/drain spacing measured at low drain voltages. The inter-electrode capacitance of the header can be determined by direct measurement on an a.c. bridge. A further effect that impedes exact correlation is the occurrence of a carrier concentration gradient in the layer.

All of the 25 μm gate length devices examined had values of I_0/I_p approaching $\infty (u=1)$

that is, the carriers in the channel were not velocity limited and the measured parameters could be predicted by the classical theory.

The classical theory failed for the 12 μm and failed completely for the 4 μm devices, for in these devices velocity saturation occurs. Table 1 below compares the measured parameters with those of the theory presented here and the Shockley classical theory. These results are typical of many batches of devices examined in this way.

Table 1

Gate length	Measured		Present Theory		Shockley Theory	
	g_m (mmhos)	C_{in} (pF)	g_m (mmhos)	C_{in} (pF)	g_m (mmhos)	C_{in} (pF)
12 μm	3·2	0·85	2·8	0·9	4·0	0·71
4 μm	4·0	0·35	5·2	0·4	9·0	0·21

The I–V characteristics shown in figure 6 are for a 4 μm gate length device. This displays some qualitative features of the theory, namely the uniformity of the transconductance with gate voltage and the difference in magnitude of the drain voltage at which saturation occurs and the cut-off voltage. In practice the voltage dropped in the source and drain series resistance causes the current to saturate at higher drain voltages than predicted by the theory for the intrinsic transistor.

7. Conclusion

The theory presented here derives expressions for transconductance and input capacitance in a junction gate field-effect transistor where carriers in the channel region become velocity limited. The main features of the theory are summarized below.

(1) As velocity saturation becomes dominant the transconductance varies as $(\sigma/L)^{1/2}$.

(2) The limiting velocity of carriers in the channel region determines the maximum operating frequency.

(3) It is possible to obtain a value of u (determined by material properties and device dimensions) for which there is an approximately linear dependence of transconductance with gate voltage. This is important in reducing distortion in large-signal operation.

(4) The degree of velocity saturation can be determined from the I–V characteristics of the devices by comparing the drain voltage at which current saturation occurs with the cut-off voltage.

The maximum operating frequency of an F.E.T. is ultimately limited by the transit time of carriers across the channel region. This implies that the argument for producing a very high frequency F.E.T. in GaAs rather than in Si simply because of its higher electron mobility is no longer valid. However, higher frequency operation will still be obtained in GaAs because of its higher limiting velocity and the higher mobility implies less degradation by parasitic series resistance. Other advantages of GaAs are the low electric fields at which velocity saturation occurs, reducing power dissipation, and the existence of semi-insulating GaAs as a substrate material, enabling devices with low capacitative strays to be fabricated. Higher transconductance (g_m) devices are also obtained, for in velocity saturation $g_m \propto \text{(mobility)}^{1/2}$.

Many advantages still exist for using GaAs as a material in which to fabricate an F.E.T. It appears that X-band operation is technically and theoretically feasible.

Acknowledgments

The authors wish to thank the directors of The Plessey Company Limited for permission to publish this note, A. Jonscher for much helpful discussion, A. M. McSwann and R. Allen for computation, J. R. Knight and his team for the supply of epitaxial layers, and P. Evans for assistance in device fabrication.

References

BUTCHER, P. N., and FAWCETT, W., 1966, *Phys. Letters*, **21**, 489.
DAS, M. B., 1966, *Proc. IEE*, **113**, 1565.
HAUSER, J. R., 1967, *Solid State Electronics*, **10**, 577.
MAGOWAN, J. A., and RYAN, W. D., 1968, *Electronics letters*, **4**, 5 March.
RICHER, I., 1963, *Proc. IEEE*, **51**, 1249.
RUCH, J. G., and KINO, E. C. S., 1967, *Appl. Phys. Lett.*, **10**, 40–2.
SHOCKLEY, W., 1952, *Proc. I.R.E.*, **40**, 1374.
WINTELER, H. R., and STEINEMANN, A., 1967, *Proc. 1966 Symp. on GaAs* 228.
WU, S. Y., and SAH, C. T., 1967, *Solid State Electronics*, **10**, 593.

Study of GaAs devices at high temperature[†]

F. H. DOERBECK, E. E. HARP and H. A. STRACK
Texas Instruments Incorporated, Dallas, Texas, U.S.A.

Abstract. GaAs n–p–n and p–n–p transistors as well as diodes and field-effect transistors were fabricated. Different GaAs starting materials and fabrication technologies were compared. A study of the leakage current and the current gain of n–p–n transistors showed that their current gain decreases at high temperatures exponentially according to an activation energy of 0·2 ev, and that the dominating parameter is the base transport factor. The critical parameter for high temperature operation is the leakage current and its stability. This applies also to p–n–p transistors. Schottky barrier field-effect transistors are limited by the high gate leakage current and the increase of the substrate conductivity.

Life test experiments with GaAs transistors and diodes showed that these devices are stable to an ambient temperature of 300°C and that they degrade during operation at 400°C. To test GaAs devices in microelectronic circuits, thick film logic gates and thin film amplifier circuits were built and operated at 300°C. Thin film resistors proved to be stable for operation at 400°C.

1. Introduction

One of the first semiconductor devices realized in GaAs was a bipolar transistor. The high electron mobility and the large bandgap appeared to favour high-frequency, high-temperature devices. Problems relating to surface states in M.O.S. field-effect transistors (Becke *et al.* 1967) and bulk states in GaAs bipolar transistors (Strack 1966) have limited devices so far to frequencies lower than theoretically predicted. It is the purpose of this paper to study GaAs as a semiconductor material for the other possible application, the high-temperature transistors and circuits field.

The upper temperature at which transistors can be operated depends on the bandgap of the semiconductor material. Based on an operating temperature of 370°K for germanium, GaAs transistors can be predicted to operate up to a temperature of 720°K. GaAs transistors have been operated at this theoretical temperature limit; however, the stability of devices at this temperature is generally poor. On the basis of our findings, we believe that a junction temperature of 620°K and an ambient temperature of 570°K are maximum safe temperatures for device operation.

The prediction of the maximum safe junction temperature is based on assumptions of the thermal generation of carriers, and does not take into account mechanisms for the generation of excess leakage currents prevailing in GaAs transistors. The main limitation is indeed not the magnitude of the current gain but the excess leakage current and the stability of the leakage current. This also means that Schottky barrier field-effect transistors cannot be considered as alternatives because of the high gate leakage current. The leakage current is related to the surface property of GaAs. It is therefore the inherent property of the material which limits the fabrication of GaAs devices which can be operated safely at temperatures near the theoretical limit.

A comprehensive study of various types of GaAs transistors, diodes and circuits will be presented. In particular, n–p–n and p–n–p GaAs transistors on bulk and on epitaxial material, as well as field-effect transistors, are discussed. The performance of GaAs devices in thick film and thin film circuits is evaluated.

2. Fabrication of GaAs devices

GaAs n–p–n transistors were fabricated from bulk and epitaxial material. The standard device had a mesa structure with a circular emitter of 3 mils diameter. For the base

[†] The research reported in this paper was in part sponsored by the NASA, Electronics Research Center under Contract NAS 12–537.

formation magnesium was diffused at 1050°C for 75 min using 0·5 mg Mg$_3$Sb$_2$ and 0·3 mg As as dopant in a quartz ampoule of 15 cm^3. The diffused layer was 7 μm deep and had to be etched back to the depth 2 μm and a sheet resistance of 400 Ω/□. Another method of achieving a thin p-layer with a low surface concentration uses a Ga–Zn alloy as diffusion source. This method was also tried, but it offered no advantage in respect to device performance and fabrication yield. The emitter of all n–p–n devices was made by a planar diffusion process. A combination of 1000 Å sputtered SiO$_2$ plus 3000 Å SiO$_2$ deposited by decomposition of tetraethylorthosilicate gave best results. As reported by Strack (1966), addition of iron during the emitter diffusion improves the high-frequency performance and high-temperature stability. The same technology was used here. Using a diffusion source consisting of 2 mg Ga$_2$S$_3$, 1 mg Fe and 5 mg As, the slices were diffused for typically 30 minutes at 925°C. AgInZn and AgInSn alloys were evaporated to form ohmic contacts, as reported by Cox et al. (1967). Mesas were then etched to form individual transistors.

The technique described was applied to different GaAs materials. Sn, S, and Ge doped epitaxial layers, grown in a halide transport reactor, were processed. Also, devices were built on Czochralski-grown GaAs doped with silicon.

GaAs p–n–p transistors were fabricated to compare their high-temperature performance with that of n–p–n devices. Starting material was bulk GaAs doped with cadmium of a concentration of 10^{17} cm^{-3}. For the base formation, Ga$_3$S$_2$ and Fe were diffused for 2 h at 925°C, similar to the emitter diffusion of n–p–n transistors, except for the diffusion time. The emitter was diffused at 625°C with 1 mg Zn and 1 mg As as dopant.

GaAs diodes were fabricated for three reasons: (1) to provide diodes for circuits, (2) to compare the high-temperature performance of diode p–n junctions with transistor p–n junctions to see whether emitter diffusion has an influence on the collector–base junction, and (3) to compare planar and mesa diodes. Mg, Cd, and Zn were diffused into different GaAs materials under conditions similar to those used for transistor fabrication.

As an alternative for bipolar devices we built and tested Schottky barrier field-effect transistors. Starting material was a thin 1–2 μm thick, n-type epitaxial layer grown on chromium doped semi-insulating GaAs substrates. For the source and drain, an AgInGe alloy was employed and evaporated molybdenum or aluminium formed the Schottky barrier gate.

3. Results

First, we will discuss the d.c. current gain, h_{fe}, of n–p–n transistors built by various fabrication processes. Figure 1 shows that for high temperatures h_{fe} of all devices decreases

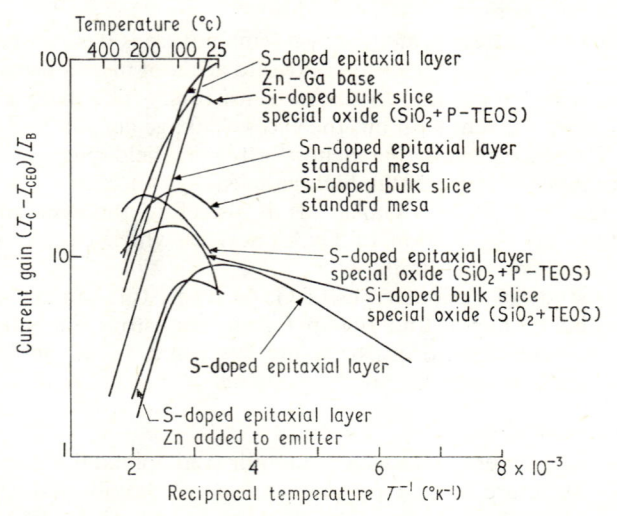

Figure 1. Current gain vs. reciprocal temperature for devices using different crystal materials and technologies.

exponentially. From the slopes of the curves one calculates an activation of 0·2 ev. The devices may be divided into two categories. Type 1 with a high h_{fe} at 25°C shows strong temperature dependence of h_{fe} even at room temperature, while type 2 has an h_{fe} which first increases with rising temperature to a maximum, then decreases, finally approaching the same slope.

To answer the question whether the emitter efficiency or the base transport factor is responsible for the temperature dependence of h_{fe}, the reverse current gain was studied. It was found that it follows the same temperature dependence. Since the impurity distribution in the vicinity of the emitter junction is completely different from that near the collector base junction and since only the base region is common to both junctions, the base transport factor must be the governing parameter. The current gain is proportional to the square of the diffusion length. Since the diffusion coefficient decreases only as $T^{3/2}$, the lifetime seems to decrease strongly with increasing temperature with an activation energy of about 0·2 ev. Strong temperature dependence is observed even at the lowest current levels, as shown in figure 2.

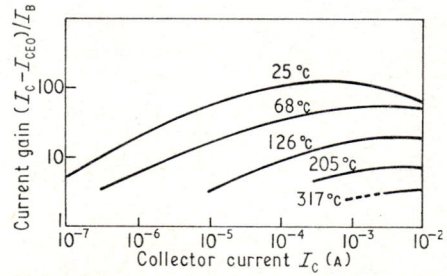

Figure 2. Common-emitter current gain vs. collector current for various temperatures (type 1).

At very low current the quasi-Fermi level of electrons is close to that of holes. The recombination of electrons is dominated by deeper levels only if the base is heavily compensated. One possibility for a base compensation is oxygen. It is possible that the advantageous effect of iron is due to a removal of the oxygen level. Sulphur from the emitter also has a compensating effect on the base, as will be discussed later.

We compare now the current gain as a function of the collector current for devices with poor (type 1) and good (type 2) high temperature performance. Figure 2 shows that this device has an $h_{fe} > 1$ in the 10^{-8} A range. A detailed study shows that h_{fe} increases at low current approximately as the square root of the collector current, as predicted for a transistor which has a recombination current component proportional to $\exp(V/2kT)$, in addition to

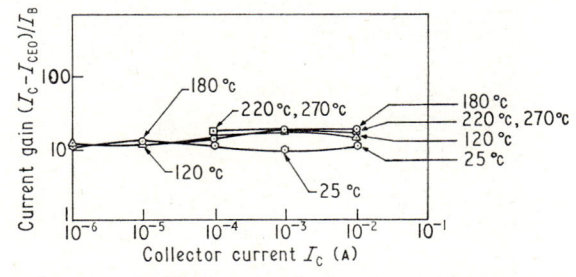

Figure 3. Common-emitter current gain vs. collector current for various temperatures (type 2).

the diffusion current. The reverse current gain also followed the same square-root dependence. The current gain behaviour of a type 2 transistor looks very different. The current gain is independent of collector current as well as of temperature in the temperature range considered, as demonstrated in figure 3. The parameter limiting the high temperature operation of these devices is the emitter leakage current I_{CE0} only.

Figure 4 shows I_{CEO} as a function of the reciprocal temperature for various n–p–n transistors. The collector junction area was 56 mils^2. The data show that the leakage current increases rapidly. In the first approximation the current depends exponentially on the reciprocal temperature with an activation energy of about one-half of the bandgap. From the circuit's point of view, the high leakage current is the limiting factor rather than the magnitude of the current gain. The high-temperature circuits designed for these devices allow for an I_{CEO} of 0·4 mA, so that 340°C is the upper temperature limit.

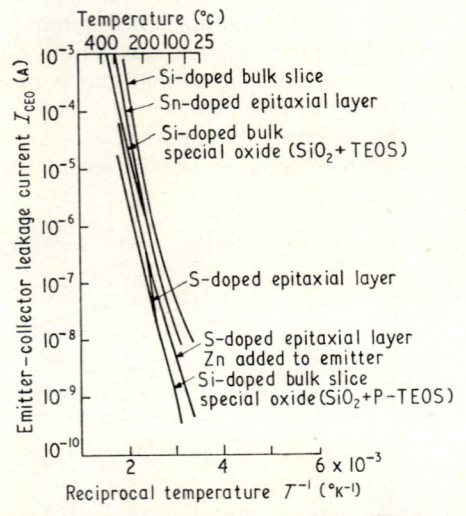

Figure 4. I_{CEO} vs. reciprocal temperature for devices using different materials and technologies.

The base-collector leakage current I_{CBO} in figure 5 is typical for an n–p–n transistor. At low reverse bias I_{CBO} has the same dependence as I_{CEO}. At higher bias and not too high a temperature, the leakage current does not change much with temperature, suggesting a tunnel current.

To answer the question whether p–n–p transistors exhibit a better high-temperature performance than n–p–n transistors, several runs of p–n–p transistors were tested. Figure 6 gives a display of p–n–p transistor characteristics at different temperatures. In addition, I_{CEO} is shown, using a different current scale. The common emitter current gain is low but it

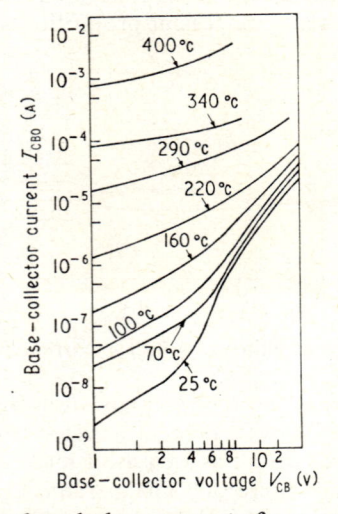

Figure 5. Collector-base leakage current of an n–p–n GaAs transistor.

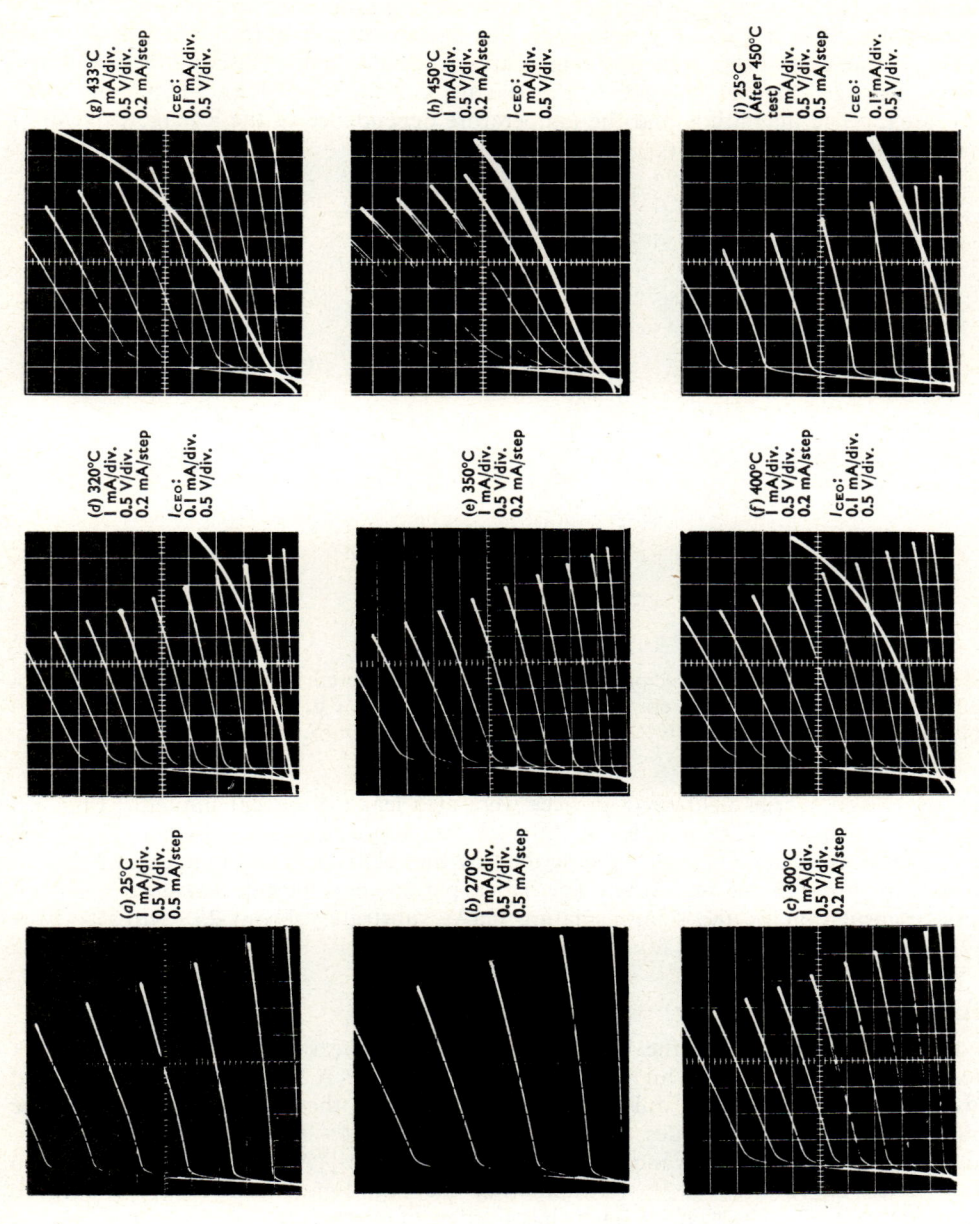

Figure 6. D.c. characteristics of a GaAs p–n–p transistor at different temperatures.

stays nearly constant up to 450°C. The emitter-collector leakage, I_{CE0}, is comparatively high at lower temperatures but also has a weak temperature dependence. P–n–p transistors with h_{fe} up to 300 were also tested. As with high-gain n–p–n transistors, the temperature dependence of I_{CE0} and h_{fe} was strong.

In addition to transistors, we evaluated GaAs diodes, fabricated in epitaxial and bulk crystal material. Cd, Zn, and Mg were employed as dopants. Figure 7 shows the leakage currents as a function of the reciprocal temperature. Planar zinc diffused diodes exhibited the lowest leakage current. At 400°C, the leakage current was 10^{-4} A for a diode with 80 mils^2 area. The magnitude and the temperature dependence of the leakage currents is comparable to that of transistors.

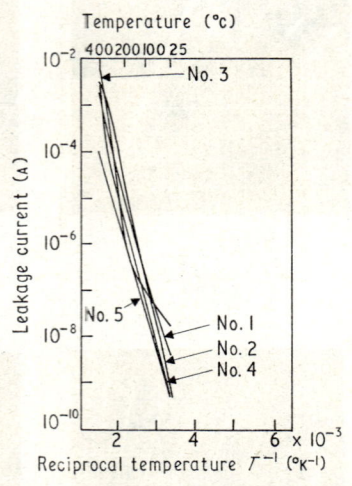

Figure 7. Leakage current vs. reciprocal temperature for different GaAs diodes. No. 1, epi (S, 10^{16} cm^{-3}) planar, Cd; No. 2, epi (Ge, 3×10^{16} cm^{-3}) planar, Cd; No. 3, bulk (Si, 10^{16} cm^{-3}) planar, Mg; No. 4, bulk (Si, 10^{16} cm^{-3}) mesa, Mg; No. 5, epi (S, 10^{16} cm^{-3}) planar, Zn.

GaAs Schottky barrier field-effect devices were also leakage current limited. Figure 8 displays the transistor characteristics at different temperatures. The temperature dependence of the transconductance can be described by the relatively weak temperature dependence of the majority carrier mobility. The critical parameter is the gate leakage current and the loss of insulation by the semi-insulating GaAs substrate. From 25°C to 250°C the substrate conductivity increases four orders of magnitude.

4. Device stability

Studies were performed on the basic limitations of device stability, the rediffusion of emitter and base regions during high-temperature operation. A Mg base layer was diffused followed by a sulphur emitter diffusion. After measuring the p–n junction depths, the samples were sealed in ampoules and then stored at 500°C. After 108 days the base–collector junction depth had advanced from 2·0 μm to 2·8 μm. An estimate of the diffusion constant gives 2×10^{-15} cm^2 s^{-1}. Extrapolating the high-temperature data for Mg to 500°C results in $D = 7 \times 10^{-20}$ cm^2 s^{-1}, which is 4 orders of magnitude lower than the observed value. One possible explanation is that the observed change is not due to a diffusion process but to an in-site conversion of neutral species into acceptor type impurities. A source of such neutral species might be sought in the formation of complexes with electrically inactive sulphur introduced during the emitter diffusion. Radiotracer measurements showed the presence of inactive sulphur atoms throughout the base and collector region. Measurements of the emitter-base junction gave no evidence for a narrowing or widening of the emitter region.

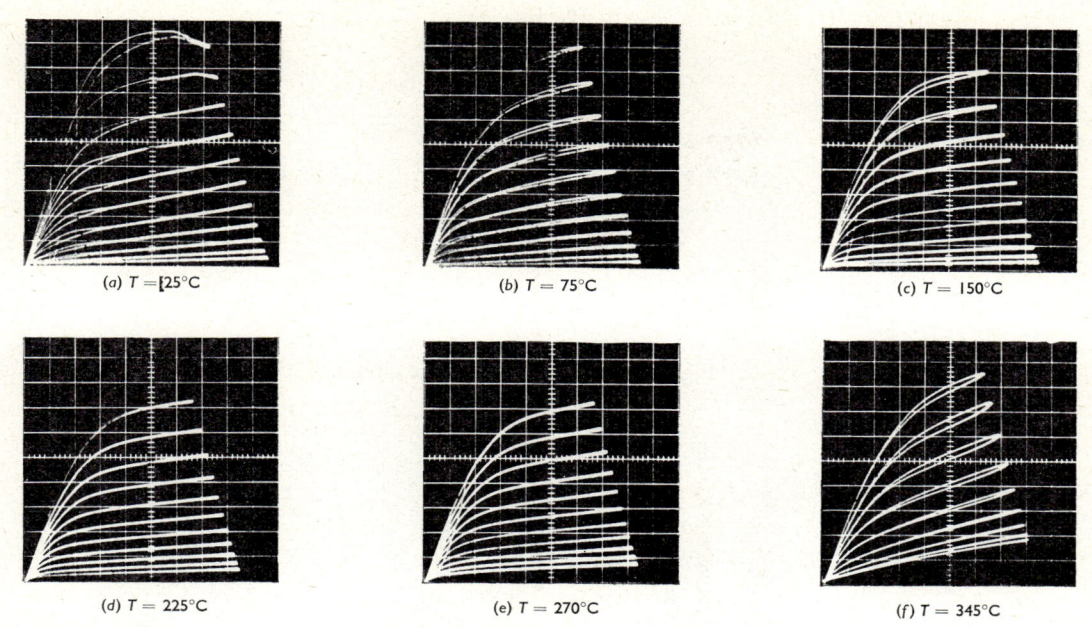

Figure 8. GaAs F.E.T. characteristics at different temperatures. Horizontal: 1 v/div; vertical: 10 mA/div. (a)–(e): 1 v/step; (f): 2 v/step.

Stability studies on actual devices were performed at 300°C and 400°C. Storage of n–p–n transistors at 300°C showed an increase of the 25°C leakage current of 10% to 30%. After operating the transistors at 300°C for more than 200 hours, many devices exhibited degradation effects in their room-temperature performance, while the current gain at 300°C did not show remarkable changes. The h_{fe} at 25°C decreased sometimes to a half of its original value. However, some devices did not degrade at all. The collector-base leakage current at 400°C was monitored as a function of time. The leakage current stayed constant during the first 20 to 50 h. During the following 100 to 200 h, I_{CBO} increased typically by a factor of 2 to 3. Experiments with GaAs diodes gave similar results. Operation at 300°C did not affect the device performance seriously, while operation at 400°C caused increases of the 25°C leakage current by one order of magnitude. No difference in life test performance could be found between mesa and planar devices. It was concluded that 400°C was beyond the maximum safe operating temperature for these GaAs devices.

5. Circuits

To test the performance and stability of microelectronic circuits using GaAs components, thick and thin film circuits were designed and built. For the NAND-gate shown in figure 9, thick-film technology was used. The resistors were fabricated by silkscreening and sintering a mixture of palladium oxide, silver, and a glass frit. Figure 10 shows the transfer characteristics of the gate at different temperatures, with GaAs transistors and diodes incorporated. The circuits were operated at 300°C for 200 h. After the test the circuits still switched; however, the room temperature characteristics were degraded.

Using standard thin-film technology, a feedback pair amplifier, as shown in figure 11, was fabricated. The Ni–Cr resistors were protected by a SiO_2 layer. During life test of the resistors at 300°C for 1000 h, without active devices incorporated, the resistor values increased typically 1%, with a maximal increase of 6%. During 500 h at 400°C the resistor increases were below 10%. With two GaAs transistors, the circuit had at 300°C a constant transimpedance of 2·4 kΩ up to 1 MHz. The input impedance at 300°C was 140 Ω and also frequency independent between 10 Hz and 1 MHz.

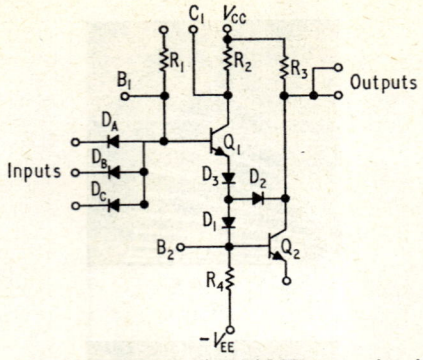

Figure 9. Basic NAND gate circuit.

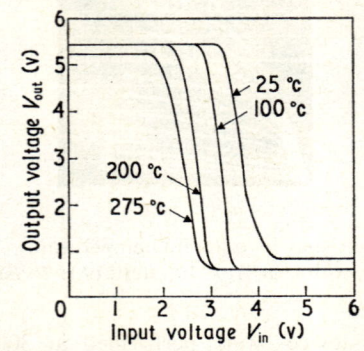

Figure 10. NAND gate transfer characteristics as a function of temperature.

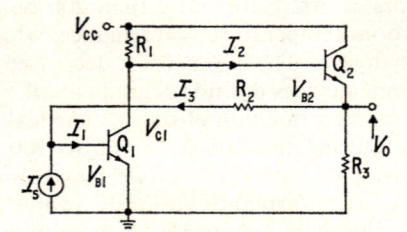

Figure 11. Feedback pair amplifier circuit.

6. Summary

A study was made on the feasibility of operating GaAs devices in high-temperature microelectronic circuits. GaAs n–p–n and p–n–p transistors and p–n diodes were fabricated on bulk and epitaxial material. Thick and thin film circuits, utilizing GaAs devices, were operated at 300°C. Performance at 400°C is poor and rapid degradation of GaAs devices occurs. The main limitation on performance is the high leakage current. This applied also to Schottky barrier field-effect transistors fabricated for comparison with bipolar devices. The following temperature ranges for active devices can be defined as follows: (*a*) below 200°C, silicon devices; (*b*) from 200°C to 300°C, competition between silicon and GaAs devices; silicon devices at the present time have a performance advantage; (*c*) from 300°C to 400°C, GaAs devices, provided the stability can be improved; (*d*) above 400°C, other materials such as GaP or SiC.

References

BECKE, H., and WHITE, J., 1967, *Electronics*, June 12, p. 82.
COX, R. H., and STRACK, H., 1967, *Solid State Electronics*, **10**, 1213–8.
STRACK, H., 1966, *Proceedings of the 1966 Gallium Arsenide Symposium* (London: Institute of Physics and Physical Society, 1967), p. 206.

PLATE XXIV—*P. L. Hower et al.: paper* 29

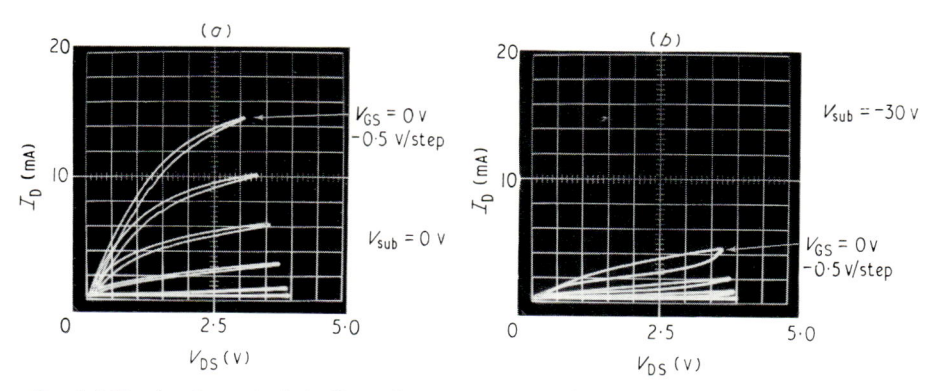

Figure 3. (a) Drain characteristic for substrate connected to the source. (b) Negative voltage applied to substrate with respect to the source.

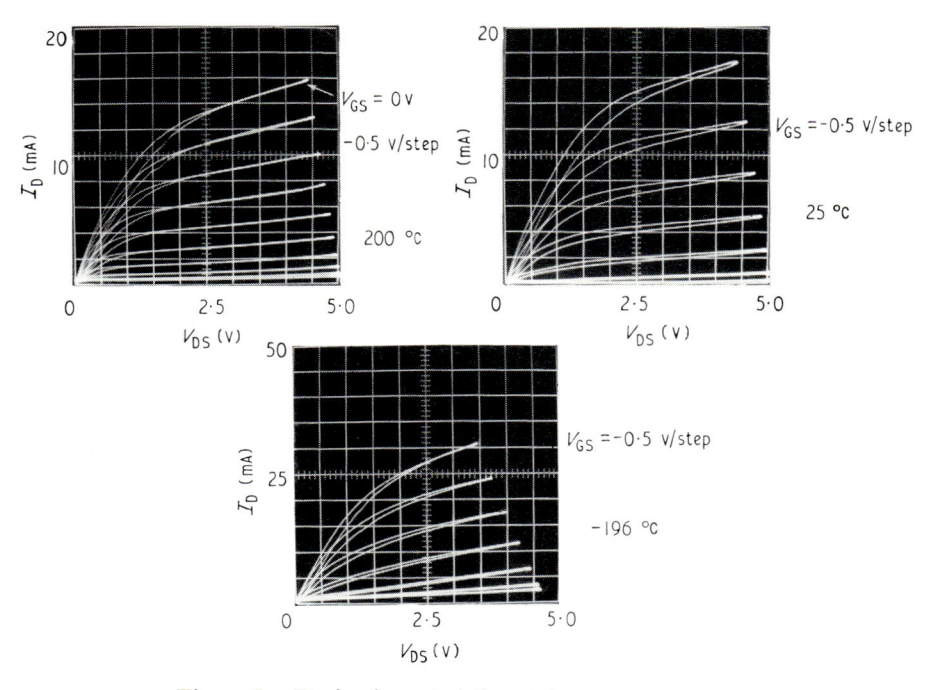

Figure 5. Drain characteristics at three temperatures.

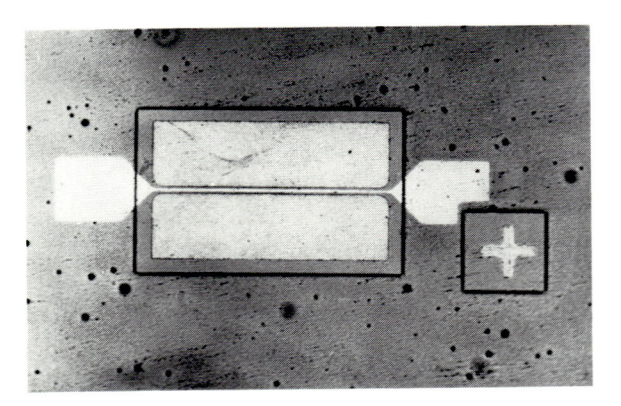

Figure 5. F.E.T. structures with 4 μm gate length.

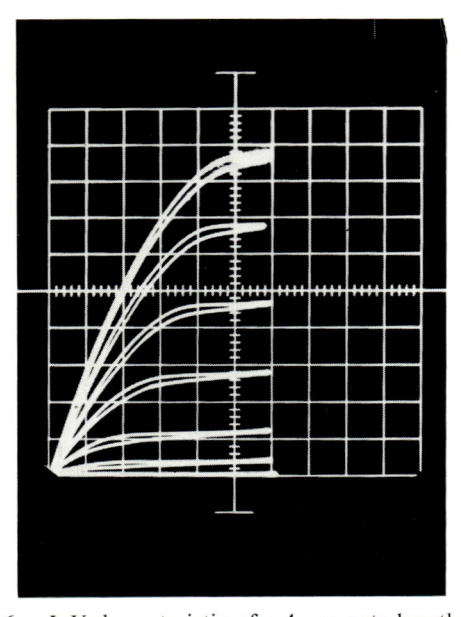

Figure 6. I–V characteristic of a 4 μm gate length device.

PLATE XXVI—*R. E. Enstrom and J. R. Appert: paper* 32

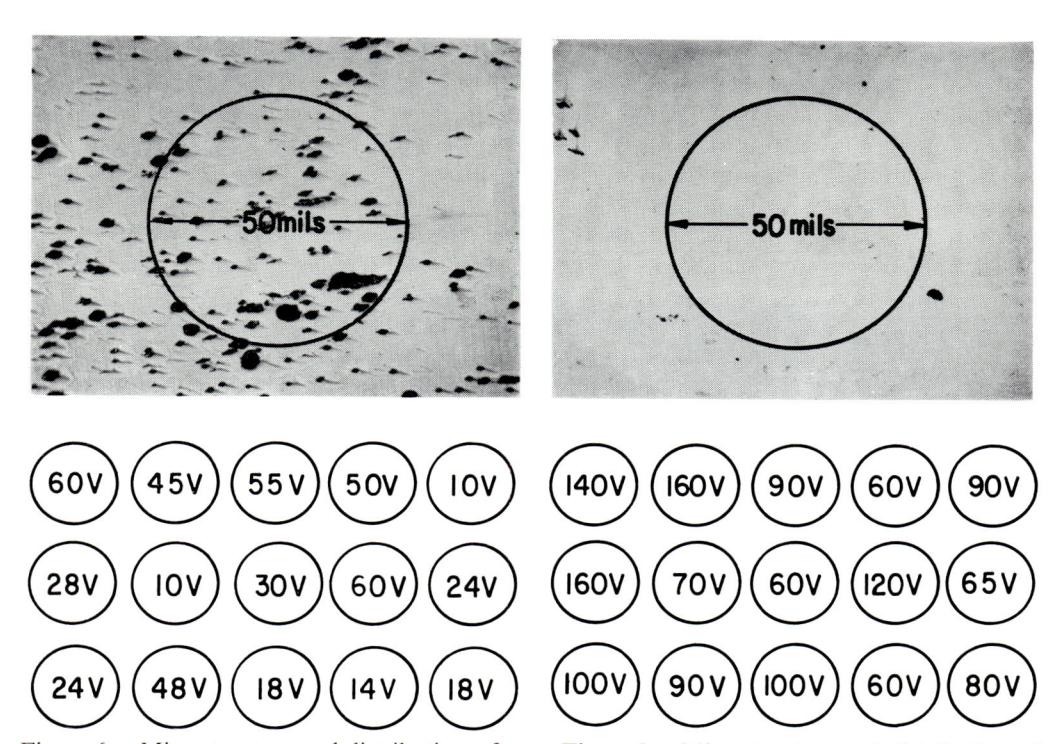

Figure 1. Microstructure and distribution of breakdown voltages for diodes on 533 substrate (Czochralski-grown, Te-doped). Deposition was made on 490 and 533 substrates simultaneously during a single run (91).

Figure 2. Microstructure and distribution of breakdown voltages for diodes on 490 substrate (Czochralski-grown, Te-doped). Deposition was made on 490 and 533 substrates simultaneously during a single run (91).

PLATE XXVII—R. E. Enstrom and J. R. Appert: paper 32

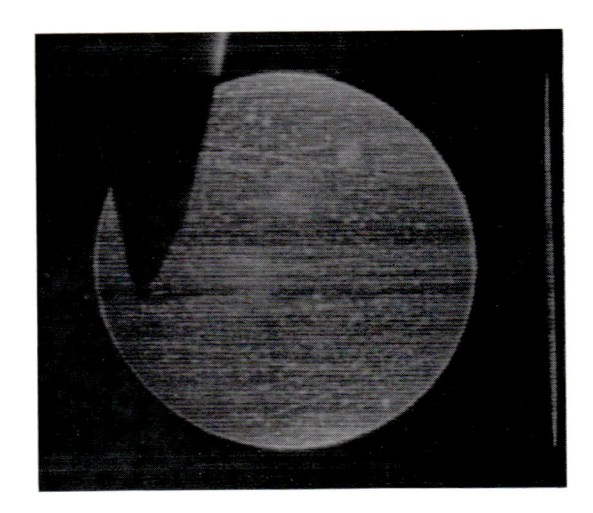

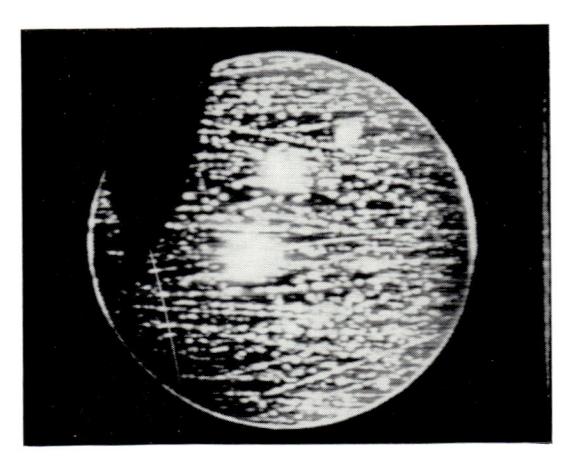

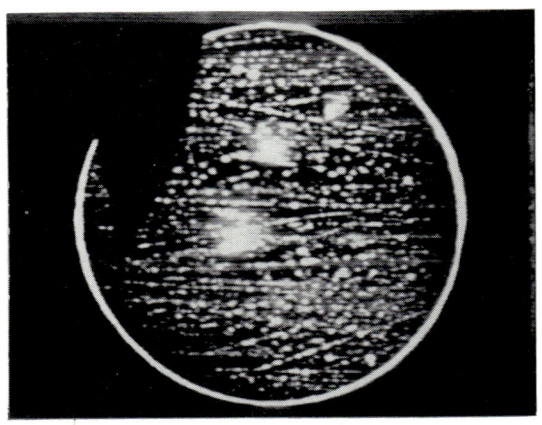

Figure 3. Scanning electron micrograph of 60-v breakdown, 50-mil-diameter diode as function of reverse bias (run 91–490). $E_{bias}=$ (top) 0 v; (centre) -3 v; (bottom) -55 v.

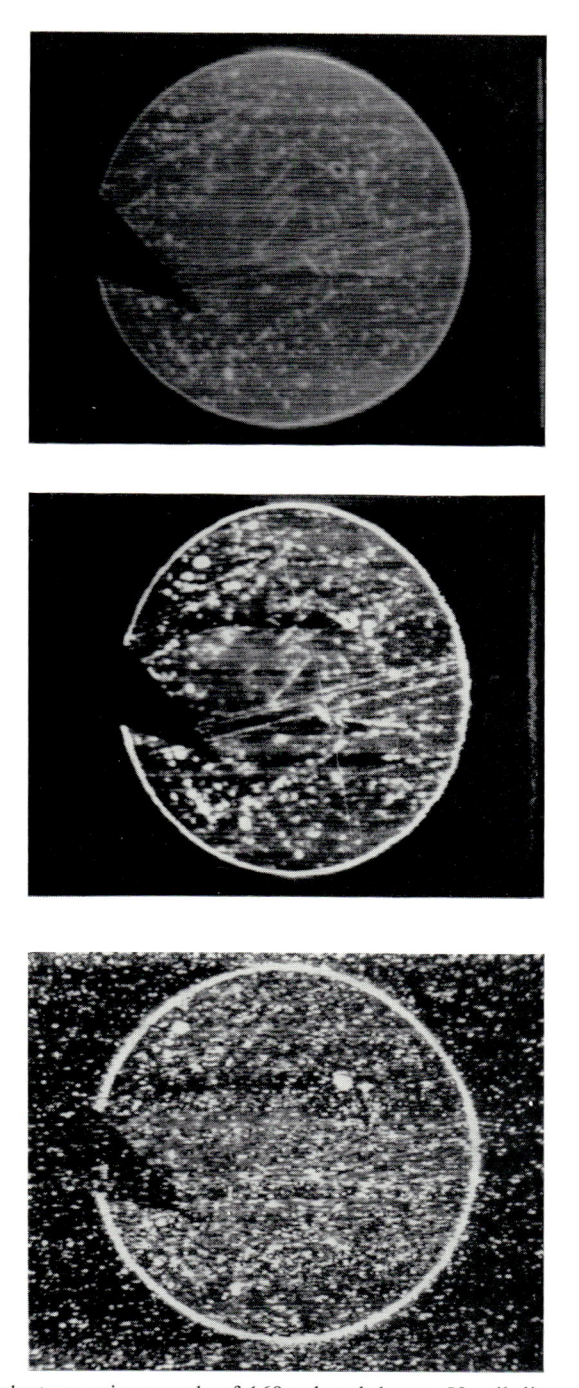

Figure 4. Scanning electron micrograph of 160-v breakdown, 50-mil-diameter diode as function of reverse bias (run 91-490). E_{bias} = (top) 0 v; (centre) −3 v; (bottom) −55 v. Random snow is spurious.

PLATE XXIX—*R. E. Enstrom and J. R. Appert*: *paper* 32

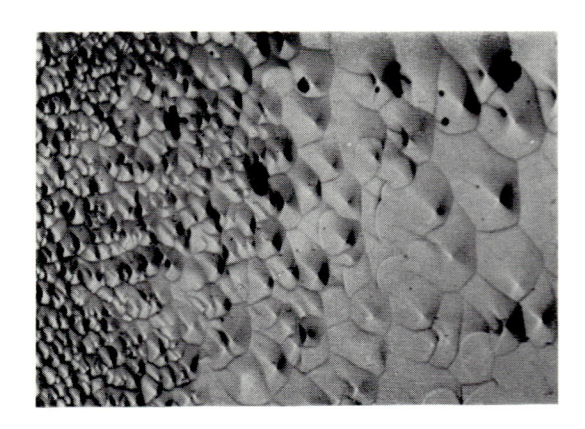

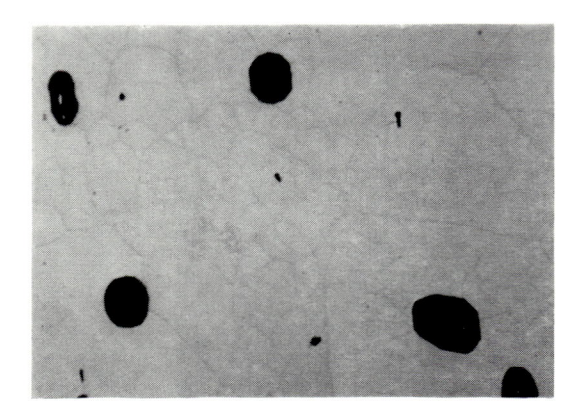

Figure 5. Appearance of wafers vapour grown at (from the top) 675°C, 725°C, and 770°C using same Czochralski-grown substrate. Magnification, × 50; photographic reduction, 5:4.

Vapour-phase growth of large-area microplasma-free p–n junctions in GaAs and GaAs$_{1-x}$P$_x$[†]

R. E. ENSTROM and J. R. APPERT

RCA Laboratories, Princeton, New Jersey 08540, U.S.A.

Abstract. Vapour phase epitaxial growth was used to prepare multi-layered structures with p–n junctions in GaAs and GaAs$_{1-x}$P$_x$ alloys ($x \leqslant 0.3$). In GaAs, a 4-layer n$^+$–n$^-$–p$^-$–p$^+$ structure was prepared, while in GaAs$_{1-x}$P$_x$ a 6-layer structure was necessary, with the additional layers required to incorporate compositional gradation from the GaAs substrate. Electron concentrations $\leqslant 1 \times 10^{15}$ cm^{-3} were achieved in the n$^-$-layer of the finished structure. Junction breakdown voltage was usually limited by microplasmas. A study of the source of the microplasmas showed that they could be mainly attributed to pits introduced during the growth process. Microplasmas associated with pits were investigated using scanning electron microscopy, and observation of the infrared emission under reverse bias. These pits were minimized by prolonged substrate etching, and by preventing reactor-generated particulates from dropping on the substrate during growth. The use of low dislocation substrates degraded reverse diode breakdown characteristics, contrary to expectations. As a result of this work diodes were prepared from layered structures, having reverse breakdown voltages as high as 475 v for a 0·050-in.-diameter mesa device and 250 v for a 0·175-in.-diameter mesa device. The latter diode operated at currents up to 50 A in the forward direction.

1. Introduction

The operation of p–n junction devices can be significantly affected by the presence of defects occurring at the junction. In particular, defects can cause the formation of microplasmas (localized avalanche breakdown) which can limit the reverse breakdown voltage of the junction. As the junction area increases, occurrence of such microplasmas becomes more probable since the microplasmas are typically randomly distributed across the junction.

In GaAs, preparation of p–n junctions free of microplasmas has been particularly troublesome. For example, in previously published work on GaAs diodes, reverse-voltage breakdown above 30 v in a 0·005-in.-diameter device was rare. More recently, breakdown voltages up to 200 v were reported in 0·020-in.-diameter mesa diodes (Bickley and McCarthy, 1967). Usually, diffusion techniques are used to prepare diodes.

An alternative method used with success to prepare p–n junctions in GaAs is vapour phase growth. In this technique, the dopants are introduced during the growth process, so that no further treatment is required to form the junction. III–V vapour deposited junction devices with excellent properties such as GaAs varactor diodes (Tietjen et al. 1966a), GaAs$_{1-x}$P$_x$ injection lasers (Tietjen et al. 1967), and in some cases, small area GaAs rectifiers (Bickley and McCarthy 1967), have been reported.

In the present work, this technique of vapour phase growth was utilized to grow large-area p–n junctions in both GaAs and GaAs$_{1-x}$P$_x$ for high temperature (300 °C) rectifier operation. The purpose was to achieve large forward currents simultaneously with high reverse breakdown voltages. To satisfy this requirement, it was necessary to grow a four-layered n$^+$–n$^-$–p$^-$–p$^+$ structure, and not just a simple p–n junction. In the case of GaAs$_{1-x}$P$_x$ alloys, a 6-layered structure was produced where compositional grading from the GaAs substrate was necessary. The junction should be microplasma-free. The research here focused first on achieving the purity necessary to sustain the desired high reverse voltage

[†] This research was supported by the Air Force Aero Propulsion Laboratory, Air Force Systems Command, Wright-Patterson Air Force Base, Ohio, under Air Force Contract AF33(615)–5352.

($n \leqslant 1 \times 10^{15}$ cm^{-3}) on a *reproducible* basis in the n$^-$-layer, and then on the characterization and elimination of microplasma-causing defects. As a result, diodes have been vapour grown with the highest breakdown voltages and forward currents yet reported for these materials.

2. Experimental procedure

Apparatus and procedures similar to those used in this work have been described in detail by Tietjen and Amick (1966b). Briefly, a horizontal-flow, quartz-tube, three controlled heat zone reactor system was used. Pd-diffused H$_2$ carrier is used with electronic-grade AsH$_3$, PH$_3$, HCl gases and metallic Ga chemical feed components. The reactants are controlled by adjusting the flows of the various gases. N$^+$ GaAs with 3×10^{18} electrons cm^{-3} is prepared by introducing H$_2$Se into the reaction tube, and p-type layers with about 3×10^{19} holes cm^{-3} are made by doping with Zn vaporized and transported in a H$_2$ stream. Sequential layers of n$^+$–n$^-$–p$^+$ or n$^+$–n$^-$–p$^-$–p$^+$ are then grown on (100) substrates without removing the sample from the apparatus during growth. Before growth, the GaAs substrates were chemically polished with a 4 : 1 : 1 (by volume) H$_2$SO$_4$, H$_2$O, H$_2$O$_2$ solution.

Ohmic contacts to the highly doped n$^+$ and p$^+$ surfaces were made by evaporating and sintering a 15 000 Å silver layer onto the surface. The silver was then covered with a 1000 Å chromium film and a 5000 Å gold film by evaporation. The chromium layer serves as a barrier to diffusion of gold through the silver (Krassner and Mayer, to be published). Selected diodes were cut out by ultrasonic abrasive technique using a Cavitron ultrasonic cutter. The device was then etched with a bromine-methanol solution to reduce surface leakage. Alternatively, 5 to 175 mil diameter mesas were etched into the multi-layer epitaxial structures without ohmic contacts, using KPR films for selective mesa etching.

Reverse-bias breakdown voltages were measured with a transistor curve tracer oscilloscope. The approximate carrier concentration in the n$^-$-layer of the n$^+$–n$^-$–p$^+$ structures was determined from point-contact breakdown voltage measurements on 3° angle lapped sections. Avalanche breakdown calibration curves were used to estimate the concentration of carriers (Weinstein and Mlavsky 1963).

3. Experimental results

3.1. *Crystal defects and microplasmas*

Low-reverse-bias-breakdown voltages and 'soft' *I–V* curves can be caused by microplasmas due to defects at the junction. These defects can include dislocations, metallic precipitates, irregularities in planarity, and inhomogeneities (Kressel 1967). Macroscopic-growth-pits on the epitaxial surface must also be considered (Tietjen *et al.* 1968). Presently these pits constitute the major imperfection in our vapour-grown layers. It is noteworthy that hillocks or pyramidal surface defects are not as predominant with the AsH$_3$ method as they are when layers are grown by the AsCl$_3$ method (Joyce and Mullin 1967).

Two approaches were taken to reduce premature breakdown caused by microplasmas. First, the origin and nature of these defects was studied and reactor techniques improved, and second, better substrate handling techniques were developed and better substrate materials were identified.

The tools used to study these defects have included: (1) anomalous x-ray transmission topographs to determine the dislocation structure in the substrate before deposition, (2) optical microscope to observe surface defects after deposition, (3) a dislocation etchant for gallium arsenide (Abrahams and Buiocchi 1965) to follow the propagation of dislocations (from the substrate into the vapour-grown layers), (4) the scanning electron microscope to observe microplasmas in reverse-biased diodes at the p–n junction, and (5) an infrared image converter to detect localized breakdown in reverse-biased etched-mesa diodes.

Concerning the origin of these defects, it was first observed that they are a function of area and substrate. In our initial studies, device structures consisting of n$^+$–n$^-$–p$^+$ layers were vapour-grown. It was found that about 75% of 5-mil-diameter diodes had reverse-bias breakdown voltages in the range of 100 to 140 V, with occasional higher values noted.

In contrast, 50-mil-diameter diodes had breakdown voltages about half as high as the 5-mil-diameter diodes in the *identical* area of the crystal, presumably because as the size of the diode increases, the probability of selecting a perfect area decreases. Some substrates gave a particularly low V_B, and simultaneously many surface defects were observed. This is shown in figures 1 and 2, and led to the main insight of the relationship between the presence of macroscopic growth pits and low V_B. The photographs show typical surfaces of wafers that have been grown simultaneously but on different substrates; the breakdown voltages for fifteen 50-mil diodes on each wafer also are shown. It may be seen that the n^+–n^-–p^+ structure with the higher growth-pit concentration leads to much lower, and more widely divergent breakdown voltages.

To further investigate the cause of low values of V_B, a series of selected 50-mil-diameter diodes, with breakdown voltages varying from 60 to 160 v, was examined with the scanning electron microscope while various values of reverse bias were applied. Figures 3 and 4 show a comparison of 60-v and 160-v diodes. As reverse bias is increased to 55 v, several large, bright spots, indicative of localized breakdown, become prominent in the 60-v diode, figure 3, whereas the 160-v diode (figure 4) remains relatively free of such sites. Similar results were found also when scanning electron microscope photographs of 120-v and 250-v diodes were compared. Thus, the scanning electron microscope results confirm that localized breakdown can be correlated with low V_B. Combined with the above result, this implies that microplasmas might correlate with growth pits.

To test this relation of growth pits with microplasmas, 50-mil diameter diodes were reverse-biased on a microscope stage and examined as a function of reverse-bias both with an infrared image converter and with infrared sensitive film. It had been shown previously that infrared radiation is emitted at microplasma sites (Constantiniscu *et al.* 1966) so that observation of a p^+-surface of a reverse-biased diode should permit a correlation to be made between microplasma sites and imperfections. Since the infrared radiation was strongly absorbed in the 10-μm-thick p^+-layer compared to the n-layer, the current had to be increased to 15 mA at 75 v to observe the radiation visually. Then, emission could be seen clearly at the site of several growth pits indicating that the pits do indeed, play a part in the reverse-bias breakdown of diodes, at least for large currents. These results were confirmed with infrared sensitive film on another 50-mil diameter diode which was reverse biased to 30 v and drew only 10 mA. The single spot on the film coincided exactly with the location of a growth pit. The small size of the spot indicates that the radiation probably originates at the bottom of the pit and that this, therefore, is the site of the microplasma. These results strengthen the association of premature-reverse-bias-breakdown and surface growth pits. Therefore, to achieve large area high-voltage-breakdown diodes, the growth pit density must be as small as possible.

3.2. Effect of growth conditions

Variations in growth conditions were studied to reduce the microplasma-density and increase V_B. Success in increasing low-current (less than 1 mA) reverse-bias breakdowns in diodes ranging from 30 to 175 mil diameter resulted.

3.2.1. Growth temperature.
The first parameter studied was the growth temperature, which is normally 725°C. Figure 5 shows that for Czochralski-grown high-dislocation-density substrates, the growth-pit density (and hillock density) decreases with increasing temperature (from 675°C to 770°C). In table 1, the reverse-bias-breakdown voltage is given as a function of temperature and diode-mesa diameter for the Czochralski-grown substrates. In the temperature range 675 to 770°C, the point contact-breakdown voltage for the n^--layer is a maximum at 725°C. Lower and higher temperatures lead to lower n^--layer breakdown voltages. Thus, the very significant increase in diode voltage with increasing temperature for all three mesa sizes directly reflects the lower pit-density at the successively higher temperatures, and indicates the dominance of the effect of crystal perfection relative to purity. The crystal perfection was increased in later runs, and therefore subsequent

Table 1. Reverse-bias-breakdown voltages as a function of temperature for Czochralski-grown substrates (run 162)

Temperature	\multicolumn{3}{c}{$(V_B$ at 0·01 mA$)/(V_B$ at 0·3 mA$)$}
	75 mil diameter	130 mil diameter	225 mil diameter
675°C	1/4, 2/6, 2/6, 2/11, 5/14	2/7, 2/6, 2/6, 1/4	2/4
725°C	18/21, 20/25, 21/24, 22/25, 60/100	14/18, 20/21, 18/21, 20/22, 17/18, 5/12, 12/28	12/22
770°C	68/80, 95/100, 36/45, 90/92, 36/38	40/54, 75/90	30/37

vapour-grown layers were prepared principally at 725°C to take advantage of the higher purity inherent with this deposition temperature.

In contrast to the above results for Czochralski-grown high-dislocation-content substrates, low-dislocation boat-grown substrates showed an opposite temperature dependence for growth-pit formation. For the latter, a growth temperature of 770°C produced many more pits than 725°C. However, as pointed out below, even when prepared at 725°C low-dislocation substrates do not produce as high breakdown voltage diodes as the high-dislocation-density Czochralski-grown substrates.

3.2.2. *Substrates.* One obvious parameter to investigate is the effect of dislocations on the breakdown voltage. In this respect, it should be noted that in separate experiments, we have shown that vapour grown GaAs layers can have fewer dislocations than are present in the substrate, in agreement with the results of others (Williams 1967).

Diodes prepared on low-dislocation-density substrates (~ 1000 pits cm^{-2}) with doping levels at 10^{16} and 10^{18} electrons cm^{-3} appear to have lower diode and n$^-$-layer point-contact voltages than those prepared on high-dislocation-density substrates as shown in table 2, Runs 188 and 191. Perhaps if a certain amount of impurity has to be accommodated into the vapour grown layer, and the impurities tend to segregate along dislocations, the higher-dislocation-density substrate may have a lower impurity density per dislocation and thus have a higher breakdown voltage.

Table 2. Effect of substrate and vapour growth procedure on 175-mil-diameter GaAs diode reverse-bias breakdown voltage

Run	Variable	$(V_B$ at 0·01 mA$)/(V_B$ at 1 mA$)$	n$^-$ layer point-contact breakdown
188	High-dislocation substrate with vapour-grown n$^+$–n$^-$–p$^-$–p$^+$ structure	60/250 60/250 55/225	150–180 v
191	Low-dislocation substrate with vapour-grown n$^+$–n$^-$–p$^-$–p$^+$ structure	25/45 65/160 25/50	130 v
190	High-dislocation substrate with vapour-grown n$^-$–p$^-$–p$^+$ structure	5/60 5/35	180–200 v

Other types of substrates have also been used, but with mixed results. For one run, the n$^+$–n$^-$–p$^+$ layer structure was vapour-grown on an n$^+$ surface prepared by liquid phase epitaxy and yielded pit-free, high breakdown-voltage diodes. However, attempts to repeat this or to grow on an n$^+$ surface prepared separately by vapour phase epitaxy have not produced the extraordinary results observed initially. Therefore, Czochralski-grown n$^+$ substrates with about 10^5 dislocations cm^{-2} were used for most experiments. However, even with this type of substrate, it has been found that some substrates give more defect-free vapour-grown layers than others, as shown in figures 1 and 2. The reasons for this are not yet completely understood.

Proper preparation of the Czochralski-grown substrates also can reduce the growth-pit density. For example, substrate polishing time and substrate position in the growth tube are particularly influential in reducing the pit density, and knowledge of this has helped us achieve the high breakdown voltage diodes. It was found that increasing the substrate chemical polishing time from 20 minutes to 60 minutes (e.g., removing 60 μm of the surface) in the 4 : 1 : 1 H_2SO_4, H_2O, H_2O_2 solution decreased the pit density in the final vapour-grown layer. Also, a 50% lower pit density and higher breakdown voltage diodes resulted from placing the substrate so that the polished surface faced downward (but simultaneously parallel to the vapour stream) during vapour growth.

A further reduction of the growth-pit density can be achieved by growing an undoped GaAs layer in the deposition tube prior to the insertion of the substrate. This pre-growth cleans the liquid Ga surface and coats the walls of the quartz deposition tube with GaAs. On the other hand, the use of a 10 mole percent (Effer 1965) KCN rinse solution after chemical polishing to eliminate possible metal contaminants (such as copper) from the surface of the substrate, neither decreased the surface imperfection density, nor increased the maximum reverse-bias breakdown voltages of 30-mil diodes compared to an untreated substrate grown back-to-back.

3.2.3. *Layer thickness and purity.* A schematic representation of the layer structures prepared for the GaAs and $GaAs_{1-x}P_x$ alloy rectifier is shown in figure 6. Initially the GaAs rectifier was prepared as an n^+–n^-–p^+ layer structure with the n^--layer about 10 μm thick and the p^+ layer about 25 μm thick. The n^--layer was increased to about 15 μm and then to 25 μm because depletion width measurements (Williams, private communication) on 50-mil diodes showed that the depletion width was 15 μm at 150 v reverse bias and 20 μm at 200 v. Thus the n^--layer should be at least 25 μm thick so that the depletion region will not punch through the n^+ layer prematurely, which would lead to a lower than expected breakdown voltage. For our best run (224), the n^--layer was grown 50 μm thick, and this has probably contributed to achieving the 475 v breakdown observed for 50-mil diodes prepared from this wafer.

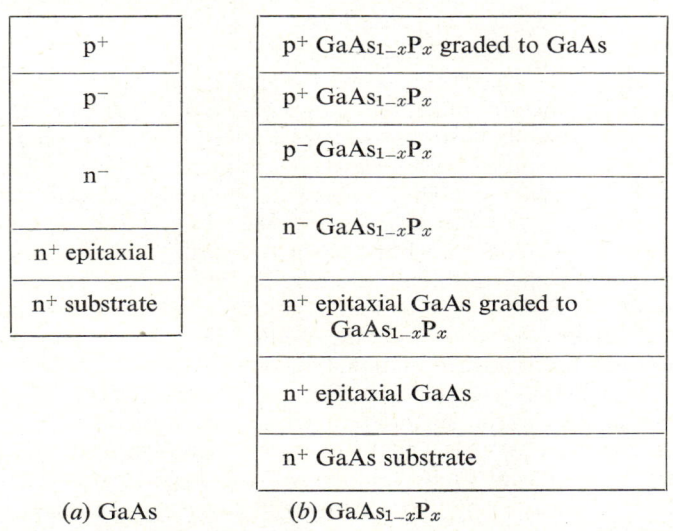

(a) GaAs	(b) $GaAs_{1-x}P_x$
p^+	p^+ $GaAs_{1-x}P_x$ graded to GaAs
p^-	p^+ $GaAs_{1-x}P_x$
n^-	p^- $GaAs_{1-x}P_x$
	n^- $GaAs_{1-x}P_x$
n^+ epitaxial	
n^+ substrate	n^+ epitaxial GaAs graded to $GaAs_{1-x}P_x$
	n^+ epitaxial GaAs
	n^+ GaAs substrate

Figure 6. Schematic representation of vapour-grown layer structure for GaAs and $GaAs_{1-x}P_x$ rectifiers.

To achieve such high breakdown voltages requires a carrier level in the n^--layer less than 1×10^{15} cm^{-3}. This was achieved on a reproducible basis by the use of clean, leak-free deposition apparatus and high-purity selected reagent gases (AsH_3, PH_3, HCl). The purity of the HCl is particularly crucial and when purer HCl was used, after run 178, the n⁻-layer point-contact breakdown voltage increased significantly. In this way, carrier levels as low as 2.3×10^{14} cm^{-3} with mobilities of 7000 cm^2 v^{-1} s^{-1} at 300°K and 53 000 cm^2 v^{-1} s^{-1} at 77°K have been achieved. More recently, higher purity GaAs with mobilities as high as 7900 cm^2 v^{-1} s^{-1} and 82 000 cm^2 v^{-1} s^{-1} at 300°K and 77°K respectively has been achieved.

It is desirable to grow GaAs diodes in subsequent runs without dismantling the system and cleaning the tube with HF–HNO₃ acid. However, Zn-doped deposits on the walls from preceding runs cause unintentional doping of the n⁻-layer in subsequent runs. To eliminate this problem, a procedure has been evolved for cleaning the completely assembled apparatus with HCl gas. This procedure has considerably facilitated these experiments, and leads to diodes comparable with those produced using the HF+HNO₃ acid cleaning normally employed, as shown in figure 7.

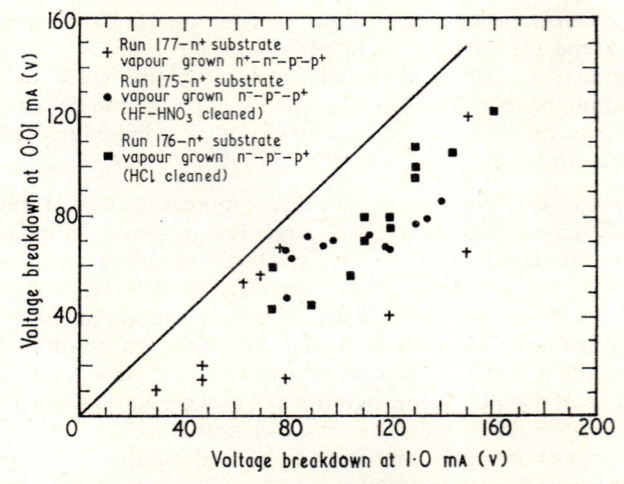

Figure 7. Reverse-bias-breakdown voltage at 0·01 and 1·0 mA for 30-mil-diameter cavitronned diodes.

3.2.4. *Growth sequence*. Generally, a vapour-grown n⁺-layer is deposited on the n⁺-substrate to keep all junctions in vapour-grown material. Wafers were also prepared by depositing the n⁻-layer directly on the n⁺-substrate to ascertain whether this would lead to higher-breakdown-voltage diodes with fewer microplasmas; the results for 30-mil-diameter diodes are compared in figure 7. The reverse-bias breakdown at 0·01 mA is plotted as a function of the breakdown at 1·0 mA for each diode; microplasma-free diodes should lie along the straight line and have high breakdown voltages. Note that the distribution for runs 175 and 176 is about the same, and generally lies higher and closer to the straight line than the distribution for Run 177 (which had an initial n⁺-vapour-grown layer). It may be concluded for *small* diodes that depositing the n⁻-layer directly on the n⁺-substrate leads to higher breakdown than incorporating an initial n⁺-vapour-grown layer between the substrate and the vapour-grown n⁻-layer. However, for the 175-mil-diameter diodes where defects are more detrimental than in the 30 to 50 mil diodes, it appears that the omission of the n⁺-layers (e.g., deposition of the vapour-grown n⁻-layer directly on the bulk-grown n⁺-substrate) leads to significantly lower diode reverse-voltages as shown by comparing run 190 with run 188 in table 2. The vapour-grown n⁺-layer was included, therefore, in our best large-area diodes.

3.2.5. *Doping*. The p–n junctions were generally formed by vapour growing a Zn-doped layer on top of the n⁻-layer. Several runs were made with a Cd-doped p-layer, but diodes prepared from these wafers all had very low, soft, breakdown voltages. Therefore the use of Cd was investigated no further and attention was focused on the Zn doping.

A p⁻-layer was incoporated into the GaAs rectifier structure to (1) reduce the Zn concentration at the p–n junction and, thereby, minimize precipitation and decoration of dislocations so that the junction would be more perfect due to elimination of metallic shorting paths that can cause premature localized breakdown (Goetzberger and Shockley 1960) and (2) grade the p–n junction to reduce the electric field for a given applied voltage. Initially a 6-μm-thick p⁻-layer was prepared with about 5×10^{17} holes cm⁻³. The level was further reduced to about 2×10^{15} holes cm⁻³ to reduce the Zn concentration at the p–n junction. This reduction in hole level and the eventual increase in thickness of the p⁻-layer to 25 μm led to significantly higher reverse bias breakdown voltage diodes, particularly for diodes greater than 100 mil diameter. A comparison of diodes with a p⁻-layer (run 168 substrate at 725 °C) and with no p⁻-layer (run 168 substrate at 760 °C) is shown in figure 8 and demonstrates that the p⁻-layer leads to about a twofold increase in breakdown voltage.

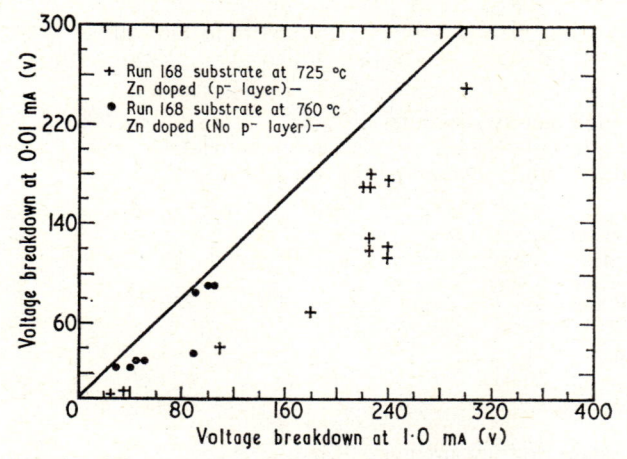

Figure 8. Reverse-bias-breakdown voltage at 0·01 and 1·0 mA for 60-mil-diameter etched-mesa diodes.

Initially, the p⁺ layer was prepared with a hole concentration of about 4×10^{19} holes cm⁻³. At this Zn-doping level, the surface was very rough due to troughs that were oriented parallel to $\langle 110 \rangle$ directions on the (100) p⁺ surface, and these were thought to be deleterious to device performance. Reducing the Zn doping temperature and thus the amount of Zn in the p⁺-layer to give 7×10^{18} holes cm⁻³ resulted in mirror-smooth surfaces; this lower doping level was therefore adopted for most of the wafers grown.

3.2.6. *GaAs$_{1-x}$P$_x$ alloys.* GaAs$_{0.85}$P$_{0.15}$ and GaAs$_{0.7}$P$_{0.3}$ alloys were prepared to achieve a higher temperature capability than is possible with GaAs. Several GaAs$_{0.7}$P$_{0.3}$ alloy wafers were prepared but, when the high-voltage (>220 v) diodes made from these wafers were found to have a higher forward resistance, attention was focused on the GaAs$_{0.85}$P$_{0.15}$ alloy. The vapour growth of GaAs$_{0.85}$P$_{0.15}$ alloy rectifiers differs from that of GaAs because of the difference in lattice parameter of the alloy and the GaAs substrate. The initial n⁺-layer must be graded from the substrate to the n⁻-layer alloy composition to reduce interfacial dislocations; this usually takes about 1 hour and results in a graded region about 25 μm thick. For convenience in alloying contacts which had been previously developed for GaAs, the p⁺-layer is graded from GaAs$_{0.85}$P$_{0.15}$ alloy to GaAs at the surface. The complete vapour-grown layer structure is shown in figure 6.

The electrical properties of GaAs$_{0.85}$P$_{0.15}$ alloy rectifiers are summarized in table 3 for different growth conditions. Runs 171, 178, and 202 were grown under similar conditions except that from 178 onwards purer HCl from Pittsburgh Chemical Corporation was used, which increased by a factor of almost 3 the n⁻-layer reverse bias breakdown voltage compared to run 171. The use of a low-dislocation content substrate, or deposition of the n⁻-layer directly on the n⁺-substrate, leads to low reverse voltages as found also for the

Table 3. Effect of substrate and vapour growth procedure on the reverse-bias breakdown voltage for 30, 60, and 175 mil-diameter GaAs$_{0.85}$P$_{0.15}$ diodes

Run	Variable	(V_B at 0·01 mA)/(V_B at 1 mA) 60 mil	175 mil
171	n$^+$–n$^-$–p$^+$	†/250 for 30 mil diodes	
178	n$^+$–n$^-$–p$^+$ Purer HCl	†/400 for 30 mil diodes	†/200, unstable
	(a) *Polished* side *up* during growth	(a) 26/75 to 97/140	
	(b) Back side of same wafer *down* during growth	(b) 22/80 to 200/220	
202	n$^+$–n$^-$–p$^+$ Purer HCl		
	(a) Back side *up*	(a) 25/70 to 130/220	
	(b) *Polished* side of same wafer *down*	(b) 60/220 to >220/>220	
182	n$^+$–n$^-$–p$^+$ Low dislocation density substrate		
	(a) *Polished* side *up*	(a) 4/20 to 13/42	10/45
	(b) Back side of same wafer *down*	(b) 4/23 to 10/33	
200	n$^+$–n$^-$–p$^+$ Deposit n$^-$-layer directly on n$^+$-substrate		†/25
219	n$^+$–n$^-$–p$^-$–p$^+$ 6-μ-thick p$^-$-layer		†/250 for 15 min.
224	n$^+$–n$^-$–p$^-$–p$^+$ 25-μ-thick p$^-$-layer 50-μ-thick n$^-$-layer	†/425 to †/475	†/250, stable

† = Breakdown voltage not measured at 0·01 mA.

GaAs diodes, table 2. On the other hand, the incorporation of a 6 μm p$^-$-layer (p ~ 1 × 10^{15} cm^{-3}) or a 25 μm p$^-$-layer with a 50 μm n$^-$-layer, runs 219 and 224 in table 3, did lead to significantly higher diode voltages for both small and large diodes. Although a 200-v diode can be achieved without a p$^-$-layer, as in run 178, the p$^-$-layer appears to be helpful in attaining a more stable reverse voltage in the large diodes. This may be related to a reduction of the influence of pits and microplasmas by grading the p–n junction and by reducing the zinc doping level that could otherwise decorate dislocations in the vicinity of the p–n junction.

The GaAs$_{1-x}$P$_x$ alloy wafers generally have smoother surfaces and fewer pits than do comparable GaAs wafers. A cross-hatch pattern of dislocations lying parallel to the [011] and [01$\bar{1}$] directions in the (100) plane of growth (Abrahams, private communication) seen on the surface of the GaAs$_{1-x}$P$_x$ alloy may serve to absorb impurities that could otherwise nucleate pits. Sixty-mil-diameter diodes on the face-down side of each wafer in table 3 consistently have higher breakdown voltages and harder I–V characteristics than 60-mil diodes on the face-up side of the same wafer. As with GaAs the face-down side of the wafer generally has a 2–3 times lower pit density and this is probably responsible for the superior electrical characteristics, compared to the face-up side. The higher density on the face-up substrate may result from particles falling onto this surface from the growth tube.

Detailed microscopic examination of the run 224 face-down wafer showed that all of the pits except for one, are less than 50 μm deep and that 75% are less than 10 μm; these pits are 3–6 times wider than they are deep. Since the p$^-$-layer and the p$^+$-layers are each 25 μm thick, the growth pits do not penetrate the p–n junction. Thus, although growth-pits have not been eliminated entirely to achieve high-voltage diodes, the number and the possible detrimental effect of these pits have been minimized.

4. Summary

High purity GaAs and $GaAs_{1-x}P_x$ alloys were prepared on a reproducible basis by using selected pure gases, clean, leak-free deposition apparatus, and the optimum substrate temperature.

Many insights were gained concerning microplasma effects and breakdown voltage. Of great importance, it was found that dislocations are probably not the primary cause of microplasmas. On the other hand, a good correlation between growth pits on the diode surface and the occurrence of microplasmas was obtained through infrared observation of localized breakdown. In turn, the occurrence of surface pits was strongly correlated with low voltage breakdown in experiments where the pit density was varied by changing substrate temperatures.

Also, growth techniques were evolved to increase large area diode breakdown voltage by reducing microplasma density. These techniques included using a 10^{18} electron cm^{-3} n^+-substrate with a moderately high rather than a low dislocation density, polishing the substrate for 60 minutes, placing the substrate face-down in the growth apparatus, using a p^--layer above the n^--layer, and growing sufficiently thick n^-- and p^--layers in the diode structure.

Acknowledgment

The authors wish to express their appreciation to L. Krassner and A. Mayer of RCA Electronic Components Division for their assistance with the evaluation of the electrical properties of these materials, and to L. R. Weisberg for helpful criticism.

References

ABRAHAMS, M. S., and BUIOCCHI, C. J., 1965, *J. Appl. Phys.*, **36**, 2855.
ABRAHAMS, M. S., RCA Laboratories, private communication, 1968.
BICKLEY, W. P., and MCCARTHY, J. P., 1967, *Gallium Arsenide*, Institute of Physics and the Physical Society, London, pp. 241–5.
CONSTANTINISCU, C. R., POPOVICH, G., and MIHAILOVIC, P., 1966, 'Radiative Recombination in GaAs p–n Junctions,' *Proceedings of Conference on Luminescence*, Budapest.
EFFER, D. J., 1965, *J. Electrochem. Soc.*, **112**, 1020.
GOETZBERGER, A., and SHOCKLEY, W., 1960, *J. Appl. Phys.*, **31**, 1821–4.
JOYCE, B. D., and MULLIN, J. B., 1967, *Gallium Arsenide*, Institute of Physics and the Physical Society, London, pp. 23–6.
KRASSNER, L., and MAYER, A., to be published.
KRESSEL, H., 1967, *RCA Rev.*, **28**, 175–207.
TIETJEN, J. J., KUPSKY, G. A., and GOSSENBERGER, H., 1966a, *Solid State Elec.*, **9**, 1049.
TIETJEN, J. J., and AMICK, J. A., 1966b, *J. Electrochem Soc.*, **113**, 724–8.
TIETJEN, J. J., PANKOVE, J. I., HEGYI, I., and NELSON, H., 1967, *Trans. Met. Soc. AIME*, **239**, 385–7.
TIETJEN, J. J., ABRAHAMS, M. S., DREEBEN, A. B., and GOSSENBERGER, H. F., 1968, *International Conference on Gallium Arsenide*, Dallas, Texas.
WEINSTEIN, M., and MLAVSKY, A. I., 1963, *Appl. Phys. Lett.*, **2**, 97–99.
WILLIAMS, F. V., 1967, *Gallium Arsenide*, Institute of Physics and the Physical Society, London, pp. 27–30.
WILLIAMS, R., RCA Laboratories, private communication, 1967.

Characteristics of GaAs based heterojunction photodetectors[†]

T. L. TANSLEY

Mullard Research Laboratories, Redhill, Surrey, England

Abstract. Heterojunctions designed as near infrared photodiodes have been prepared by interfacing ternary derivatives of gallium arsenide, namely gallium arsenide-phosphide and gallium-indium arsenide, with gallium arsenide substrates. A conventional $H_2/AsCl_3/PCl_3$ transport system was used. Reported in this paper are the detailed measurement of forward characteristics and the formulation of an intraband tunnelling model capable of accounting for them. Qualitatively the predictions of the theory account for the general similarity of heterojunction forward characteristics reported in the literature and a fair degree of quantitative agreement is found for the instances cited, the discrepancies being discussed.

1. Introduction

Since the weight of evidence now points to the fact that little or no bulk injection of minority carriers takes place at a p-n heterojunction the only viable application of such structures appears to be in the sensitive and fast detection of radiation in the waveband prescribed by the choice of materials. Of interest in both device and physical terms are materials of general type $A_y^{III} B_{1-y}^V C_x^{III} D_{1-x}^V$, although one degree of freedom is usually removed from this formulation in practical work, attention being devoted to the III–V ternary compounds. Specifically the gallium arsenide derivative alloys $Ga_yIn_{1-y}As$ and $GaAs_xP_{1-x}$ have been chosen for the present study and interfaced with GaAs to form rectifying heterojunctions. The complete range of solid solutions is available by well established open-tube vapour epitaxy techniques with material forbidden energy gaps between about 0·33 ev (3·7 μm, InAs) and about 2·4 ev (0·5 μm, GaP). In device terms this affords the possibility of creating photodetectors with responses anywhere in this range of wavelengths (Tansley 1967). In physical terms we have the facility to prepare heterojunctions with various degrees of dissimilarity between their constituents (tending to the homojunction case in the limit) from which the singular properties of the heterostructure might hopefully be isolated.

Of fundamental importance in gaining an understanding of the mechanisms involved in the operation of heterojunction devices are the nature of the interface band profile and the way in which this interacts with photons, photogenerated carriers, and dark-current carriers. The original work of Anderson (1960) proposed a band structure in which differences in band gap and work function led to a profile including discontinuities in both conduction and valence band edges. A list of fifty-four possible detailed variations was subsequently given by Van Ruyven (1964). A number of workers have found this band model capable of explaining experimental observations, although the effects of interface states frequently could not be discounted. The interaction of incident photons with such a structure has been reported in several investigations and, with the exception of isotype heterojunctions, the results have been shown capable of interpretation in terms of the Anderson band model.

In the case of current transport problems most attention has been devoted to the study of forward characteristics. The two models which have proved most successful both involve quantum mechanical tunnelling as the major, current-limiting, effect in forward bias current–voltage characteristics. The intraband model proposed by Riben (1965) and others involves

[†] Parts of this work are to be included in a thesis submitted to the University of Nottingham, England, in partial fulfilment of the requirements for the degree of Ph.D.

diagonal band-to-band tunnelling at the interface through a multistep array of recombination states distributed across the interface region. The intraband model first mooted by van Ruyven involved tunnelling through the forbidden energy gap spike into a 'sink' which may be either an array of interface states or the band notch, the assumption being that tunnelling into rather than recombination/emission from the sink is the limiting process. This model was rejected by Riben on the grounds that the majority of the tunnelling current flux was transmitted near the top of the spike and would therefore be expected to show a diffusion type temperature dependence of the form $\exp-(g/kT)$ rather than the form $\exp(T/T_0)$ commonly encountered for a wide variety of heterojunction pairs (Newman 1965). In this paper the intraband model will be developed rigorously to show its quantitative applicability to the prediction of heterojunction forward characteristics and a degree of insensitivity to a number of interface parameters which may explain the similarity of the results seen in a number of pairs.

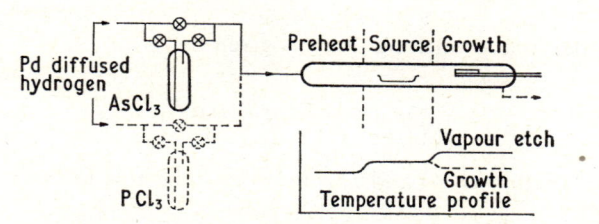

Figure 1. Schematic diagram of growth apparatus.

2. Experimental

A conventional open tube system, as indicated in figure 1, has been employed for the growth of both types of material on GaAs substrates with $p \sim 10^{18}$ cm^{-3}. In the case of Ga(AsP) palladium-diffused hydrogen was passed through arsenic trichloride and phosphorous trichloride bubblers in parallel, mixed and reacted with 6N gallium held in a spectrosil boat at about 800°C. In the case of (GaIn)As a single arsenic trichloride source was used and an alloy of indium and gallium held in the boat. Homogeneous monocrystalline layers were then deposited at 700°C on polished p-type gallium arsenide substrates which had been submitted to a light vapour etch immediately before growth. As-grown material on (100) substrates was n-type with doping densities in the range 1×10^{16}–6×10^{17} and mobilities in the range 3500–5000 cm^2 V^{-1} s^{-1} as determined by Van der Pauw (1958)—Hall measurements. The composition parameter x for GaAs$_x$P$_{1-x}$ alloys was varied between 1·0 and 0·7 by adjustment of the relative rates of flow to the two bubblers, fine adjustment being obtained by directing gas flow through lengths of stainless steel capillary tube. For Ga$_y$In$_{1-y}$As values of y from 1·0 to 0·85 were obtained by variation of the boat alloy composition.

The layers were characterized by measurements with control samples grown on semi-insulating substrates placed close to the diode substrates during growth. The layers were diced and contacts made by the evaporation of gold–zinc and gold–tin alloys (for p and n type materials respectively) with subsequent alloying-in. The diodes were soldered to transistor headers with wires thermocompression bonded to the upper surfaces.

Specimens were placed in a cryostat and current–voltage characteristics at a set of temperatures recorded digitally with a data-logging system.

3. Theory

Figure 2 shows the four principal band profiles possible with doping asymmetry included. In general we can assume that (1) one type of carrier is in plentiful supply at the interface recombination sites in such an instance and that (2) current transport is limited by the rate at which the other can tunnel into this region. For the case of electron tunnelling at energy ϵ above the bottom of the conduction band the relative parameters are introduced in figure 3. If $T(\epsilon)$ is the tunnelling probability at this level and $J(\epsilon)\, d\epsilon$ the incident electron flux between

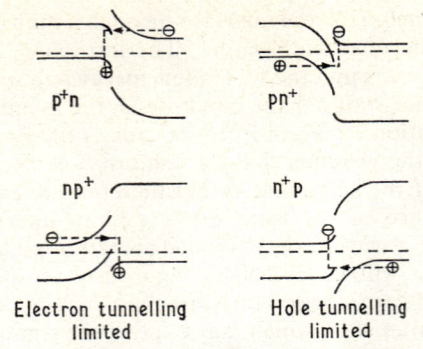

Figure 2. The four principal band profiles.

ϵ and $\epsilon + d\epsilon$, then the total tunnelling flux J is given by

$$J = \int_0^B T(\epsilon) J(\epsilon) \, d\epsilon \qquad (1)$$

and we require the functions $T(\epsilon)$ and $J(\epsilon)$ in order to be able to evaluate this integral.

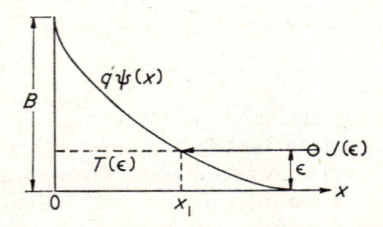

Figure 3. Tunnelling current parameters.

3.1. *Tunnelling probability*

The WKB method can be applied to the calculation of tunnelling probability, T, for problems in which slowly varying potentials are involved. The result for an electron wave striking the potential barrier illustrated at $x = x_1$, and penetrating to $x = 0$ is

$$T(\epsilon) = \exp\left\{ -\sqrt{\frac{8m^*}{\hbar^2}} \int_0^{x_1} \sqrt{(q\psi(x) - \epsilon)} \, dx \right\} \qquad (2)$$

where $\psi(x)$ is the potential at distance x from the interface and the remaining nomenclature is standard.

In expressing $\psi(x)$ as an explicit function of x we must choose a form for the ionized impurity distribution in the depletion region. For an abrupt junction

$$q\psi(x) = B \left[1 - x \sqrt{\frac{q\rho_0}{2K\epsilon_0 B}} \right]^2. \qquad (3)$$

One may begin to investigate at this point whether the result will be critically dependent on the shape of the potential barrier by replacing equation (3), representing an ideal case, by a different function. Probably fictitious but mathematically convenient is a function of the form

$$q\psi(x) = B \exp -x \sqrt{\frac{\rho_0 q}{K\epsilon_0 B}} \qquad (4)$$

which would characterize a retrogressive exponential doping profile. However, (3) and

(4) are not very different and we shall see in due course that the result is only weakly dependent on the exact choice of doping profile.

After integration and a little manipulation, equation (2) for the abrupt junction case becomes

$$T(\epsilon) = \exp\left\{-\sqrt{\frac{2m^*}{\hbar^2}}\sqrt{\frac{2K\epsilon_0}{q\rho_0}}.B.\left[\sqrt{(1-\xi)} - \xi \ln\left(\frac{1+\sqrt{(1-\xi)}}{\sqrt{\xi}}\right)\right]\right\} \quad (5)$$

where $\xi = \epsilon/B$. Its equivalent in the exponentially graded case is

$$T(\epsilon) = \exp\left\{-\sqrt{\frac{2m^*}{\hbar^2}}\sqrt{\frac{K\epsilon_0}{q\rho_0}}.B.\left[\sqrt{(1-\xi)} - \sqrt{\xi} \arctan\sqrt{\frac{1-\xi}{\xi}}\right]\right\}. \quad (6)$$

3.2. Incident flux

The incident electron flux depends on the distribution of electron states, their occupation and the thermal velocity of the occupants. The conduction band distribution of electron states in these materials is well approximated by

$$N(\epsilon)\,d\epsilon = \frac{4\pi}{\hbar^3}(2m^*)^{3/2}\epsilon^{1/2}\,d\epsilon. \quad (7)$$

The occupation is given by the Boltzmann approximation to the Fermi function for non-degenerate material

$$F(\epsilon) = \exp-\left(\frac{E_F + \epsilon}{kT}\right). \quad (8)$$

For simplicity the electron mean free path (thermalization length) is assumed to be long compared with the depletion region width so that the energy distribution of thermal velocities of electrons striking the barrier is taken at its bulk value

$$v(\epsilon) = \sqrt{\frac{2\epsilon}{m^*}}. \quad (9)$$

Equations (7), (8), and (9) are then combined to give the total incident electron flux as a function of ϵ (assuming normal incidence of all electrons, see section 3.3).

$$J(\epsilon)\,d\epsilon = N(\epsilon)\,F(\epsilon)\,v(\epsilon)\,d\epsilon. \quad (10)$$

3.3. Calculation

Equation (1) has been calculated on an Elliot 503 computer to yield a set of heterojunction forward characteristics with temperature as additional parameter. Only two disposable variables appear in the calculation. The choice of B represents the barrier height and variations in B the applied bias, whilst choice of ρ_0 decides the value of zero-bias depletion width. A number of detailed modifications have been made to the programme to test its sensitivity to situations likely to arise in real structures. In particular the one-dimensional case described above, which implicitly assumes that all electrons at a given energy are normally incident on the interface, has been extended to the real situation in which the direction of motion of electrons striking the barrier is distributed over 2π steradians. The results quoted are for such a case.

4. Results

4.1. Results from the theory

Figure 4 shows the forward characteristics of a heterojunction, calculated from the model, in the range of barrier heights from 1·0 ev to 0·4 ev for both abrupt and exponentially graded cases. The barrier width for the abrupt case is 0·03 μm at $\epsilon = 0$ and $B = 1$ ev and the exponential barrier, since it cannot be defined in this way, is scaled to approximately fit the parabolic potential at the $1/e$ point. Figure 5 compares the temperature dependences of the characteristics at $B = 0·9$ ev.

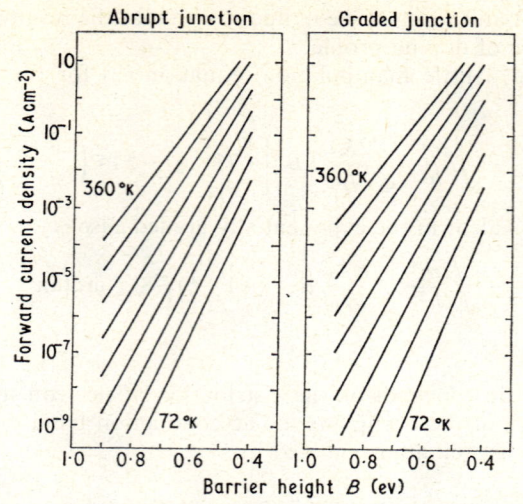

Figure 4. Computed forward I–V characteristics for a heterojunction. The temperatures are from $kT = 6$ mev ($72°$K) to 30 mev ($360°$K) in 3 mev steps.

Immediately apparent is the qualitative agreement with the empirical expression of Newman

$$J \alpha \exp\left(\frac{V}{V_0} + \frac{T}{T_0}\right)$$

found to be typical of heterojunction forward characteristics. Furthermore this behaviour is predicted not only for the ideal abrupt junction, but also for the mathematical fiction of an exponential junction. In other words, this type of characteristic may be expected irrespective of minor deviations from the ideal case.

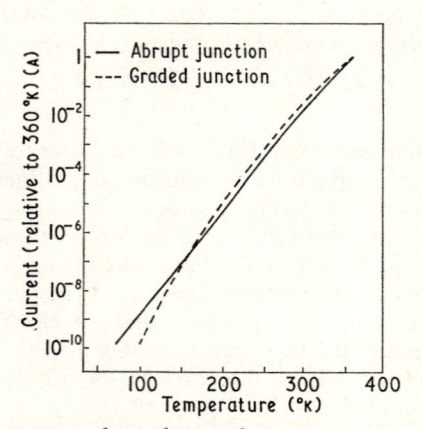

Figure 5. Temperature dependence of current at 0·9 ev barrier height.

The constants V_0 and T_0 decrease with increased barrier width, that is the characteristics become steeper and further apart, and a degree of convergence becomes apparent at large widths (see for example figure 4).

4.2. *Experimental results*

Compared in figure 6 at a set of temperatures between $230°$K and $360°$K are the theoretical and experimental results for a GaAs–Ga(AsP) p$^+$n heterojunction in which the ternary

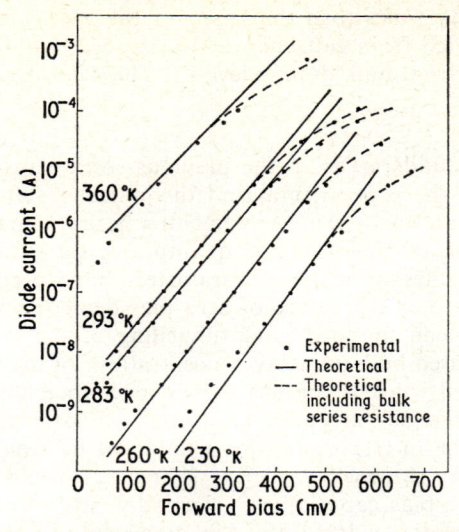

Figure 6. Forward characteristics of a Ga(AsP)–GaAs heterojunction.

epilayer has a gallium phosphide content of approximately 20 mol percent. An abrupt barrier has been used and this was supported by an inverse square law evident in capacitance–voltage plots. The absolute values of measured and calculated current density were used, these agreeing well. The best theoretical fit to this result was found to require the postulation of a zero-bias barrier height of 0·9 ev and width of 0·05 μm. The diode was formed by the growth of a layer with $N_d \simeq 4 \times 10^{17}$ on a nominal $N_a = 10^{18}$ Zn doped substrate from which a zero voltage barrier height of 0·93 ev and barrier width of 0·059 μm may be calculated assuming an abrupt junction. These values are in good agreement with the theoretical values used to match the forward characteristics. At high current densities a bulk resistance term has been included (figure 6).

Figure 7 repeats this procedure over a wider temperature range for a Ga(InAs)–GaAs n–p$^+$ heterojunction with about 5 mol percent InAs in the solid-solution component. The fit is obtained with a zero-bias barrier height of 0·65 ev and width of 0·024 μm, the absolute values of measured and calculated current densities again being close. This time, however, the barrier width calculated from measurements of bulk impurity concentration

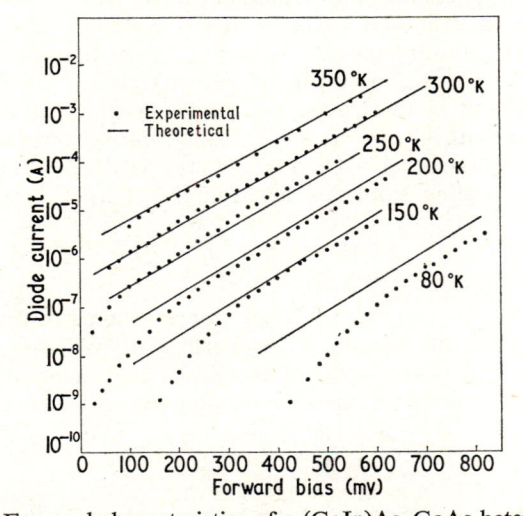

Figure 7. Forward characteristics of a (GaIn)As–GaAs heterojunction.

is larger by a factor of five times than that used in the computation to give a good fit. The barrier width calculated from capacitance data agreed, within experimental accuracy, with the value calculated from bulk doping levels. This is discussed in section 5.

5. Discussion

From the comparisons undertaken in the previous section two significant points arise. Firstly, in many respects the experimental and theoretical results show good correlation, especially at higher temperatures. This agreement is both qualitative in its broad support for the empirical equation of Newman and quantitative (at least for the GaAs–Ga(AsP) case) where numerical values of material parameters have been introduced. Secondly, there is a discrepancy between the values of zero-bias depletion width predicted from the forward characteristics when an intraband tunnelling model is applied, and the value calculated from the measured bulk impurity concentrations of the (GaIn)As epilayer and its substrate. This calculated value is supported by capacitance data. One could explain a narrowing of the depletion region in a number of ways. Firstly, autodoping might generate inhomogeneity close to the interface and this would not be seen by either van der Pauw techniques or by taking the slope of the C^{-2} vs V line. It would, however, be manifest in the absolute value of zero-bias capacitance. Secondly, the effect of an array of charged interface states is to reduce the width of the depletion region in the material whose ionized impurities bear the same sign whilst increasing the other (this is clear from space charge neutrality considerations). Calculations show that an interface density of about 10^{12} cm^{-2} would be necessary to account for the discrepancy noted in section 4. This value is approximately one-tenth of the number of dangling bonds expected at a Ga$_{0.95}$In$_{0.05}$As–GaAs interface, as calculated by the method of Holt (1966). Such a narrowing mechanism would be accompanied by a commensurate decrease in barrier height if the charge was strongly localized at the interface, although this is progressively less true if the charge spreads into the depletion regions. In any case the foregoing remarks on capacitance evidence apply equally well to this situation.

We remark then that the barrier width used in the calculations to describe the forward characteristics is not in good agreement with its expected value for the (GaIn)As–GaAs heterojunction. Although such a discrepancy could easily be explained in terms of interface charge it is not supported by capacitance data. The Ga(AsP)–GaAs heterojunction shows good agreement with theory.

6. Conclusions

A model of general applicability to the prediction of forward characteristics of rectifying heterojunctions has been proposed. The model is general in that it requires, as it is a single-step process, no arbitrary choice of energy and temperature distribution of intermediate stages in order to fit experiment as was the case for Riben. For the two real heterojunctions considered in this paper the model provides an adequate description of forward characteristics with a good degree of quantitative agreement in one case and a significant discrepancy in the other. Qualitatively the model provides a theoretical basis for the observation of a puzzling similarity between the characteristics of a wide variety of heterojunction pairs.

Acknowledgments

I have been indebted to a number of people for assistance in the preparation of diodes reported in this paper, notably Messrs. R. A. Ford, T. E. Williams, and R. J. Tree. I am also grateful to Dr. S. J. T. Owen of the University of Nottingham and many colleagues at Mullard Research Laboratories for their advice and encouragement.

References

ANDERSON, R. L., 1960, *I.B.M. Jnl. Res. Dev.*, **4**, 283–7.
HOLT, D. B., 1966, *J. Phys. Chem. Solids*, **27**, 1053–67.

NEWMAN, P. C., 1965, *Electronics Letters*, **1**, 265.
RIBEN, A. R., 1965, Ph.D. Dissertation, Department of Electrical Engineering, Carnegie Institute of Technology.
TANSLEY, T. L., 1967, *Phys. Stat. Sol.*, **23**, 241–52.
VAN DER PAUW, L. J., 1958, *Philips Res. Repts.*, **13**, 1–9.
VAN RUYVEN, L. J., 1964, *Thesis* T.H. Eindhoven.

The GaAs photocathode[†]

L. W. JAMES, J. L. MOLL and W. E. SPICER

Stanford University, Stanford, California, U.S.A.

Abstract. A p^{++} type GaAs single crystal when cleaved in an ultra-high vacuum and coated with caesium has a low enough work function to permit photo-excited electrons from all energies within the conduction band to escape, giving a high quantum yield at all photon energies greater than the band gap. The actual yield obtained can be accurately predicted in terms of the diffusion lengths of electrons in the Γ and X minima, and the escape probabilities of electrons reaching the caesiated surface. The diffusion lengths and escape probabilities have been measured for crystals with various doping densities in an effort to determine the optimum doping. The escape probabilities, which depend on the work function, have been studied as a function of surface treatment. By applying additional alternating layers of caesium and oxygen to the surface, the work function is lowered appreciably below the bottom of the conduction band, giving very high yields near threshold and indicating the possibility of obtaining high yields with a lower threshold by using a narrower band gap material.

The details of the process used to obtain these high yield results repeatedly will be presented. Several properties of the completed GaAs–Cs–O photocathode have been studied. The effects of heating and cooling through liquid-nitrogen temperature on its operation has been determined. Conditions affecting the stability of the photocathode have been studied.

1. Introduction

Heavily doped p-type GaAs cleaved in an ultra-high vacuum and coated with caesium has a low enough work function to permit photo-excited electrons from all energies within the conduction band to escape, giving a high quantum yield at all photon energies greater than the band gap, and producing a high efficiency photocathode (Scheer and Van Laar 1965). The application of additional oxygen–caesium layers increases the efficiency over that obtained with caesium alone.

High resolution measurement of emitted electron energy distributions provides a valuable tool for studying the operation of the GaAs–Cs–O photocathode in detail. In this paper the results obtained using this technique will be discussed, principally in terms of obtaining the maximum detector sensitivity for low photon energies.

2. Photocathode preparation

Photocathodes are prepared in a high-vacuum cleaving chamber shown schematically in figure 1. A single crystal of commercially available boat-grown p^+ GaAs 1 cm square

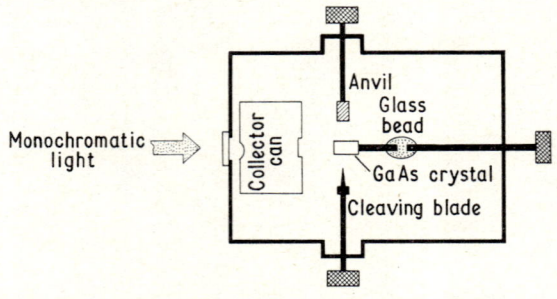

Figure 1. Ultra-high-vacuum cleaving chamber in which photocathodes are prepared and measured.

[†] Work supported by the U.S. Army, Ft. Belvoir, and by the Advanced Research Projects Agency through the Center for Materials Research at Stanford University.

by $1\frac{1}{2}$ cm long is mounted on a movable rod, aligned such that the $\langle 110 \rangle$ face (the cleavage plane) faces the window on the front of the chamber. After pumping the chamber down to a pressure of 10^{-11} torr, the crystal is moved into position between the tungsten carbide blade and the annealed copper anvil. Pressure is applied between the blade and anvil until the crystal cleaves, giving a mirror-like surface with a few cleavage lines.

Caesium is applied to the freshly cleaved surface from a caesium chromate channel source while monitoring the photocurrent produced by a dim white light. As caesium is applied, the photocurrent increases approximately exponentially until a peak sensitivity is reached. Additional caesiation reduces the sensitivity slowly. The additional caesium is unstable on the surface; it will come off in a day or two at room temperature. That amount of caesium which gives the optimum sensitivity, and which is stable on the surface, is referred to as a 'layer'. No measurement of the amount of caesium contained in a layer is possible in our experiment.

For those photocathodes treated with additional oxygen–caesium layers, after applying the first layer of caesium (as described above), oxygen is leaked into the chamber at a partial pressure of 2×10^{-8} torr for a period of 20 minutes. During this time the photocurrent decreases. The oxygen supply is then turned off, and the chamber is allowed to pump back to a low pressure. There is an immediate small increase in photocurrent when the pressure is reduced, and a further one over about a one-hour period. After the photocurrent has stabilized, caesium is again applied until a peak in sensitivity is reached. This process gives an additional 'oxygen–caesium layer', and may be repeated as many times as desired to obtain multiple oxygen–caesium layers, referred to as $(O+Cs)^n$ for n additional layers.

After preparation, the photocathode is moved into the collector can where measurements are made.

3. Theory of operation near threshold

Figure 2 shows a band diagram for a p^+ crystal of gallium arsenide coated with a layer of caesium. The bands are bent near the surface and the work function is lowered sufficiently for the vacuum level to be below the bottom of the conduction band in the bulk of the material. Near threshold the absorption coefficient for light is so small that only a few per cent of the light is absorbed in the band-bending region, and almost all photo-excitation takes place in the bulk of the material. The hot-electron scattering length is also short compared with the optical absorption length, so that photo-excited electrons thermalize in a conduction band minimum, then diffuse to the band bending region where they are accelerated toward the surface and emitted. Thus, we may consider the photo-emission near threshold as a series of processes: photo-excitation, thermalization, diffusion, and escape through the band-bending region and surface layers into the vacuum. We will consider each of these processes in order in more detail.

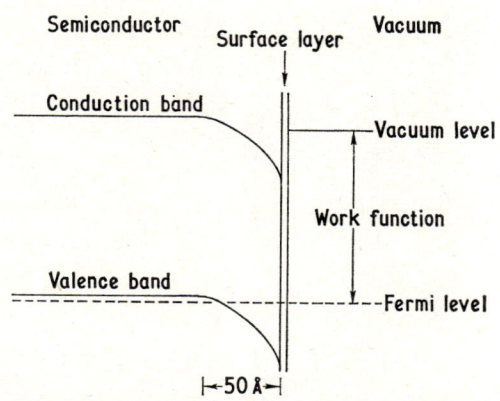

Figure 2. Band-bending diagram showing the effects of a layer of caesium applied to a p^+ GaAs surface.

Figure 3 shows a simple band structure for GaAs near the band gap. Photo-excitation in this material requires conservation of k-vector and energy, giving vertical transitions between states in the valence band and states in the conduction band which differ in energy by $h\nu$, where $h\nu$ is the photon energy. For low photon energies, such as shown at 'a' in figure 3,

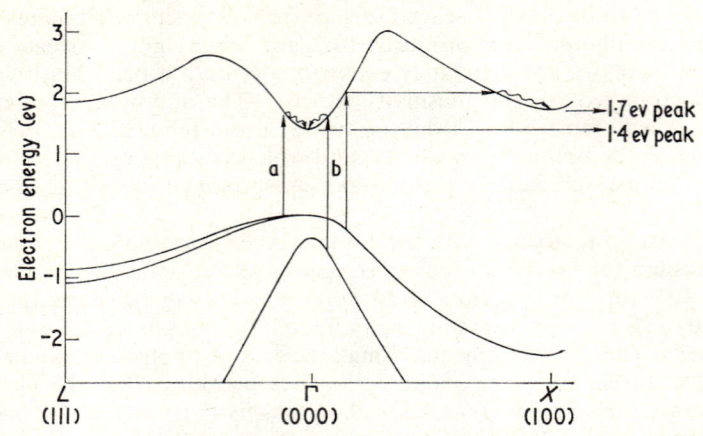

Figure 3. GaAs band structure near the energy gap showing examples of photo-excitation, scattering, and thermalization in the Γ and X minima. For photon energies below 1·7 ev ('a'), all electrons thermalize in the Γ minimum. Above 1·7 ev ('b'), some electrons are excited to a high enough energy to thermalize in X.

all photo-excitation will be to final states lower in energy than the X minimum and thermalization will occur into the Γ minimum. For higher photon energies, such as shown at b, some excitation will be to energies above 1·7 ev and some to energies below 1·7 ev. An electron excited above 1·7 ev will rapidly scatter into X and thermalize there due to the much higher density of states in X. The fraction which is excited to energies greater than 1·7 ev may be calculated from the band structure and will be defined as F_X. These electrons are assumed to travel only a very short distance through the crystal before thermalizing in X. The remaining fraction of excited electrons, F_Γ, are assumed to rapidly thermalize in the Γ minimum. F_Γ and F_X are shown in figure 4.

Examination of the experimental energy distribution curves shows that for photon energies from threshold at 1·4 ev (near-infrared) to 2·3 ev (blue-green) almost all emitted electrons are thermalized in either the Γ or X minima, while for higher photon energies a significant number of higher energy, unthermalized, electrons may be seen in the distribution. Also,

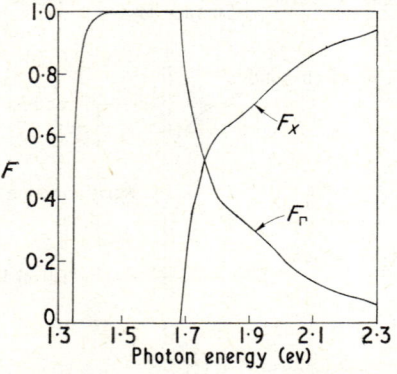

Figure 4. Fraction of photo-excited electrons which thermalize in each minimum, calculated from the GaAs band structure.

above 2·3 ev, α increases to a point where excitation in the band-bending region may no longer be neglected. Below 2·3 ev we need to consider only those electrons generated in the bulk crystal, and these are assumed to be thermalized in either the Γ or X minima, so we may solve for electron transport in terms of the coupled diffusion equations for these minima.

$$-D_\Gamma \frac{\partial^2 n_\Gamma}{\partial y^2} + \frac{n_\Gamma}{\tau_{\Gamma V}} = \frac{n_X}{\tau_{X\Gamma}} + I(1-R)\, F_\Gamma\, \alpha e^{-\alpha y} \quad (\Gamma \text{ equation})$$

$$-D_X \frac{\partial^2 n_X}{\partial y^2} + \frac{n_X}{\tau_{X\Gamma}} = I(1-R)\, F_X\, \alpha e^{-\alpha y} \qquad (X \text{ equation})$$

where y is the distance into the crystal.

The first term in each equation is the diffusion term where D is the diffusion coefficient. The second term is the rate at which carriers are lost from each minimum. $\tau_{\Gamma V}$ is the relaxation time for recombination from the Γ minimum to the valence band (or to traps). $\tau_{X\Gamma}$ is the relaxation time for scattering from the X minimum to the Γ minimum. Notice that $n_X/\tau_{X\Gamma}$ is a rate of generation term in the Γ equation as well as a rate of loss term in the X equation. The last term is the rate of generation by photo-excitation where I is the incident light intensity, R is the reflectivity, and α is the optical absorption coefficient. We may solve these equations for the current density flowing into the band-bending region, giving

$$J_X = \frac{qI(1-R)\, F_X}{1 + 1/\alpha L_X}$$

and

$$J_\Gamma = \frac{qI(1-R)}{1 + 1/\alpha L_\Gamma} \left[F_\Gamma + \frac{F_X}{1 + \alpha L_X} \right] \qquad \text{for} \qquad L_\Gamma \gg L_X$$

where the diffusion lengths are given by

$$L_X = \sqrt{(D_X \tau_{X\Gamma})}$$

$$L_\Gamma = \sqrt{(D_\Gamma \tau_{\Gamma V})}.$$

Of that current flowing into the band-bending region, a certain fraction, given by the escape probability P, will be emitted into the vacuum. P will be a function of both surface treatment and electron energy. The photoelectric quantum efficiency, known as yield, is then given for each minimum by

$$Y_X = \frac{P_X J_X}{qI(1-R)} = \frac{P_X F_X}{1 + 1/\alpha L_X}$$

$$Y_\Gamma = \frac{P_\Gamma J_\Gamma}{qI(1-R)} = \frac{P_\Gamma}{1 + 1/\alpha L_\Gamma} \left[F_\Gamma + \frac{F_X}{1 + \alpha L_X} \right].$$

Everything is known in these equations except the diffusion lengths, L_X and L_Γ, and the escape probabilities, P_X and P_Γ. The X and Γ yields may be obtained experimentally from the energy distribution curves. Examining the yield equations, we see that the magnitude of the yield vs. photon energy curves is determined by the escape probability, while the shape of the curves is determined by the diffusion length. Thus P_X and L_X may be determined uniquely from the experimental X yield curve, and P_Γ and L_Γ may be obtained from the Γ yield curve. L_X is limited by optical phonon scattering from X to Γ, and is measured to be 0.03 ± 0.01 μm at room temperature independent of doping. L_Γ is limited by recombination, and varies with doping and trap density. Measured values (obtained by a least squares fit to experimental Γ yield data) for boat-grown material are shown in table 1. The Γ yield is proportional to $(1 + 1/\alpha L)^{-1}$; therefore, to obtain the best yield, especially in the important near infrared region where α is small, we desire a long diffusion length, indicating the use of a lightly doped material which is free of deep traps.

Table 1. Measured Γ diffusion lengths for boat-grown Zn doped material

Carrier concentration cm^{-3}	Diffusion length μm 300°K	100°K
1×10^{19}	1·6±0·2	
3×10^{19}	1·2±0·2	1·0±0·3
4×10^{19}	1·0±0·2	

So far we have considered in detail all steps in the photo-emission process up to escape through the band-bending region and surface layers. No quantitative theory has been developed for the calculation of escape probability, but through analysis of high resolution energy distribution curves for several samples, we have been able to derive empirical relationships for escape probability. Figure 5 shows an energy distribution curve for a photon energy of 1·6 ev with the photocathode at liquid-nitrogen temperature. The sharp

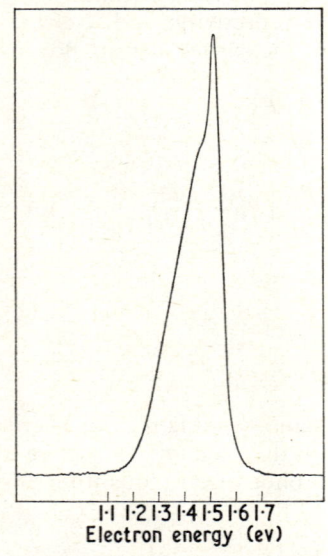

Figure 5. High resolution electron energy distribution curve for a photon energy of 1·6 ev at 77°K showing the peak of electrons thermalized in the Γ minimum and the scattering tail of electrons which are scattered in the band-bending region and surface layers.

peak corresponds to electrons thermalized in the Γ minimum, which cross the band-bending region and are emitted without scattering. The low-energy tail consists of electrons scattered in the band-bending region. The low-energy end of the tail is the value of the work function; no electrons with energy lower than the work function are emitted. Thus we can measure independently the work function and the Γ and X escape probabilities for each sample. The escape probability is expected to be a function of the difference between the electron energy and the vacuum level. Figure 6 shows this function for a surface treatment of a single layer of caesium. The function is very steep near threshold. For electron energies near the vacuum level, a 10 millivolt decrease in work function will triple the Γ escape probability (tripling the yield near threshold). This extreme sensitivity to small changes in work function probably accounts for the wide range of sensitivities obtained by some workers under seemingly identical preparation conditions.

For caesium coated GaAs the band gap and the work function are both 1·4 ev. In order to have the vacuum level lower than the conduction band, we must have the Fermi level lower than the valence band,† requiring the use of heavily doped, degenerate material. This is in conflict with the desire for long diffusion lengths, so a compromise is necessary. The yield obtained using a sample with a carrier concentration of 4×10^{19} (near the optimum compromise doping with one layer of caesium) is shown in figure 9.

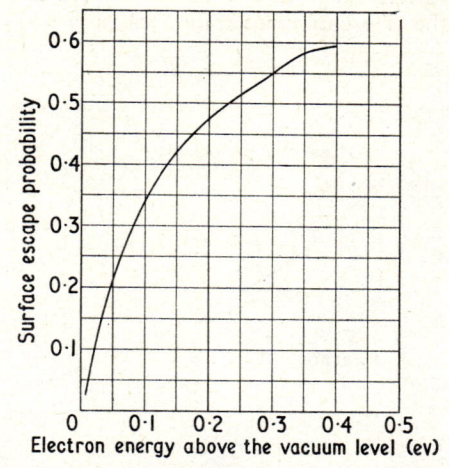

Figure 6. Surface escape probability vs. electron energy above the vacuum level, measured for a 1×10^{19} cm^{-3} sample. Slight differences could be expected for different doping due to the differing width of the band-bending region.

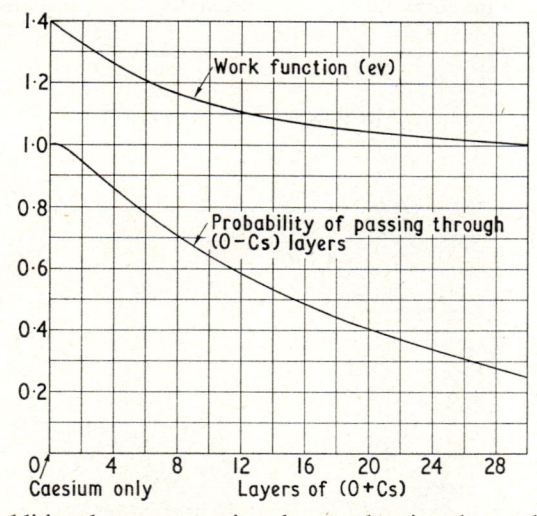

Figure 7. Effects of additional oxygen–caesium layers, showing the work function lowering and the electron absorption as measured experimentally.

In order to avoid this compromise, improve the sensitivity, and decrease the critical dependence on work function position, it is desirable to lower the work function. This may be done by applying additional oxygen–caesium layers to the surface as described earlier. The top curve in figure 7 shows the measured work function vs. the number of additional oxygen–caesium layers applied. However, there will be some absorption of electrons in the oxygen–caesium layers. This absorption was measured by comparing the hot electron (2·5 ev) escape probabilities before and after applying 30 additional (O+Cs) layers, and is

† The work function, which is the Fermi level to vacuum level spacing, is fixed.

shown in the bottom curve of figure 7. The actual escape probability is then the product of the surface escape probability and the probability of passing through the (O+Cs) layers. Using figures 6 and 7 we may determine the optimum surface treatment. Figure 8 shows the Γ and X escape probabilities for a 1×10^{19} cm^{-3} sample calculated using figures 6 and 7. If we desire optimum sensitivity near threshold, we should apply six additional (O+Cs) layers. This will reduce the X escape probability, causing a slight decrease in ultraviolet

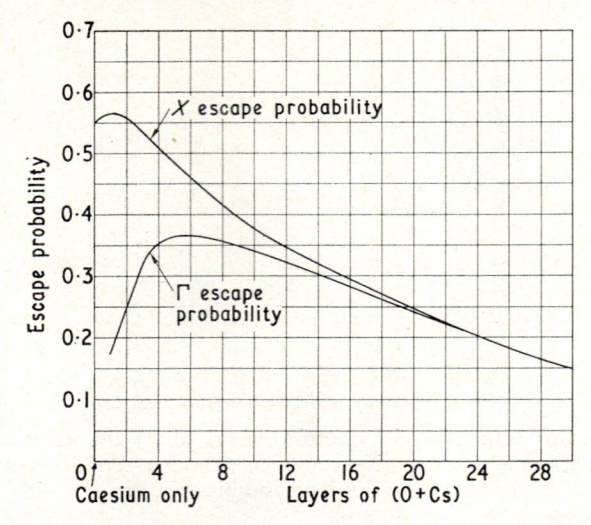

Figure 8. Γ and X escape probability for a 1×10^{19} cm^{-3} sample calculated using figures 6 and 7.

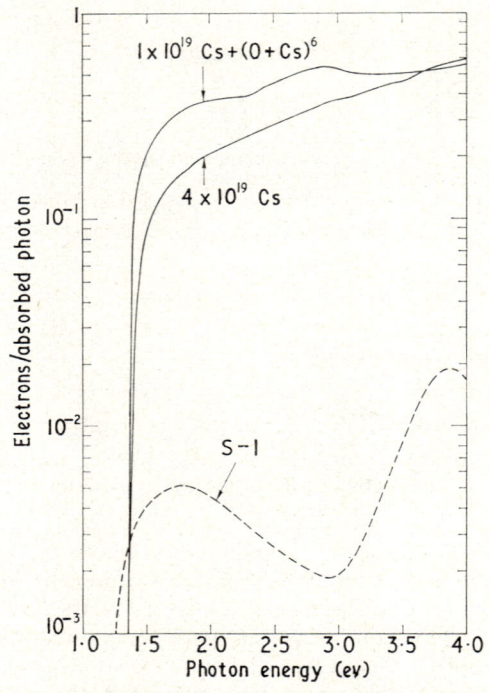

Figure 9. Absolute quantum yield curves shown for optimum caesium-only treatment (4×10^{19} Cs), optimum oxygen–caesium treatment (1×10^{19} Cs+(O+Cs)^6), and, for comparison, a commercial S–1 photocathode.

sensitivity where the X yield becomes dominate. The experimental yield obtained using this optimum surface treatment is shown in figure 9. This optimized photocathode has a sensitivity of 1000 μA lumen^{-1}, and has a yield at 1·5 ev which is almost two orders of magnitude better than the S–1 photocathode (also shown in figure 9 for comparison).

4. The future for practical photocathodes

It should be emphasized that the results given in this paper were obtained on cleaved surfaces under ultra-high-vacuum conditions. Two problems become apparent in processing the photocathode under less ideal conditions of poor vacuum or uncleaved surfaces. In some cases the work function is larger than shown in figure 6 for the same surface treatment, and in other cases a 'barrier' is apparently present between the GaAs and the first caesium layer, lowering the escape probability for a given work function. More work is required adequately to understand these effects.

Stability of the GaAs–Cs photocathode is also a problem. Deterioration in sensitivity is thought to occur with adsorption of oxygen or other contaminates on the caesium surface causing an increase in work function. At 10^{-11} torr, the sensitivity of a caesium-only treated photocathode is reduced by about 15% in a two-week period. The oxygen–caesium treated photocathode shows less deterioration due to the much lower sensitivity to small changes in work function. In both cases, practically full sensitivity may be restored by applying a small amount of additional caesium.

Heating the photocathode to 75°c causes caesium to leave the surface, raising the work function about 0·25 ev, and drastically reducing the yield. Optimum sensitivity may be restored by reapplying caesium, but high temperature operation is obviously prohibited.

Cooling the photocathode to liquid-nitrogen temperatures causes an increase in threshold and an increase in Γ escape probability due to the increase in the band gap to 1·5 ev. The dark current due to thermal generation in the band-bending region should be completely negligible at liquid-nitrogen temperatures.

The materials work currently in progress on III–V mixed alloys points to additional possibilities in photocathode development. If high sensitivity in the visible light range is desired and near infrared sensitivity is of no importance, a wider band gap material such as GaAs$_x$P$_{1-x}$As should give higher Γ escape probabilities and be easier to fabricate. If a lower threshold than 1·4 ev is desired, a smaller band-gap material such as I$_n$P (Bell and Uebbing 1968) or Ga$_x$In$_{x-1}$As (Uebbing and Bell 1968) may be used with oxygen–caesium surface layers at the expense of reduced escape probability.

5. Conclusions

Through a physical understanding and measurement of the details of the photo-emission process in caesiated p$^+$ GaAs, we have been able to determine the doping and surface treatment necessary for optimum sensitivity, and have been able to make photocathodes demonstrating this optimum sensitivity.

References

BELL, R. L., and UEBBING, J. J., 1968, *Appl. Phys. Lett.*, **12**, 76–8.
SCHEER, J. J., and VAN LAAR, J., 1965, *Solid State Comm.*, **3**, 189–93.
UEBBING, J. J., and BELL, R. L., 1968, *Proc. IEEE*, **56**, 1624–5.

Photon emission during avalanche breakdown in GaAs

M. H. PILKUHN and G. SCHUL

Physics Institute, Frankfurt University, Germany

Abstract. Emission spectra from different types of GaAs avalanche diodes were investigated in the temperature range between 80 and 340°K. At high temperatures, a broad emission band typical of avalanche breakdown dominates the spectra of all diodes. The spectral shape of this band is analysed in detail and discussed in view of theoretical models concerning photon emission during avalanche breakdown. It was found that temperature has no influence on the shape of the broad emission band. In addition, two near edge emission lines may be seen, particularly at low temperatures. These lines are similar to the photoluminescent lines of the n- and p-material. The quantum efficiency of the broad emission band is independent of temperature as long as the reverse bias I–V characteristic is typical of avalanche breakdown. It increases with decreasing temperature in those cases where avalanche breakdown and tunnelling are both present. The relative efficiency of the near edge emission lines always increases as the temperature is lowered.

1. Introduction

In recent years, technological progress in the methods of p–n junction fabrication has led to the development of improved GaAs avalanche diodes. As in silicon, the avalanche breakdown in GaAs may be interpreted by an impact ionization process, which can be distinguished from Zener tunnelling by the shape and temperature dependence of the I–V characteristic. Ionization rates have been studied in GaAs by photomultiplication experiments (Kressel and Kupsky 1966; Logan, Chynoweth, and Cohen 1962) leading to values for ionization energies and mean free paths. A theoretical computation of breakdown voltages (Sze and Gibbons 1966) agrees well with experimental values found for abrupt junctions (Weinstein and Mlavsky 1963) and for those with linear gradients (Kressel and Blicher 1963).

There is relatively little information available about the photon emission associated with avalanche breakdown. Very broad emission spectra were reported for Si (Chynoweth and McKay 1956), Ge (Chynoweth and Gummel 1960), SiC (Kholuyanov 1962) and GaP (Gershenzon and Mikulyak 1961). However, many of these measurements have been limited to energies larger than the bandgap or to particular temperatures. The broad emission band has been interpreted through combined intraband and interband transitions of hot carriers (Wolff 1960). It has also been proposed that bremsstrahlung of hot carriers interacting with charged impurities is the cause of the broad emission (Figielski and Torun 1962).

In GaAs, the photon emission from reverse biased p–n junctions was mainly studied at low temperatures (Michel, Nathan, and Marinace 1964). As a dominant feature, sharp emission lines have been reported in the neighbourhood of the bandgap, similar to those observed in photoluminescence and typical of regular minority carrier recombination. In this paper, the emphasis lies on the study of the broad emission band which we find to be dominant at high temperatures.

2. Experimental

Two different types of p–n junctions were studied under reverse bias: (1) Zn-diffused diodes (n-substrate doping levels in the range between 7×10^{16} and 3×10^{18} cm^{-3}), and (2) diodes prepared by liquid phase epitaxy. Under reverse bias, the diffused diodes with low substrate doping level showed 'hard' I–V characteristics with a positive temperature coefficient ($dV_b/dT > 0$) of the breakdown voltage, V_b, which is typical of avalanche multiplication. The highly doped epitaxial diodes had 'soft' reverse bias I–V characteristics and a negative temperature coefficient, dV_b/dT, typical of Zener tunnelling. There were also mixed cases of combined avalanche and Zener breakdown.

The reverse bias emission was viewed either through the n- or p-layer of the junction. The diodes were mounted on a heat sink in a variable temperature Dewar which allowed measurements between 77 and 340°K. Great care was taken to calibrate the monochromator and photodetector (cooled S–1 photomultiplier) in order to get reliable response corrections for the emission spectra. The calibration was made with a known black-body radiation source.

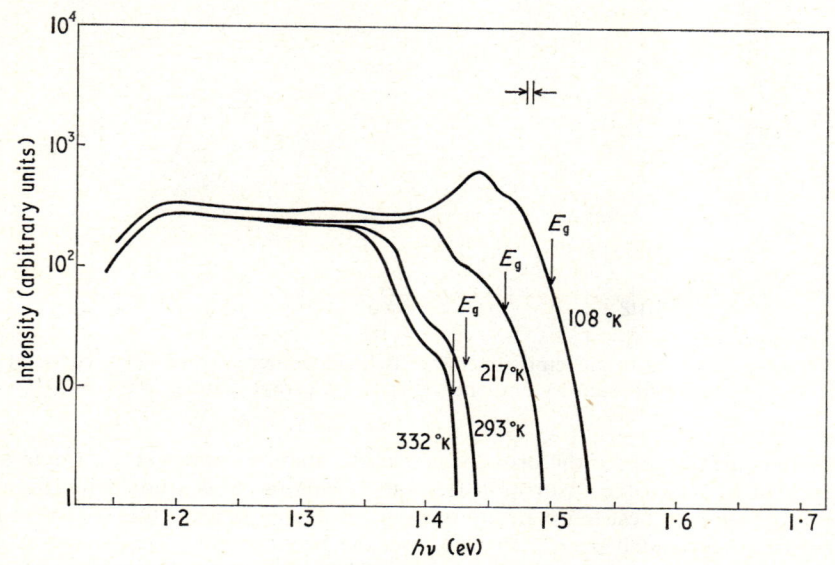

Figure 1. Reverse bias emission spectra of a GaAs avalanche diode at various temperatures. The corrected intensity has been plotted in a logarithmic scale and in arbitrary units against photon energy. The arrows indicate the position of the energy gap. Current density: 2·5 A cm^{-2}.

3. Results and discussion

Spectra for various temperatures are shown in figure 1 for a Zn-diffused diode whose I–V characteristic is typical of avalanche breakdown. The corrected intensity has been plotted in a logarithmic scale. It can be seen that a very broad emission band dominates the spectra at high temperatures. There is only slight structure on the high energy side of this band at photon energies near the bandgap. This structure becomes stronger as the temperature is lowered and eventually develops into additional emission lines. The shift of these lines to higher energy reflects the increase of the bandgap with decreasing temperature. This general description of the emission spectra applies to all diodes investigated. Especially, the shape of the broad emission band at high temperature and for $h\nu < E_g$ changed very little from diode to diode; however, the relative intensity of the near edge lines did vary.

We first discuss the spectral shape of the broad emission band neglecting near edge lines. In figure 1, it may be seen that this shape does not vary at all with temperature between 100 and 340°K. This result is not unexpected, if one assumes that the temperature of hot carriers in the avalanche plasma does not change much with the lattice temperature.

The broad emission band has a high energy cut-off at photon energies somewhat below the bandgap. This cut-off appears to be predominantly due to re-absorption. The energy where it starts coincides with the beginning of the absorption edge (Hill 1964; Turner and Reese 1964). Since the absorption coefficient becomes very high at $h\nu > E_g$ in GaAs, no corrections for re-absorption were tried above the energy gap. Qualitatively it was found, however, that weak re-absorption (diode emission viewed through thin n-layers) resulted in a gradual high energy cut-off. Furthermore, if the emission was viewed through a highly doped p-layer instead of an n-layer, the cut-off was shifted to lower energies reflecting the shift in absorption edge.

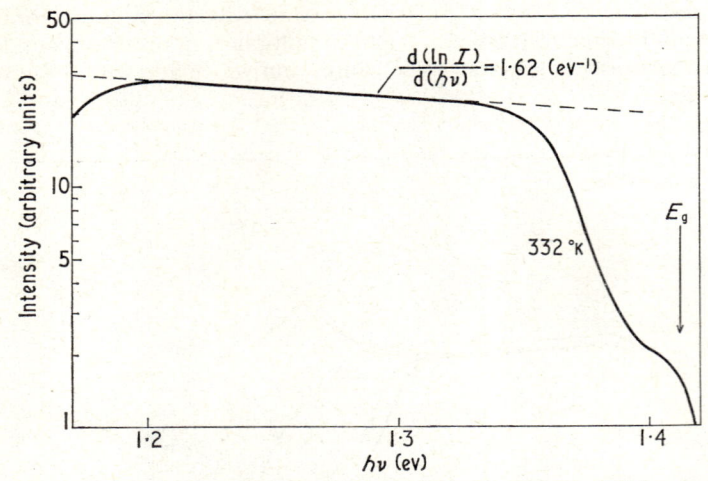

Figure 2. Analysis of the broad emission band at high temperature (332°K—curve of figure 1) in an expanded intensity and energy scale. Current density: 2·5 A cm^{-2}.

The most significant part of the broad spectrum to analyse is the photon range below the high energy cut-off where re-absorption is small. This range is studied for the spectrum corresponding to the highest temperature of figure 1, and in figure 2 this spectrum has been replotted with an expanded scale. In this case, the broad band is almost undisturbed by additional lines. In the energy range between 1·2 and 1·34 ev, the intensity appears to follow the relation

$$I \propto \exp(-\beta h\nu).$$

This relation would be expected from the bremsstrahlen theory of Figielski and Torun, and an electron temperature, T_e, may be derived from the slope of the linear region in figure 2:

$$T_e = \frac{1}{k} \cdot \frac{d(h\nu)}{d(\ln I)} = 7100°K.$$

This value for an electron temperature is not too different from values obtained by a crude estimate using the breakdown field, F_b, the mean free path, l, and the LO phonon energy, $(\hbar\omega)_{ph}$:

$$T_e = \frac{1}{k} \frac{(qF_b l)^2}{3(\hbar\omega)_{ph}}.$$

If a mean free path of 43 Å (Kressel and Kupski 1960), a breakdown field of 5×10^5 v cm^{-1}, and an optical phonon energy of 36 mev is assumed, the electron temperature comes out to be about 5000°K. A certain disagreement is found for the elemental semiconductors Si and Ge too, where the spectral shape of the broad band was analysed at $h\nu > E_g$ (Figielski and Torun 1962). In those cases, improved bremsstrahlen theories assuming non-Maxwellian distribution functions for hot carriers were tried for better agreement with experimental data (Shewchun and Wei 1965, Kamieniecki 1964). We have not made corrections for non-Maxwellian distributions in the case of GaAs.

Below 1·2 ev, there is a low energy intensity drop. It is clearly noticeable in all diodes, and it is not caused by an insufficient response correction. The low energy intensity drop is also not caused by free-carrier absorption, because a correction for re-absorption does not make it disappear.

Recent investigations of the high temperature spectra of GaP avalanche diodes (Pilkuhn 1968) show that the broad band is very similar in shape to that reported here for GaAs. Beside a nearly linear portion in the $\ln I$ vs. $h\nu$ plot at energies $h\nu < E_g$, a low energy intensity drop is observed, occurring at about 1·8 ev, i.e. at $0·8 \times E_g$, as in GaAs. In fact, if a dimensionless energy co-ordinate $h\nu/E_g$ is used, the broad spectra for GaP and GaAs are very similar to one another. This indicates that the emission does not depend on the band structure of the material, which seems to rule out the intra/interband transition model of Wolff. Although the bremsstrahlen theory appears to be favoured through this experimental comparison, difficulties remain: in particular, it is difficult to understand the low energy intensity drop at $h\nu < 0·8 \times E_g$. Increased phonon emission during the interaction of the less energetic charge carriers with ionized impurities may be a possible explanation.

We shall now discuss the additional emission lines observable near the energy gap. These lines have been studied by Michel, Nathan, and Marinace at low temperatures and attributed to recombination of charge carriers which have assumed lattice temperature after diffusing away from the hot plasma region. In figure 1 it may be seen that there are two near edge lines. They can be correlated with the photoluminescence lines of the n- and p-material, and also with the forward bias emission lines. We may conclude that the recombination of carriers which are at lattice temperature during avalanche breakdown occurs both on the n- and p-side of the space charge region. Instead of a diffusion away from the plasma, an ionization through hot carriers in the n- and p-regions at the edge of the space charge layer may be suggested as an alternative model.

In the highly doped and epitaxial diodes, the near edge line in reverse bias is mostly the one at lower energy which resembles the photoluminescence of p-material and the forward bias emission. Contrary to the broad emission band, the relative intensity and spectral shape of the additional lines were found to depend on the structure of the p–n junction. For a given diode, the shape of the emission spectrum, i.e., also the intensity of near edge lines relative to the broad band, did not change with current density. It did change, however, with temperature: we have found in all diodes that the near edge lines increase in intensity relative to the broad band when the temperature is lowered. This may either mean that a larger portion of carriers cools down to lattice temperature, or that the efficiency of their recombination process increases with decreasing temperature.

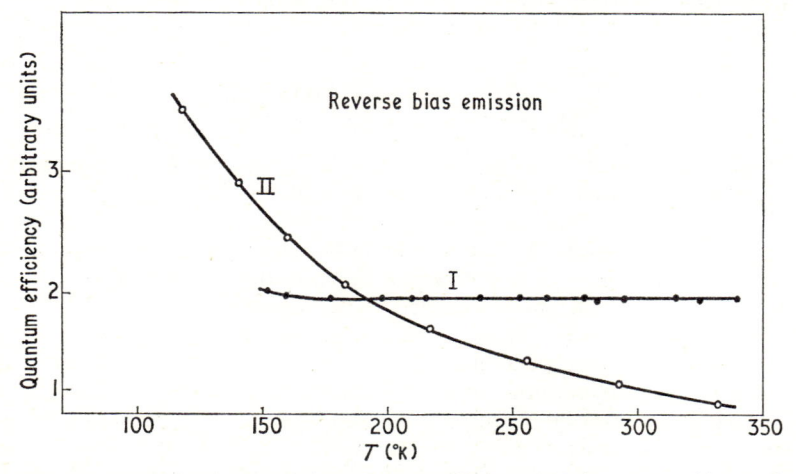

Figure 3. Temperature dependence of the quantum efficiency of the reverse bias emission from a GaAs avalanche diode. Curve I refers to the broad part of the spectrum, curve II to the additional near edge lines. Quantum efficiencies are plotted in a linear arbitrary scale. Current density: 2·5 A cm^{-2}.

In diodes where the forward bias spectra contained low energy lines due to transitions to deep levels, these transitions were also observed as additional lines in the reverse bias spectra at low temperatures.

We have also investigated the quantum efficiency of the reverse bias emission as a function of temperature. Since the emission intensity was found to increase linearly with current density, a relative quantum efficiency was defined as the ratio of light intensity to current. Results for an avalanche diode are depicted in figure 3. Curve I refers to the quantum efficiency of the broad emission band, measured at constant current density. One can see that in the temperature range between 150 and 340°K the quantum efficiency remains constant. The slight rise at $T < 150°K$ is possibly due to decrease in free carrier absorption. This result, together with the fact that the broad band does not vary its shape with temperature, demonstrates that the photon emission from an avalanche plasma in GaAs is not correlated with the lattice temperature.

Curve II in figure 3 is the relative quantum efficiency of the additional near edge lines. It increases with decreasing temperature, as already mentioned.

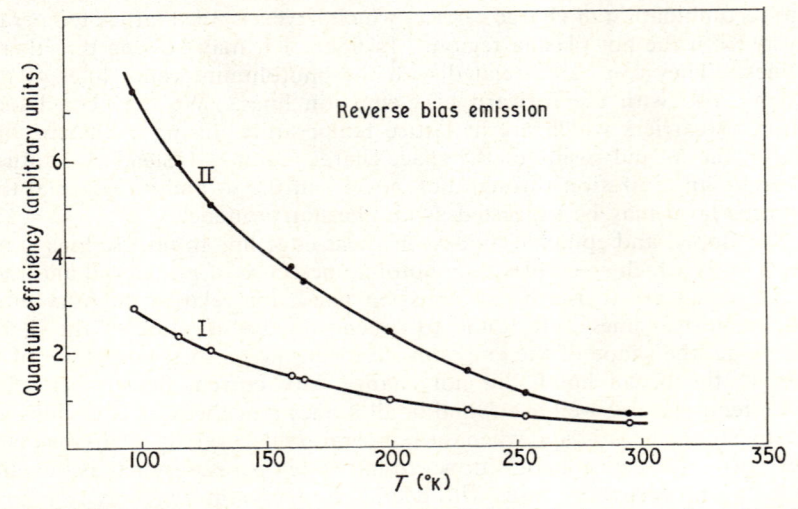

Figure 4. Temperature dependence of the quantum efficiency in the case of combined Zener and avalanche breakdown. The emission of an epitaxial diode ($N_d = 10^{18}$ cm^{-3}) was viewed through the n-layer. As in figure 3, curves I and II refer to broad emission and near edge emission lines. Current density: 16·7 A cm^{-2}.

In highly doped diodes with soft I–V characteristics, i.e., where avalanche breakdown and tunnelling are both present, the quantum efficiency of the broad emission was found to increase with decreasing temperature which is shown in figure 4, curve I. This behaviour can be interpreted by the different influence of temperature on Zener tunnelling and avalanche breakdown, resulting in a larger relative portion of the avalanche current at lower temperature. If the Zener current is assumed to be non-radiative and if the experiment is done at constant current density, the light intensity must increase with decreasing temperature. Here, the temperature dependence of the quantum efficiency of the broad emission band may serve to get information about the relative portions of Zener and avalanche current. However, care must be taken to avoid additional low energy emission lines due to recombination of 'cold' carriers through deep levels. Curve II in figure 4 represents the quantum efficiency of the additional near edge lines which increases strongly when the temperature is lowered, as in figure 3.

In conclusion, the broad emission which is observed during avalanche breakdown appears to be independent of temperature, junction structure, and band structure (if the results obtained for GaAs are compared with those obtained for GaP). This favours a theory like the bremsstrahlen theory of Figielski and Torun. However, more work is needed to fit theoretically the detailed spectral shape of the broad band and to remove disagreements like those mentioned in this paper. The analysis should be done over a larger energy range and preferably for more semiconductors than discussed here. The additional near edge lines which are seen in GaAs appear to be dependent on temperature and junction structure.

Acknowledgment

We are particularly grateful to Prof. H. Beneking and Dipl.-Ing. W. Vits of the Aachen Technical University and to the IBM Laboratories, Boblingen, for supplying GaAs diodes.

References

CHYNOWETH, A. G., and GUMMEL, H. K., 1960, *J. Phys. Chem. Sol.*, **16**, 191.
CHYNOWETH, A. G., and MCKAY, K. G., 1956, *Phys. Rev.*, **102**, 369.
FIGIELSKI, T., and TORUN, A., 1962, *Internat. Conference on the Physics of Semiconductors, Exeter*, p. 863.
GERSHENZON, M., and MIKULYAK, R. M., 1961, *J. Appl. Phys.*, **32**, 1338.
HILL, D. E., 1964, *Phys. Rev.*, **133**, A866.
KAMIENIECKI, E., 1964, *phys. stat. sol.*, **6**, 877.
KHOLUYANOV, G. F., 1961, Fiz. Tverd. Tela, **3**, 3314. Soviet Physics–Solid State, **3**, 2405.
KRESSEL, H., and BLICHER, A., 1963, *J. Appl. Phys.*, **34**, 2495.
KRESSEL, H., and KUPSKY, G., 1966, *Int. J. of Elect.*, **20**, 535.
LOGAN, R. A., CHYNOWETH, A. G., and COHEN, B. G., 1962, *Phys. Rev.*, **128**, 2518.
MICHEL, A. E., NATHAN, M. I., and MARINACE, J. C., 1964, *J. Appl. Phys.*, **35**, 3543.
PILKUHN, M. H., to be published.
SHEWCHUN, J., and WEI, L. Y., 1965, *Solid State Electronics*, **8**, 485.
SZE, S. M., and GIBBONS, G., 1966, *Appl. Phys. Lett.*, **8**, 111.
TURNER, W. J., and REESE, W. E., 1964, *J. Appl. Phys.*, **35**, 350.
WEINSTEIN, M., and MLAVSKY, A. J., 1963, *Appl. Phys. Lett.*, **2**, 97.
WOLFF, P. A., 1960, *J. Phys. Chem. Solids*, **16**, 184.

Author Index

Abrahams, M. S.   55
Allred, W. P.   66
Antell, G. R.   160
Appert, J. R.   213
Ashley, K. L.   123
Barnett, A. M.   136
Beneking, H.   96
Bittman, C. A.   187
Bowers, H. C.   136
Caraballès, J. C.   28
Casey, H. C.   141
Cohen, L.   153
Cumming, G.   66
Day, G. F.   22
Diguet, D.   28
Dobson, C. D.   36
Doerbeck, F. H.   205
Drago, F.   153
Dreeben, A. B.   55
Dyment, J. C.   83
Enstrom, R. E.   213
Fenner, G. E.   131
Franks, J.   36
Galginaitis, S. V.   131
Goodwin, A. R.   36
Gossenberger, H. F.   55
Greene, P. E.   18
Harp, E. E.   205
Hooper, W. W.   187
Hower, P. L.   187
Hwang, C. J.   83
Ilegems, M.   3
James, L. W.   230
Jensen, H. A.   136
Kang, C. S.   18
Kawazura, S.   167
Kim, H. B.   110
Kinoshita, J.   22
Koyama, J.   167
Kumabe, K.   167
Kung, J.   66

Lebailly, J.   28
Lehrer, W.   187
Lindley, W. T.   43
Meikleham, V. F.   136
Mohn, E.   101
Moll, J. L.   230
Mooney, J. B.   22
von Münch, W.   77
Ohara, S.   167
Oku, T.   116
Pearson, G. L.   3
Pilkuhn, M. H.   238
Reid, F. J.   59
Riley, T. J.   173
Ripper, J. E.   91
Robinson, L. B.   59
Schul, G.   238
Shaw, D. W.   50
Shortt, B.   153
Silversmith, D. J.   141
Socci, R.   153
Sogo, T.   116
Solomon, R.   11
Spicer, W. E.   230
Spitzer, W. G.   66
Stein, W. W.   22
Steinemann, A.   73
Stillman, G. E.   43
Strack, H. A.   123, 205
Susaki, W.   116
Tansley, T. L.   222
Tietjen, J. J.   55
Tremere, D. A.   187
Turner, J. A.   195
Urban, M.   153
Vits, W.   96
Wilson, B. L. H.   195
Winteler, H. R.   73
Wolfe, C. M.   43
Zuleeg, R.   181